Preface to the Third Edition

This project began nearly 25 years ago, and the first edition was published in 1981. Since then, plant science and ecology have undergone radical revolutions, but the need to understand the environmental physiology of plants has never been greater. On the one hand, with the excitement generated by molecular approaches, there is a real risk that young plant scientists will lack the necessary understanding of how whole plants function. On the other hand, there are global problems to tackle, most notably the consequences of climate change; those who are charting possible futures for plant communities need to have a good grasp of the underlying physiology. Environmental physiology occupies a vital position as a bridge between the gene and the simulation model.

Although this edition retains the basic structure and philosophy of previous editions, the text has been completely rewritten and updated to give a synthesis of modern physiological and ecological thinking. In particular, we explain how new molecular approaches can be harnessed as *tools* to solve problems in physiology, rather than rewriting the book as a primer of molecular genetics. To balance the molecular aspects, we have made a positive decision to use relevant examples from pioneering and classic work, drawing attention to the foundations of the subject. New features include a more generic approach to toxicity, explicit treatment of issues relating to global climate change, and a section on the role of fire. The text illustrations, presented according to a common and improved format, are complemented by colour plates. Even though the rewriting of the book has been a co-operative enterprise, AHF is primarily responsible for Chapters 2, 3 and 7 and RKMH for Chapters 4, 5 and 6. We thank Terry Mansfield, Lucy Sheppard, Ian Woodward, Owen Atkin and Angela Hodge for their helpful comments on individual chapters in draft.

Environmental physiology is a rapidly expanding field, and the extent of the literature is immense. Our aim has not been to be comprehensive and authoritative but to develop principles and stimulate new ideas through selected examples, and we remain committed to a policy of full citation to facilitate access to key publications. Where the subject area is in rapid flux, we have attempted to provide a balanced review, which will inevitably be overtaken by events; and we have consciously focused attention on studies in North America, Europe and Australia, because of our personal experience of these areas, and because we hope that this will give a greater coherence to the examples chosen. Although much excellent work is published in languages other than English, we have not relied heavily on this, since the book is intended primarily for students to whom such literature is relatively inaccessible. Finally, we would recommend the advanced monograph *Physiological Plant Ecology*, edited by M.C. Press, J.D. Scholes and M.G. Barker (1999; Blackwell Science, Oxford) as a useful complement to this book.

This third edition is a celebration of a quarter of a century of working together towards a common goal from different viewpoints and experiences. We are very grateful to our editor, Andy Richford, whose vision, encouragement and persistence have kept us to the task.

A.H. FITTER
R.K.M. HAY

Environmental Physiology of Plants

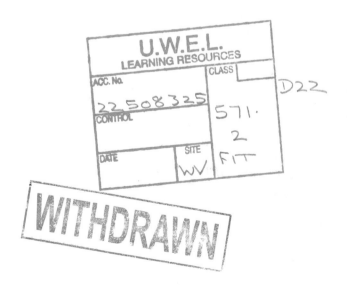

Third Edition

Acknowledgements

We are grateful for permission from the following authorities to use materials for the figures and tables listed:

Academic Press Ltd., London – Figs. 3.23, 4.9, 5.10;
Annals of Botany – Fig. 1.1a;
American Society of Plant Physiologists – Figs. 4.2, 6.3;
Blackwell Science Ltd., Oxford – Figs 2.8, 2.27, 3.7, 4.11, 4.12, 5.6, 5.15, 6.8, 6.9, Tables 4.4, 6.7;
BIOS Scientific Publishers Ltd., Oxford – Fig. 4.17;
Cambridge University Press – Table 5.2;
Ecological Society of America – Fig. 5.13;
Elsevier Science, Oxford – Fig. 4.4;
European Society for Agronomy – Fig. 4.10;
HarperCollins, London – Fig. 2.14;
J. Wiley and Sons, New York – Fig. 5.16;
Kluwer Academic Publishers, Dordrecht – Fig. 3.2;
Munksgaard International Publishers and Dr Y. Gauslaa – Fig. 5.9;
National Research Council of Canada – Fig. 5.11;
New Phytologist and the appropriate authors – Figs. 2.26, 2.29, 3.26, 3.37, 4.19, 5.1, 5.14, 6.7, 6.10;
Oikos – Table 6.4;
Oxford University Press – Figs. 2.25, 4.8, 5.2;
Pearson Education, Inc., New Jersey – Fig. 4.6;
Physiologia Plantarum – Fig. 6.2;
Professor I.F. Wardlaw – Fig. 5.5;
Professor M.C. Drew – Fig. 6.6;
Royal Society of Edinburgh and Dr G.A.F. Hendry – Table 6.5;
Royal Society of London and Professor K. Raschke – Fig. 4.15;
Springer Verlag (Berlin) and the appropriate authors – Figs. 1.7, 2.18, 5.4, 5.8, 5.12, 5.17, 6.11 Tables 5.5, 6.6;
Urban & Fischer Verlag, Jena – Fig. 6.5;
Weizmann Science Press of Israel and Professor Y. Gutterman – Table 4.2.

To
Rosalind and Dorothea

Contents

1

Introduction

1. Plant growth and development

This book is about how plants interact with their environment. In Chapters 2 to 4 we consider how they obtain the necessary resources for life (energy, CO_2, water and minerals) and how they respond to variation in supply. The environment can, however, pose threats to plant function and survival by direct physical or chemical effects, without necessarily affecting the availability of resources; such factors, notably extremes of temperature and toxins, are the subjects of Chapters 5 and 6. Nevertheless, whether the constraint exerted by the environment is the shortage of a resource, the presence of a toxin, an extreme temperature, or even physical damage, plant responses usually take the form of changes in the rate and/or pattern of growth. Thus, environmental physiology is ultimately the study of plant growth, since growth is a synthesis of metabolic processes, including those affected by the environment. One of the major themes of this book is the ability of some successful species to secure a major share of the available resources as a consequence of rapid rates of growth (the concept of pre-emption or 'asymmetric competition'; Weiner, 1990).

When considering interactions with the environment, it is useful to discriminate between plant growth (increase in dry weight) and development (change in the size and/or number of cells or organs, thus incorporating natural senescence as a component of development). Increase in the size of organs (development) is normally associated with increase in dry weight (growth), but not exclusively; for example, the processes of cell division and expansion involved in seed germination *consume* rather than generate dry matter.

The pattern of development of plants is different from that of other organisms. In most animals, cell division proceeds simultaneously at many sites throughout the embryo, leading to the differentiation of numerous organs. In contrast, a germinating seed has only two localized areas of cell division, in meristems at the tips of the young shoot and root. In the early stages of development, virtually all cell division is confined to these meristems but, even in very short-lived annual plants, new meristems are initiated as development proceeds. For example, a root system may consist initially of a single main axis with an apical meristem but, in time, primary laterals will emerge, each with its own meristem. These can, in turn, give rise to further branches (e.g. Figs. 3.20, 3.22). Similarly, the shoots of herbaceous plants can be resolved into a set of modules, or phytomers, each comprising a node, an internode, a leaf and an

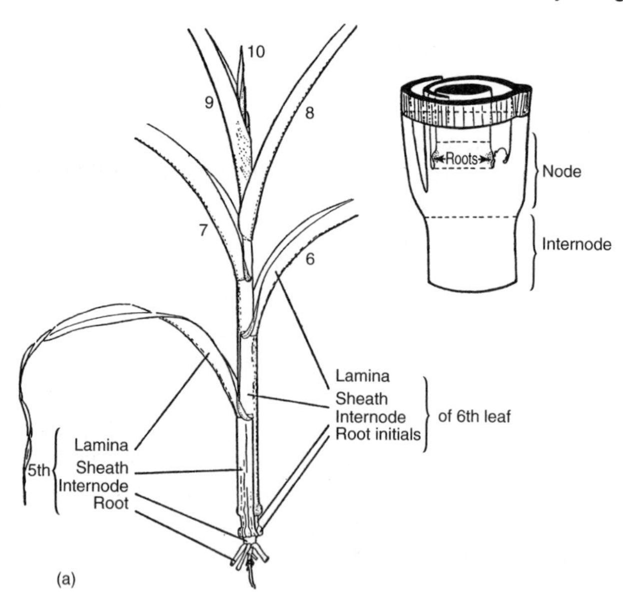

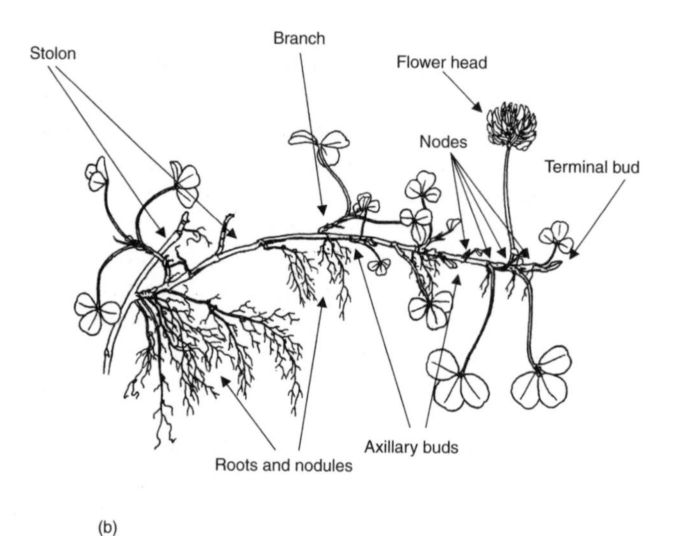

Figure 1.1

Two variations on the theme of modular construction. (a) Maize, where the module consists of a node, internode and leaf (encircling sheath and lamina; see inset diagram for spatial relationships). The axillary meristem normally develops only at one or two ear-bearing nodes, in contrast to many other grasses, whose basal nodes produce leafy branches (tillers). Nodal roots can form from the more basal nodes. Note that in the main diagram, the oldest modules (1–4) are too small to be represented at this scale, and the associated leaf tissues have been stripped away (adapted from Sharman, 1942). (b) White clover stolon, where the module consists of a node, internode and vestigial leaf (stipule). The axillary meristems can generate stolon branches or shorter leafy or flowering shoots, and extensive nodal root systems can form (diagram kindly provided by Dr M. Fothergill).

axillary meristem (Fig. 1.1). Such branching patterns are common in nature (lungs, blood vessels, neurones, even river systems); in each case, the 'daughters' are copies of the parent branches from which they arose.

The modular mode of construction of plants (Harper, 1986) has important consequences, including the generalization that development and growth are essentially indeterminate: the number of modules is not fixed at the outset, and a branching pattern does not proceed to an inevitable endpoint. Whereas all antelopes have four legs and two ears, a pine tree may carry an unlimited number of branches, needles or root tips (Plate 1). Plant development and growth are, therefore, very flexible, and capable of responding to environmental influences; for example, plants can add new modules to replace tissues destroyed by frost, wind or toxicity. On the other hand the potential for branching means that, in experimental work, particular care must be exercised in the sampling of plants growing in variable environments: adjacent pine trees of similar age can vary from less than 1 m to greater than 30 m in height, with associated differences in branching, according to soil depth and history of grazing (Plate 1). Such a modular pattern of construction, which is of fundamental importance in environmental physiology, can also pose problems in establishing individuality; thus, the vegetative reproduction of certain grasses can lead to extensive stands of physiologically-independent tillers of identical genotype.

Even though higher plants are uniformly modular, it is simple, for example, to distinguish an oak tree from a poplar, by the contrasting shapes of their canopies. Similarly, although an agricultural weed such as groundsel (*Senecio vulgaris*) can vary in size from a stunted single stem a few centimetres in height with a single flowerhead, to a luxuriant branching plant half a metre high with 200 heads, it will never be confused with a grass, rose or cactus plant. Clearly recognizable differences in form between species (owing to differences in the number, shape and three-dimensional arrangement of modules) reflect the operation of different rules governing development and growth, which have evolved in response to distinct selection pressures. For example, the phyllotaxis of a given species is a consistent character whatever the environmental conditions. The rules of 'self assembly' (the plant assembling itself, within the constraints of biomechanics, by reading its own genome or 'blueprint') are still poorly understood (e.g. Coen, 1999; Niklas, 2000).

Where the environment offers abundant resources, few physical or chemical constraints on growth, and freedom from major disturbance, the dominant species will be those which can grow to the largest size, thereby obtaining the largest share of the resource cake by overshadowing leaf canopies and widely ramifying root systems – in simple terms, trees. Over large areas of the planet, trees are the natural growth form, but their life cycles are long and they are at a disadvantage in areas of intense human activity or other disturbance. Under such circumstances, herbaceous vegetation predominates, characterized by rapid growth rather than large size. Thus, not only size but also rate of growth are influenced by the favourability of the environment; where valid comparisons can be made among similar species, the fastest-growing plants are found in productive habitats, whereas unfavourable and toxic sites support slower-growing species (Fig. 1.2).

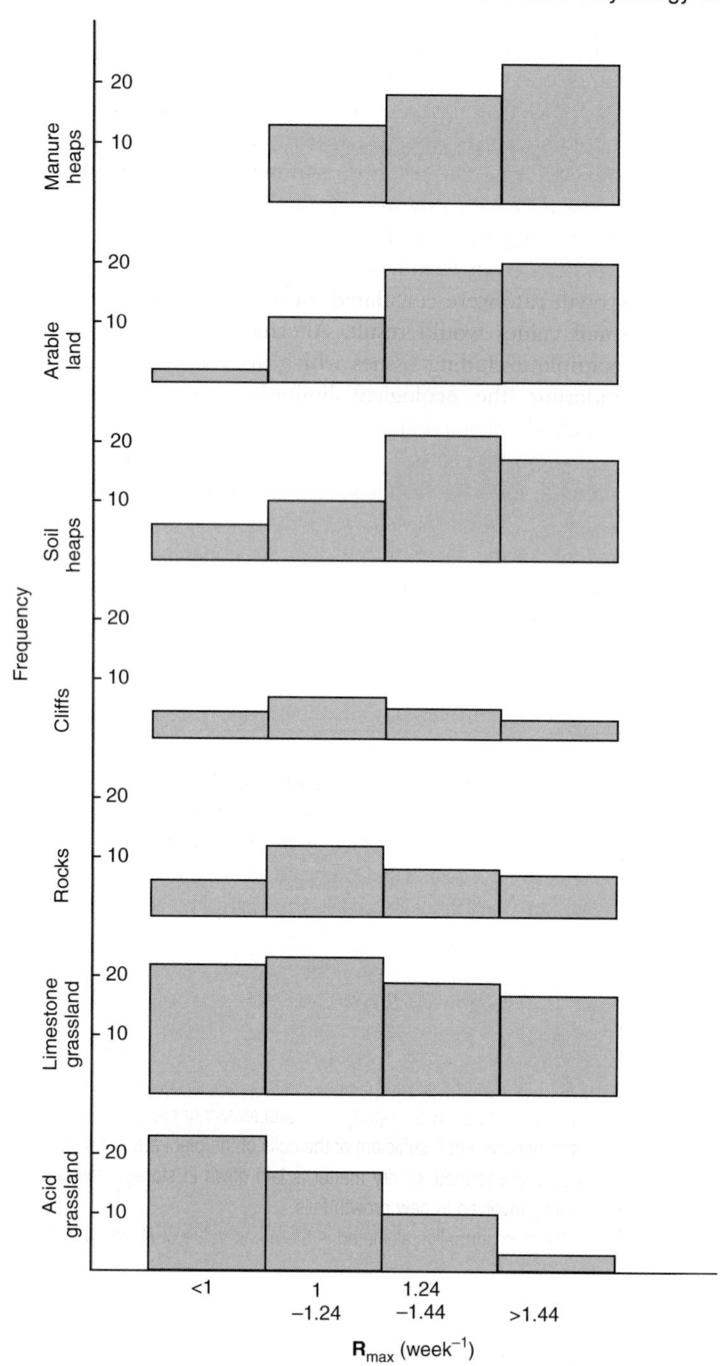

Figure 1.2

Frequency distribution of maximum growth rate $\mathbf{R}_{max}$ of species from a range of habitats varying in soil fertility and degree of stress (data from Grime and Hunt, 1975). Frequencies do not add up to 100% because not all habitats are included. Manure heaps and arable land are the most fertile and disturbed of the habitats represented.

The assumption (Box 1) that the growth rate of a plant is in some way related to its mass, as is generally true for the early growth of annual plants, is dramatically confirmed by the growth of a population of the duckweed *Lemna minor* in a complete nutrient solution (Fig. 1.4). The assumption is, however, not tenable for perennials. For example, the trunk of an oak tree contributes to the welfare of the tree by supporting the leaf canopy in a dominant position, and by conducting water to the crown, but most of its dry matter is permanently immobilized in dead tissues, and cannot play a direct part in growth. If relative growth rate were calculated for a tree as explained in Box 1, then ludicrously small values would result. Alternative approaches have been proposed, for example excluding tissues which are essentially non-living, but these serve to underline the ecological limitations of the concept. All plants use the carbohydrate generated by photosynthesis for a range of functions, such as support, resistance to predators and reproduction, with the result that growth rate is lower than the maximum potential rate; indeed such a maximum would be achieved by a plant consisting solely of meristematic cells. It is no accident that the fastest growth rate measured in an extensive survey by Grime and Hunt (1975) was for *Lemna minor*, a plant comprising one leaf and a single root a few millimetres long; or that the unicellular algae, the closest approximations to free-living chloroplasts, are the fastest-growing of all green plants.

1. Relative growth rate and growth analysis

The measure of growth used in Fig. 1.2 is relative growth rate (**R**), a concept introduced to describe the initial phase of growth of annual crops (Blackman, 1919; Hunt, 1982). Use of **R** assumes that increase in dry weight with time (t) is simply related to biomass (W) and, therefore, like compound interest, exponential (i.e. the heavier the plant, the greater will be the growth increment):

$$\mathbf{R} = 1/W \,.\, \mathrm{d}W/\mathrm{d}t = \mathrm{d} \ln W/\mathrm{d}t$$

Calculated in this way, **R** represents, at an instant of time, the rate of increase in plant dry weight per unit of existing weight per unit time. If growth were truly exponential, **R** would be constant, and a fixed property or characteristic of the plant; in reality, this is the case only for short periods when sufficient of the cells of the plant are involved in division. Once specialized organs are formed, or dry matter is laid down in storage, the proportion of plant dry weight directly involved in new growth falls.

What is normally calculated is the mean value of **R** over a period of time:

$$\bar{\mathbf{R}} = (\ln W_2 - \ln W_1)/(t_2 - t_1)$$

This equation is useful when comparing the growth of plants of different size, but since growth is usually exponential only in the very early stages, the values of **R** obtained are continually changing, and usually declining.

An alternative approach to growth analysis, pioneered by Hunt and Parsons (1974) involves fitting curves to dry weight data obtained at a series of time intervals, and calculating instantaneous values of **R** at intervals along the curves. Figure 1.3 below illustrates the characteristic steady decline in **R** as the growing season proceeds, calculated in this way.

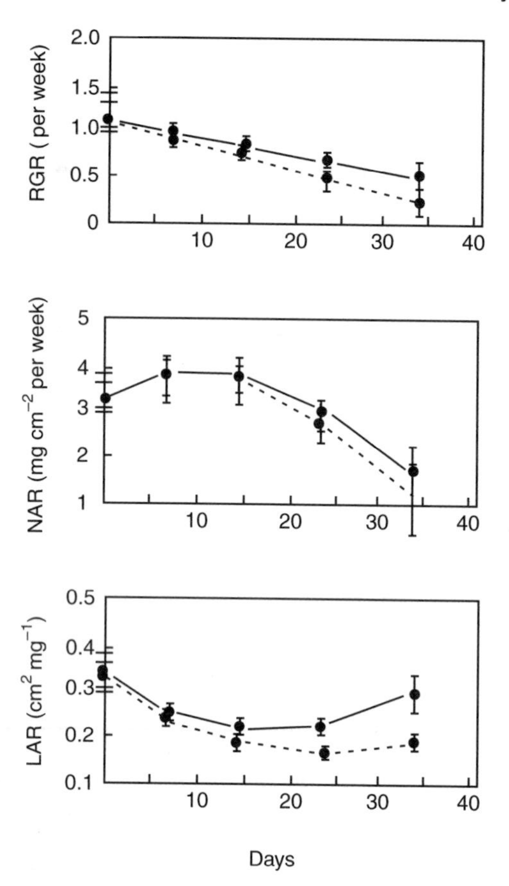

Figure 1.3

Relative growth rate (RGR) and its components, net assimilation rate (NAR) and leaf area ratio (LAR), of plants of *Phleum pratense* cv. Engmo grown at a constant temperature (15 °C) and 8h (- - -) or 24h (—) daylength. The error bars indicate confidence limits (from Heide *et al.*, 1985).

Values of **R** should be calculated on a whole plant basis, including below-ground biomass, but, for practical reasons, most estimates are based on above-ground tissues only, and should be referred to as shoot **R**.

As explained in Chapter 2, growth analysis can be extended to provide more powerful tools in the interpretation of plant growth, by resolving **R** into net assimilation rate and leaf area ratio (leaf weight ratio × specific leaf area) (see Fig. 1.3).

During the later stages of growth of a plant stand or crop, if the interception of solar radiation by the canopy is complete, the increase in dry matter with time will tend to be linear, unless growth is limited by another environmental factor such as water or nutrient stress. Absolute growth rate can then be used:

$$\mathbf{A} = W_2 - W_1/t_2 - t_1$$

A is widely used in crop physiology, where the emphasis is on the maximization of interception of solar radiation. Over a given time interval, it can be resolved into: intercepted solar radiation and radiation use efficiency (g dry weight gained per unit of radiation) (Hay and Walker, 1989).

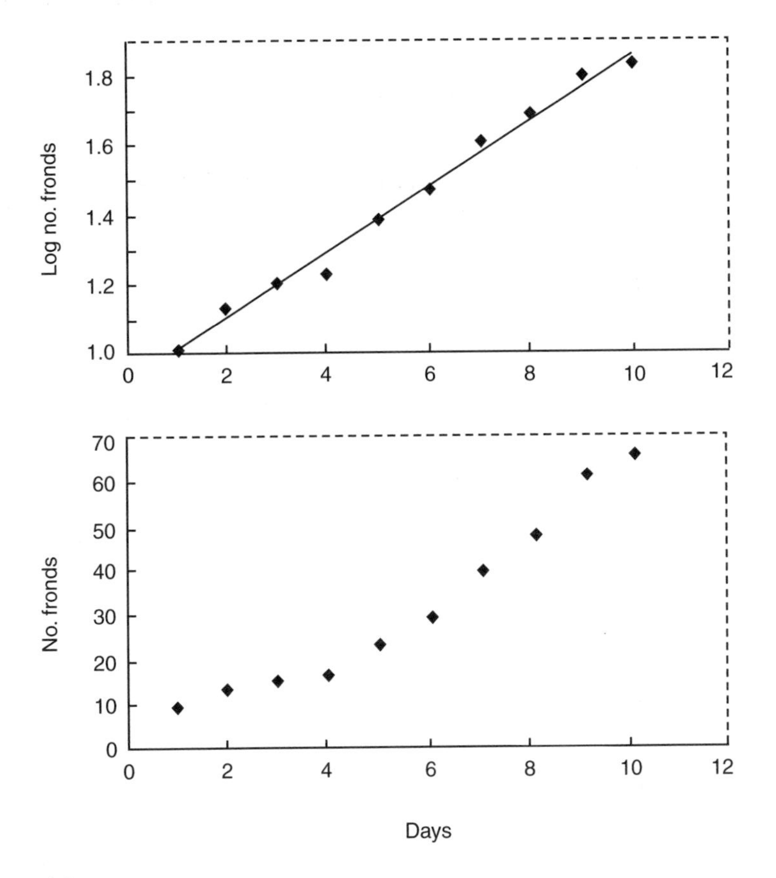

Figure 1.4

Growth of duckweed (*Lemna minor*) in uncrowded culture. The growth rate (based on frond numbers since frond dry weight remains constant) is 0.20 d^{-1} and is represented by the slope of the plot of *ln* numbers against time (d *ln N*/d*t*) (from data of Kawakami *et al.* (1997). *J. Biol. Educ.* **31**, 116–118).

Relative growth rate can, therefore, be used as an indicator of the extent to which a species is investing its photosynthate in growth and future photosynthesis (the production and support of more chloroplasts), as opposed to secondary functions, such as defence, support, reproduction, and securing supplies of water and mineral nutrients. In many habitats, usually unfavourable or toxic ones, growth can actually be disadvantageous; here the emphasis is on survival, and priority is given to the securing of scarce resources, or protection from grazing or disease. These characteristics, which are features of plants from deeply shaded (Chapter 2), very infertile (Chapter 3), very dry (Chapter 4), very hot or cold (Chapter 5) or toxic (Chapter 6) environments, are termed conservative.

2. The influence of the environment

Research on the physiology of plants is normally conducted under controlled conditions, where the environment is engineered to remove all constraints to growth: under such conditions, the growth rate of control plants is 'optimal' or 'maximum' (highest inherent rate), and the influence of environmental factors can be assessed in terms of their ability to depress growth rate. Comparisons among species reveal that there can be a tenfold variation in maximum growth rate (Fig. 1.2), largely because of variation in the proportions of photosynthate re-invested in photosynthetic machinery. Thus, fast-growing annual plants direct most of their photosynthate successively into above-ground leaves, flowers and fruit. In contrast, the temperate umbellifer pignut (*Conopodium majus*) scarcely progresses beyond the emergence of the cotyledons in the first year of growth, with surplus photosynthate being stored in an underground storage organ (the 'pignut'); in the next season, the stored resources enable it to produce leaves and reproductive structures rapidly in early spring. Taking the conventional approach, very low relative growth rates would be recorded in the first season, because dry matter is being invested in storage rather than leaves, which could create more biomass. Here there is an important interaction between development and growth.

Even under non-limiting conditions, therefore, species vary markedly in their use of resources, and in their patterns of growth and development. In natural habitats, such conditions are rare, and the supplies of the different resources for life are, typically, unbalanced. For example, the uppermost leaves of the C_3 leaf canopy in Fig. 1.5(a) would be unable to make full use of even moderate photon flux densities (>500 $\mu mol\ m^{-2}\ s^{-1}$ photosynthetically active radiation or *PAR*) because of limitations in the supply of CO_2 from the atmosphere (around 360 $\mu l\ l^{-1}$). Although normally light-saturated at higher photon flux densities, the leaves could achieve higher rates of photosynthesis if the CO_2 concentration were higher. In contrast, the photosynthetic rate of the C_4 leaves in Fig. 1.5(b) reached a plateau at 150 $\mu l\ CO_2\ l^{-1}$ under low light (300 $\mu mol\ m^{-2}\ s^{-1}$ *PAR*) but much higher rates at CO_2 concentrations above 150 $\mu l\ l^{-1}$ could be achieved with increased supplies of PAR. The rates of flux of CO_2 required to satisfy the light-saturated rates of photosynthesis in Fig. 1.5(a) could be achieved only if the stomata were fully open, but this would lead to rapid loss of leaf water, exposure to water stress, and a reduction in influx of CO_2 as a consequence of stomatal closure (Chapter 4). Thus, under different combinations of factors, rates of photosynthesis and growth can be limited by solar radiation, CO_2 supply, water relations, or even the mineral nutrient status of the leaf.

In some habitats, limitation of plants or plant communities by a specific environmental factor can be demonstrated by the increases in growth observed when the factor is alleviated; the rate rises to the point where some other factor becomes limiting (e.g. Fig. 1.5). However, it is probably more common for two or more factors to contribute simultaneously to the limitation, and only when both or all are alleviated will there be a response (e.g. Figure 1.6). Such interactions ensure that the adaptive responses made by plants to their environment are complex.

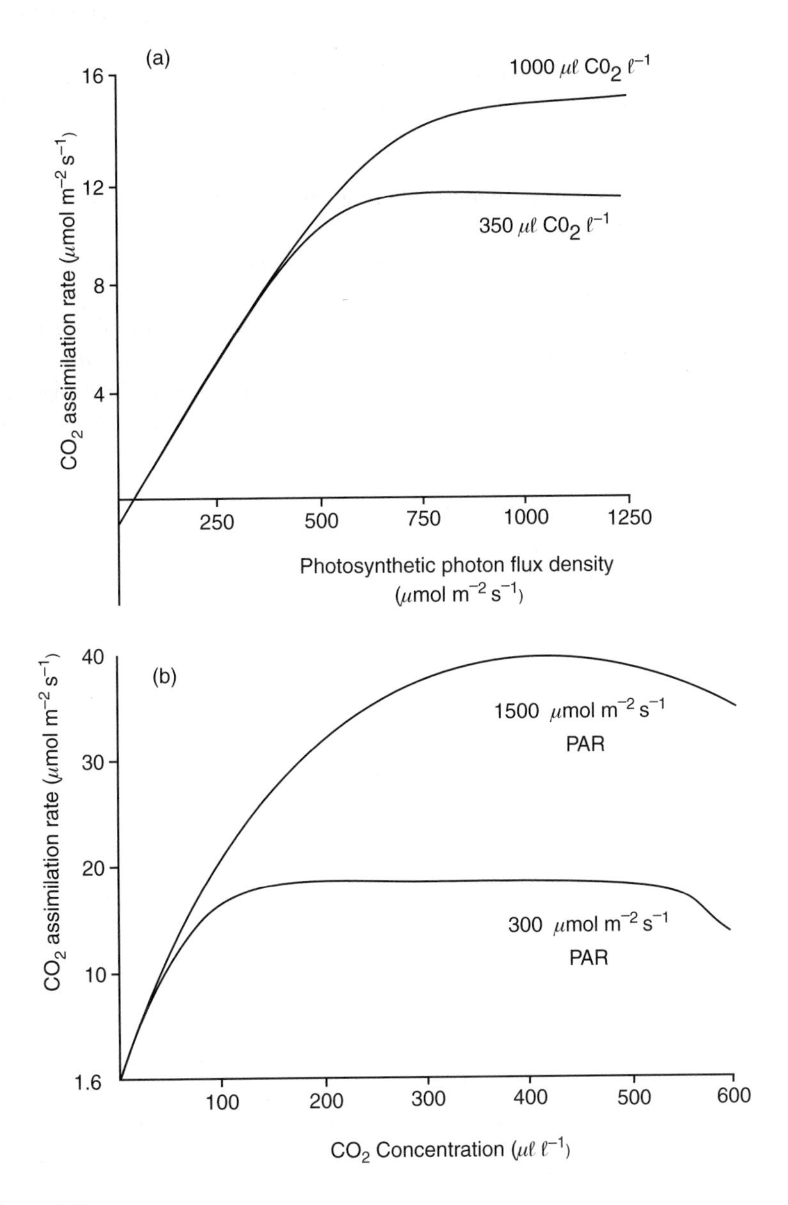

Figure 1.5

Responses of the rate of net photosynthesis to variation in environmental conditions: model responses of the leaves of (a) a temperate C_3 species (e.g. wheat), and (b) a tropical C_4 species (e.g. sorghum), to variation in the supply of photosynthetically active radiation (PAR) and CO_2. Note that assimilation at high concentrations of CO_2 will be affected by stomatal closure (see p 149). The contrasting physiologies of C_3 and C_4 plants are considered in detail in Chapters 2 and 4.

Understanding the environmental physiology of a plant can be particularly difficult where the responses to different factors are in conflict. For the leaves illustrated in Fig. 1.5, the maintenance of an adequate supply of CO_2 to the chloroplasts requires the stomata to be fully open, thereby exposing the leaf to the risk of excessive water loss. It is likely, therefore, that there has been strong selection for optimization of stomatal function: balancing the costs and benefits of stomatal opening (Cowan, 1982). Chapter 7 includes an exploration of the extent to which the concepts of economics and accountancy (investment of resources etc.) can be used to evaluate the costs and benefits of complex plant responses.

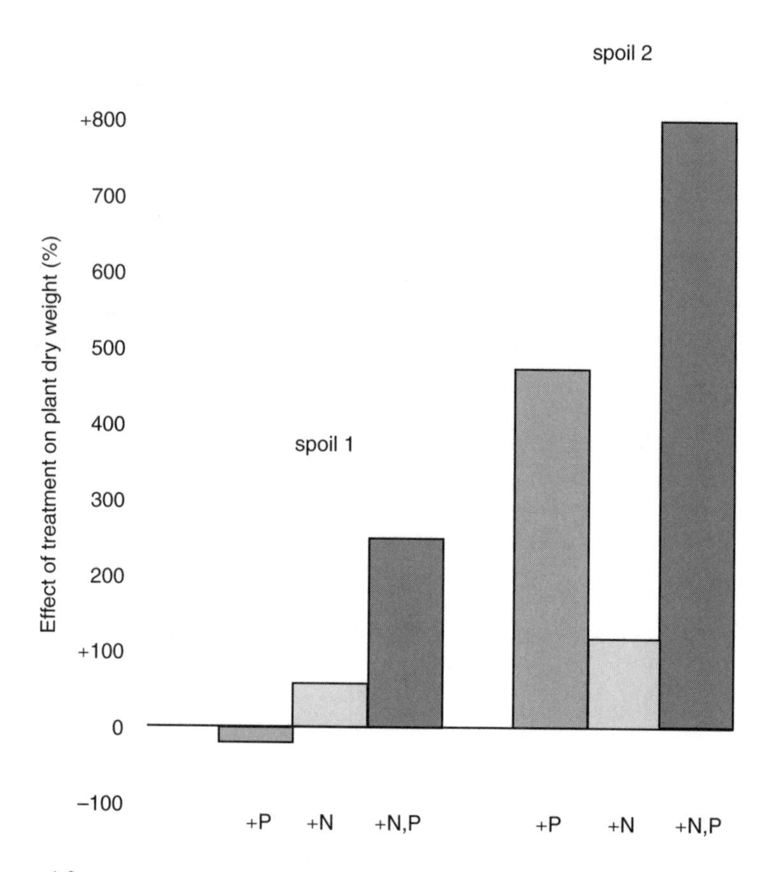

Figure 1.6

Differing patterns of interaction between the effects of phosphorus and nitrogen supply on the growth of plants: responses of plants of *Lolium perenne* growing in pots of extremely nutrient-deficient colliery spoil (receiving 2.5 kg ha^{-1} each of N and P) to the addition of further N (25 kg ha^{-1}), P (25 kg ha^{-1}) or both. The growth of the control plants in Spoil 1 (0.40 g/pot) was higher than that in Spoil 2 (0.07 g/pot). In each medium, both nutrients were required for the full stimulation of growth and, in Spoil 1, the application of P alone actually depressed growth (from data of Fitter, A. H. and Bradshaw, A. D. (1974). *J. Appl. Ecol.* **11**, 597–608).

The analysis of responses can also be complicated when the limiting factors vary with time. For example there are considerable diurnal variations in temperature, supply of solar radiation and leaf water status, even in temperate areas, but such variations reach an extreme form in tropical montane environments (Fig. 5.15; Plate 14) where the plants can experience 'winter' and 'summer' each day: night temperatures are so low that frost resistance is necessary but, from sunrise, irradiance and temperature rise sharply, such that photosynthesis can be limited by photoinhibition (see p. 57), CO_2 supply, or water and mineral deficiencies (owing to frozen soil). By mid-day, under very high radiant energy flux, the stomata will close, restricting the uptake of CO_2, and exposing the leaves to potentially damaging high temperatures. On other days, low cloud can result in conditions where photosynthesis is limited by the supply of solar radiation.

Variability of this scale demands enormous flexibility of the physiological systems of plants, at timescales from the almost instantaneous upwards. In any habitat, there will be significant fluctuations within the lifetime of any individual plant. Where the fluctuation is sufficiently predictable, it may be dealt with by rhythmic behaviour (for the many diurnal fluctuations) or by predetermined ontogenetic changes, such as the increase in dissection of successive leaves of seedlings emerging from shaded into fully-illuminated conditions (Chapter 2). The timing of such ontogenetic changes and the duration of the life-cycle may be highly plastic (see Box 2), and represent major components of adaptation to temporal fluctuation. Thus the environmental control of autumn-shedding of leaves by temperate deciduous trees is confirmed by the retention of functional leaves under artificially extended photoperiods.

Damage and plant response

Most habitats are potentially hazardous to plants; for example, as noted in Chapter 4, exposure to water stress is a routine experience for terrestrial plants. The resulting damage can vary from reduced growth caused by physiological malfunction, to the death of all or part of the plant tissues, but there are striking differences, among species and among populations, in the degree of damage sustained in a given habitat. By definition, all species that survive in a habitat must be able to cope with the range of environmental variation within it, but a rare event, such as an unseasonable frost or extreme drought, can cause the extinction of species that are otherwise well-adapted to the habitat. In other words, the niche boundary of these species will have been exceeded (Fig. 1.7), and large differences in the ability to survive such events can be predicted.

The occurrence of significant damage implies a lack of resistance to the relevant environmental factor. Resistance can be conferred by molecular, anatomical or morphological features, or by phenology (the timing of growth and development); it is a fundamental component of the plant's physiology and ecology, and differences in resistance are responsible for all major differences in plant distribution. The critical feature is that such resistance is constitutive: a particular enzyme will be capable of operating over a certain range of temperature, or concentration of toxin, outside of which damage will occur (e.g. Table 5.5; Figs. 6.2, 6.11). Resistance can be viewed as a factor in

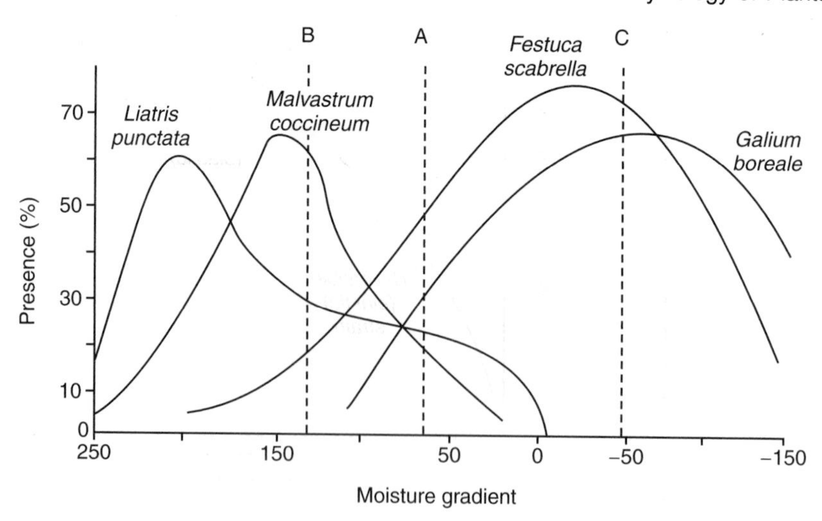

Figure 1.7

Niche relationships of four prairie species in relation to a moisture gradient (the *x*-axis is a statistical representation of the gradient). At A, all four species can co-exist but deviation towards B (drier) will lead to the extinction of *Galium boreale* and towards C (wetter) to the loss of *Liatris punctata* and *Malvastrum coccineum* (from data of Looman, J. (1980). *Phytocoenologia*, **8**, 153–190).

homeostasis, permitting the plant to maintain its functions in the face of an environmental stimulus, without apparent physiological or morphological changes. Outside the limits of resistance, the plant will sustain obvious damage.

Adaptive responses are the fine control on such constitutive resistance to damage. They involve a shift of the range over which resistance occurs, and such shifts can be reversible (usually metabolic/physiological, e.g. Figure 5.6) or irreversible (usually morphological, e.g. Figs 2.13 and 2.14). Both traits (resistance, and the potential for the adaptation of resistance) are permanent features of the genotype, having evolved under the particular selection pressures of the habitat. Thus, although resistance is a fixed feature of the phenotype, individual plants or populations of a species can appear and behave quite differently according to the degree of adaptation evoked by the environment. It has become customary to use the terminology of physics in the analysis of adaptation (Box 2).

2. Stress and strain

The application of a mechanical force (compression or tension stress) to a solid body causes deformations that can be reversible (elastic strain) or irreversible (plastic strain) when the stress is withdrawn. Thus a copper wire, or an elastic band, under increasing tension first undergoes reversible stretching, followed by irreversible stretching and, ultimately, failure (Fig. 1.8).

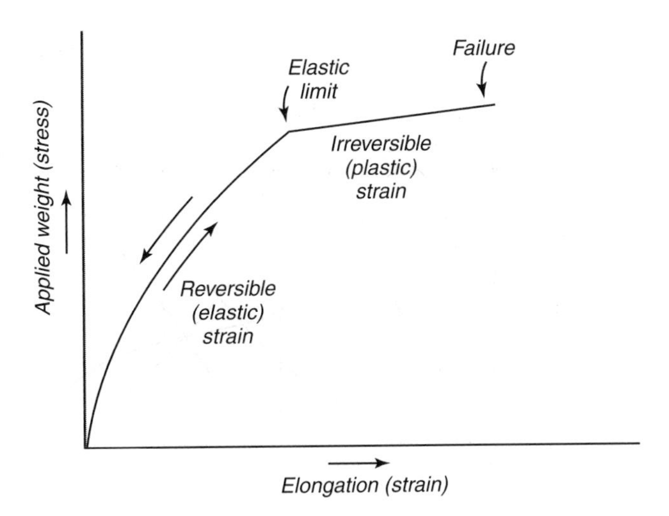

Figure 1.8

The effect of increasing the weight applied to an elastic band on its length. Up to the elastic limit, removal of the weight will permit the band to return to its original dimensions. Once the band has passed its elastic limit, permanent damage will be sustained, which can not be repaired by removal of the stress.

Levitt adapted these concepts of stress and strain to aid in the interpretation of plant responses (strain) to the application of environmental stresses (e.g. Levitt, 1980), but, although the term 'stress' is now widely used in plant physiology, 'strain' is rarely encountered. The term 'plasticity' is also widely used, and sometimes abused (e.g. referring to reversible metabolic changes), but the term 'elasticity' is used only in the strict physical sense (as in the characterization of cell expansion; Chapter 4).

Shading stress (reduction in irradiance) can induce reversible/'elastic' changes (i.e. strains) in the light compensation point and the photosynthetic efficiency of the leaves of woodland plants (e.g. Figs. 2.16, 2.17). These changes are the 'fine control' of the constitutive resistance of such species, whose photosynthetic apparatus is already geared to low irradiance, and they facilitate exploitation of the very variable light environment of the forest floor. Applying the same stress to a weed or crop plant, which is not intrinsically resistant to low irradiance, will induce irreversible (plastic) morphological responses (notably internode extension). The non-adaptiveness of such responses (i.e. the unlikelihood of being able to grow through the tree canopy) is considered in Chapter 2.

For shading, both stress and strain can be quantified independently, in terms of irradiance (stress) and photosynthetic parameters, or internode length (strain). It is, therefore, possible to construct stress/strain diagrams analogous to those used in physics (e.g. Fig. 2.16). Similar quantification is possible for temperature (e.g. Figs. 5.2, 5.11), concentration of toxins (e.g. Figs. 6.2, 6.5), and supply of oxygen, but not in studies of water relations because of the difficulty of expressing the degree of water stress in terms of environmental rather than plant parameters (Chapter 4).

Nevertheless, the analogy between physics and physiology fails ultimately because of the more dynamic attributes of plants: they are alive and have the capacity to *replace* irreversibly damaged tissues by new growth (addition of new modules).

Phenotypic plasticity of morphology (i.e. irreversible changes in response to environmental cues, Box 2) is a universal feature of plants; outstanding examples, such as the heterophylly of water buttercups (*Ranunculus aquatilis*) (feathery submerged leaves, entire aerial leaves; Fig. 2.14) and of certain eucalypts (juvenile rounded shade-tolerant leaves succeeded by drought-resistant strap-like leaves) are well known. Although *reversible* changes in phenotype are also ubiquitous, they are less obvious. Examples include changes in the concentration of enzymes, particularly inducible enzymes such as nitrate reductase (Chapter 3); and behavioural responses such as the opening and closing of flowers and compound leaves (Chapters 2, 4), or the diurnal tracking of the sun, either to maximize or minimize interception of solar radiation (Chapters 2, 5, 6). Even the ability to enter, or not, into a symbiotic relationship, such as a mycorrhiza (Chapter 3), can be viewed in terms of plasticity. These phenomena may be direct responses to the environment (irradiance, temperature, nutrient supply) or the consequence of endogenous rhythms which can continue without being reset by an environmental cue.

Each individual plant, therefore, has access to a range of responses to environmental fluctuation. Clearly, all must have an ultimate molecular basis, but it is possible to classify them according to whether the molecules deliver the adaptation directly, or act by creating structures or behavioural patterns which are adaptive. Resistance to injury is most easily classified in this way: either the metabolically-active molecules are themselves resistant to stress (e.g. the enzymes of thermophilic bacteria), or they are protected from damage by other molecules, special structures, or patterns of behaviour. The tools of molecular genetics are now being deployed to elucidate plant response at the molecular level (e.g. the effects of heavy metal ions on aquaporins, Fig. 6.3; evaluation of the roles of heat shock, and low temperature response, proteins, Chapter 5). A wide range of possible responses is reviewed in Table 1.1; the types of response shown by a given plant depend upon the way in which the environmental stimulus is presented.

3. Evolution of adaptation

Plants that survive in their habitats are clearly adapted; to that extent, the term adaptation is redundant. Körner (1999a), for example, suggests that alpine conditions are not stressful to alpine plants. Their physiology and ecology are so closely attuned to the harsh conditions that they survive better under those conditions than under the apparently more favourable conditions at low altitude, a fact well-known to gardeners. Nevertheless, plants are never perfectly adapted; for example, photosynthetic processes show widely differing levels of adaptation to high temperature (Table 5.5). This apparent mal-adaptation may arise from several causes: because the plant lacks the genetic variation required to produce a better 'fit' to the environment ('phylogenetic constraint'); because in practice other steps in a metabolic or developmental pathway are more sensitive to the environment and more critical to plant survival; or because the environment is spatially and temporally heterogeneous (and unpredictably so), and the character in question is well-suited to some other set of conditions. Selection acts differentially, at the level of the individual plant, and not at the level of the organ, response or process. The various components of an organism

Table 1.1
A classification of responses to environmental stimuli

Response		In response to variation in	
		Time	Space
Changes in amounts of molecules	*e.g. Rates of protein synthesis and degradation*	• Antioxidant production in response to stress (Chapter 6) • Levels of intracellular solutes in cryo- and osmo-protection (Chapters 4 and 5) • Unsaturated fatty acids in membranes in response to cold (Chapter 5)	• Rubisco activity in shaded and unshaded leaves (Chapter 2) • Carrier molecules for ion transport (Chapter 3) • Secretion of protons and organic compounds that modify the rhizosphere (Chapters 3 and 6)
Changes in types of molecules	*e.g. Patterns of gene expression*	• Reduced frequency of –SH groups in enzymes of cold-hardened and drought-resistant plants (Chapters 4 and 5) • Defence compounds against pathogens (Chapter 7) • Switch from C_3 to CAM photosynthesis following drought (Chapters 2 and 4)	• Synthesis of photoprotective pigments in full sun (Chapter 2) • Expression of symbiosis genes in roots colonized by rhizobium or mycorrhizal fungi (Chapter 3)
	e.g. Post-translational activation/inactivation of enzymes, e.g. by phosphorylation	• Circadian control of CAM (Chapters 2 and 4) • Inactivation of alternative oxidase by oxidation of –SH groups to –SS-bonds (Chapter 5) • Activation/inactivation by protein kinases	• Movement of light-harvesting complexes from PSII to PSI (Chapter 2)
Behavioural responses	*e.g. Movements*	• Leaf and stomatal movements (Chapters 2 and 4) • Sun-tracking by flowers (Chapters 2 and 5)	• Foraging responses of stolons and roots to nutrients (Chapter 3)
Developmental responses	*Phenotypic plasticity*	• Resource allocation changes in response to shading (Chapter 2) or drought (Chapter 4)	• Aquatic heterophylly (Chapter 2) • Aerenchyma production in waterlogged soil (Chapter 6)

may be well or poorly adapted in a mechanistic sense, but it will have the opportunity to reproduce only if the sum of the components is sufficiently suited to the environment, and marginally more suited than that of its competitors. In this context, it is important not to push the concept of 'optimization' of the physiology of a population of plants too far.

There is abundant evidence that when plant populations are exposed to novel environmental conditions they evolve more adapted genotypes. This is well known for plants on metal-contaminated soils and those exposed to air pollution (Chapter 6, p. 281), and where fertilization creates distinct nutritional environments (Chapter 3, p. 128), among others. The selection pressures involved can be very large and, consequently, such evolutionary differentiation can occur over very short distances, as little as a few centimetres. Such patterns demonstrate that selection pressures are large enough to counter the effects of gene flow. It is less obvious that similar selection pressures act where there are no such imposed environmental patterns. However, Nagy (1997) showed that natural populations are exposed to high levels of stabilizing selection. He crossed two subspecies of *Gilia capitata* (Polemoniaceae) and planted the resulting F2 hybrids into the habitats of the two subspecies. Their offspring had a common evolutionary response across a range of characters: they evolved in the direction of similarity to the subspecies that was native to the site. In other words, the native character states were adaptive.

Even though selection may promote differentiation of genetically distinct populations (ecotypes) on adjacent, but environmentally contrasting, sites, there are strong forces discouraging this process. All environments are heterogeneous: they show variation in both space and time, and commonly on scales that are small relative to the size of plants (see Fig. 3.6, p. 85). Consequently, an individual plant or, in a clonal species, a genotype may experience very contrasting conditions. Short-lived plants may be less likely to experience temporal fluctuation, but they tend to have wide seed dispersal and therefore their offspring may encounter very different habitats. Long-lived plants are bound to experience temporal variation and are commonly large, thus increasing their exposure to spatial heterogeneity.

In many, perhaps most, environments, fitness will be maximized by characters which allow the organism to track environmental fluctuations and patchiness, rather than those which render it suited to one particular set of factors. Indeed, in some habitats, survival may depend on the ability to survive occasional extreme events. Thus, although resistance to stress is of central importance, phenotypic plasticity of processes and structures will contribute strongly to the fitness of particular individuals by extending the environmental range over which the plant can survive. This is particularly true of plants since they are sessile, and liable to experience greater temporal variation than more mobile animals; it is elegantly illustrated by a study by Weinig (2000) on velvetleaf *Abutilon theophrasti*, a common weed in north America. Plants from fields in continuous corn grew slightly taller at the seedling stage than those from corn–soybean rotations or weedy fields. This can be interpreted as an ecotypic differentiation to the more severe competition for light experienced by the seedlings in continuous corn fields. However, for all populations, the elongation stimulated by shading was much greater than the differences in height between the populations. In other words, plasticity is a more effective response to this stress, and the existence of plasticity tends to suppress the development of ecotypes. Importantly, the degree of plasticity was also different between fields; plants from continuous corn fields showed less plasticity in the elongation of later internodes. This response was also interpreted as adaptive because velvetleaf cannot grow taller than corn, so that later elongation will have a cost but offer no benefit. These results emphasize that plasticity is a trait that undergoes selection.

4. Comparative ecology and phylogeny

By its nature, then, environmental physiology is about the adaptation of plants to existing habitats, and their ability to survive wider amplitudes of environmental factors within the existing range of phenotypes. Nevertheless, it would be perverse to study the interactions between environmental and physiological processes without considering the evolutionary framework in which these changes might have come about. The subject is essentially a comparative one: in order to see the diversity of physiological responses that has evolved, it is essential to examine a wide range of species growing in a variety of habitats. The study of comparative ecology and phylogeny, which have been integral components of plant ecology since its foundation, has recently assumed an enhanced role with the development of new taxonomic and molecular tools (Ackerly, 1999).

The adoption of similar morphologies and physiologies, by distantly-related species, growing in similar habitats, at widely-separated locations, provides strong evidence that such characteristics are adaptive. As already noted, the giant rosette habit, with associated frost and drought tolerance, which is found in Old and New World tropical montane zones (Chapter 5), is a spectacular example of such convergent evolution. The taxonomic distances among these species indicate that the set of adaptations has evolved independently at different times. Similarly, there are marked similarities among the xerophytic and fire-resistant species in the plant communities of mediterranean zones in Europe, the Americas, South Africa and Australia (Mooney and Dunn, 1970) (Chapters 4 and 5).

Taking the opposite approach, study of the divergent morphologies and physiologies of closely related species, for example by reciprocal transplanting, has also provided evidence for the adaptive nature of characters. For example, the Death Valley transplant experiment described in Chapter 4 demonstrated the differing metabolic adaptations of species of the genus *Atriplex*; and the varying morphologies of *Encelia* species have facilitated understanding of the roles of colour and pubescence in thermal and water relations (Chapters 4 and 5).

Deployment of crassulacean acid metabolism (CAM, see p. 59, 176) permits plants to continue to assimilate CO_2 without opening their stomata during daylight hours, thus reducing water loss by transpiration, and the reduction in assimilation caused by midday closure of stomata under water stress (see p. 153). The adaptive value of CAM in dry habitats is confirmed by the divergent photosynthetic characteristics of closely related tropical trees of the genus *Clusia* (Fig. 1.9). In well-watered *C. aripoensis* (moist montane forest),

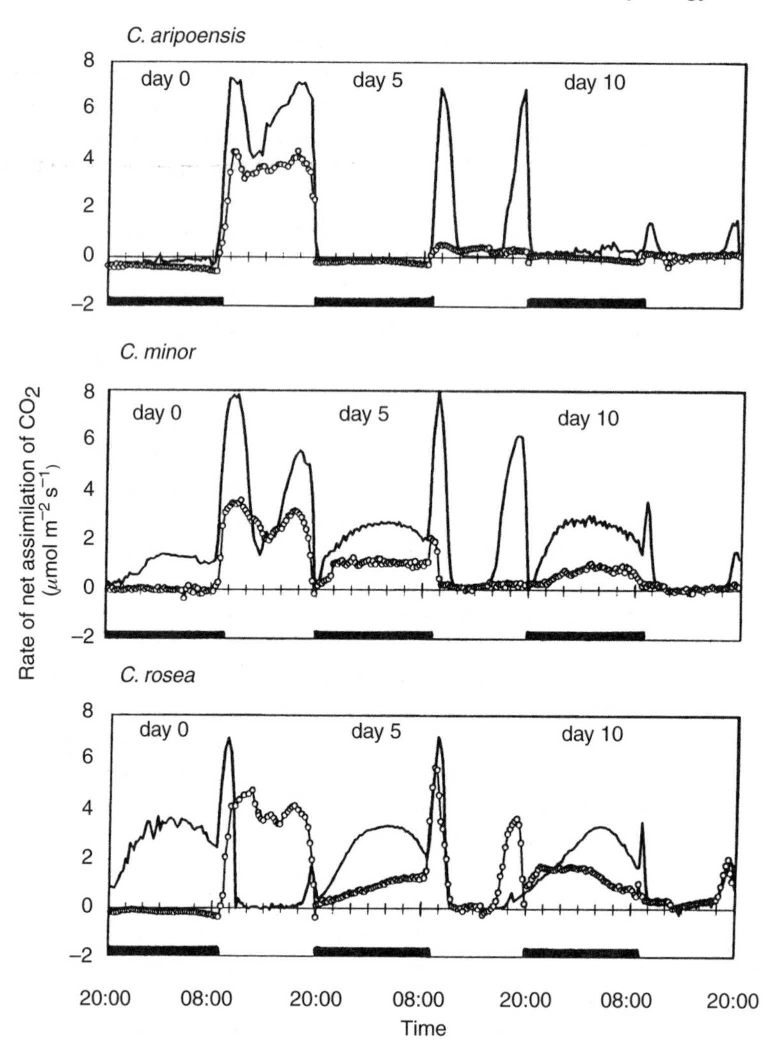

Figure 1.9

Rates of net CO_2 assimilation in young (O) and mature (——) leaves of young trees of species of *Clusia*, native to Trinidad, raised under identical conditions in a growth chamber. The measurements were made under well-watered conditions (day 0) and after 5 and 10 days without further irrigation. The solid bar on the *x*-axis indicates the duration of darkness. See text for full description (reprinted from Borland, A.M., Tecsi, L.I., Leegood, R.C. and Walker R.P. Inducibility of crassulacean acid metabolism (CAM) in *Clusia* species, *Planta* **205**, © 1998, with the permission of Springer Verlag, Berlin.)

assimilation by the C_3 pathway was restricted to daylight hours; drought caused prolonged midday closure of stomata, and eventually brought about a very limited induction of CAM on day 10 (weak CAM-inducible). In *C. minor* (drier lowland deciduous forest), mature and young leaves fixed CO_2 by both the C_3 and CAM pathways from day 0, with the contribution from the C_3 pathway decreasing progressively as water stress intensified (C_3-CAM intermediate). In *C. rosea* (dry rocky coast), mature leaves relied almost completely on night-time fixation of CO_2 by the CAM pathway, showing the characteristic spike of assimilation after dawn, from day 0 (constitutive CAM). However, young leaves used the C_3 pathway on day 0, relying more on CAM on days 5 and 10. Integration of the areas under the curves for mature leaves shows that the two species deploying CAM assimilated progressively more CO_2 than *C. aripoensis* as the drought intensified.

Similarly, by using more precise techniques for estimating the evolutionary distance between species, it is now possible to show that the C_4 photosynthesis syndrome (p. 59) has evolved independently many times in dry environments, even in closely related species (Ehleringer and Monson, 1993). The application of advanced statistical techniques now permits identification of ancestors which first acquired certain characteristic traits (Ackerly, 1999). The key development is the availability of reliable phylogenies based on molecular genetic information. The classification of plants is fundamentally based on morphology. When constructing classifications, taxonomists have, traditionally, given greatest weight to characters of the reproductive system, on the argument that these are the most stable in evolution. Whereas leaf, stem or root features might evolve rapidly, say within a genus, in response to a local selection pressure, this would less often be true of reproductive characters, because such a change would make it less likely that an individual would be able to exchange gametes with another. This conservatism was held to be especially true of certain fundamental features of plant reproductive systems, such as the number of carpels and the shape of the flower. However, there was no reliable and independent way of checking these assumptions, beyond the rather inadequate fossil record, especially when considering evolution within low-level taxonomic groups such as genera or families. Equally seriously, there was no way of estimating the rate of evolution within particular groups. This situation has been overturned by the availability of gene sequence data. It is now possible to examine a wholly independent data set of, for example, the sequence of the large subunit of the photosynthetic enzyme Rubisco (rbcL: Chase *et al.*, 1993). Species that have diverged recently in evolution will have more similar gene sequences than those that diverged further back in time.

These techniques allow a detailed analysis of the evolutionary patterns within groups of plants. For example, Ackerly and Donoghue (1998) examined the evolution of a number of characters in the genus *Acer* (sycamore and maples) using a molecular phylogeny of the genus. They were able to show that some characters had evolved early in the history of the genus: one example was the angle of bifurcation between the shoots that grow out below the apical meristem. In Japanese maples (section *Palmata*) this angle is very large (>65°) because the apical meristem dies each season. The resulting tree architecture is quite distinct from that of other maples, which have narrow bifurcation angles (45–60°). In contrast, leaf size is a character that appears to change frequently, presumably as species evolve in distinct environments where powerful selection pressures apply, and closely related species may therefore have very different-sized leaves. On this basis, it could be said that shoot architecture determines the ecological niche of maples, whereas leaf size is determined by it.

In the chapters that follow, we discuss a range of adaptations of physiological processes to environmental conditions. Underlying all of these discussions is the assumption that they are the result of natural selection. It should, however, not be assumed that selection acts on such processes; it acts on organisms, on entire phenotypes. A plant with the most exquisitely optimized phenotype with respect to water use efficiency may not survive to reproduce if, for whatever reason, it has an ineffective defence against pathogens or grazing animals. The entire genotype will be lost, whereas another phenotype, apparently inferior in terms of adaptation, may have greater fitness in practice. Equally, it would be a mistake to assume that selection acts on a single function or structure: the non-glandular hairs on a leaf may contribute to plant fitness by altering its energy balance (Chapters 2, 4, and 5) or by deterring herbivores, or for both reasons. However, if another trait renders the plant vulnerable to stress, any improvement in the matching of the physiology of the hair-carrying leaf to its environment will be in vain.

Part I

The Acquisition of Resources

2

Energy and Carbon

1. Introduction

Photosynthesis is fundamental to plant metabolism, and the acquisition of radiant energy and CO_2 is critical to the ecological success of a plant. Radiation that is photosynthetically active (*PAR*) roughly corresponds to visible light, but both represent only a small part (*c.* 400–700 nm) of the full solar radiation spectrum (Fig. 2.1), and plants are also sensitive to other wavelengths: for example far-red radiation (far-red 'light' is a convenient misnomer) of wavelength *c.* 700–800 nm strongly influences morphogenesis (Smith, 1995). Radiation affects organisms by virtue of its energy content and is active only if absorbed. Thus, ultraviolet 'light' is strongly absorbed by proteins and can cause damage; blue light is absorbed by carotenoid pigments and chlorophyll, red light by chlorophyll, and both red and far-red by phytochrome. The existence of pigments, therefore, is basic to any response and most plants appear green simply because most plant pigments absorb green light weakly.

At longer wavelengths one can no longer think in terms of pigments (which of course strictly refer to only the visible range), since long-wave radiation is absorbed by all plant tissues, with consequent heating. The energy budgets of plant organs are discussed in Chapter 5; they are of great importance in regulating the temperature of plants, particularly in extreme climates. In many situations there is a conflict between the need to intercept light for photosynthesis and the resulting increases in leaf temperature. Energy loss, by convection and evaporation, then becomes paramount; consequently there may be benefits from both changes in leaf morphology which increase convective loss, and changes in transpiration rate which increase evaporative loss of energy, despite their often deleterious effect on the absorption and utilization of radiant energy for photosynthesis.

Because of this dual effect of solar radiation – in supplying the energy for metabolism and in influencing the temperature of plants – responses to sunlight may have no photosynthetic or photomorphogenetic basis. For example, flowers in Arctic regions, such as *Dryas integrifolia* and *Papaver radicatum*, are saucer-shaped and follow the sun, acting rather in the manner of a radio telescope, so concentrating heat on the reproductive organs in the centre of the flower and attracting pollinating insects to these 'hot spots'. A temperature differential of 7 °C or more is frequently attained between flower and air, and a temperature of 25 °C has been recorded (Kevan, 1975; cf. Chapter 5).

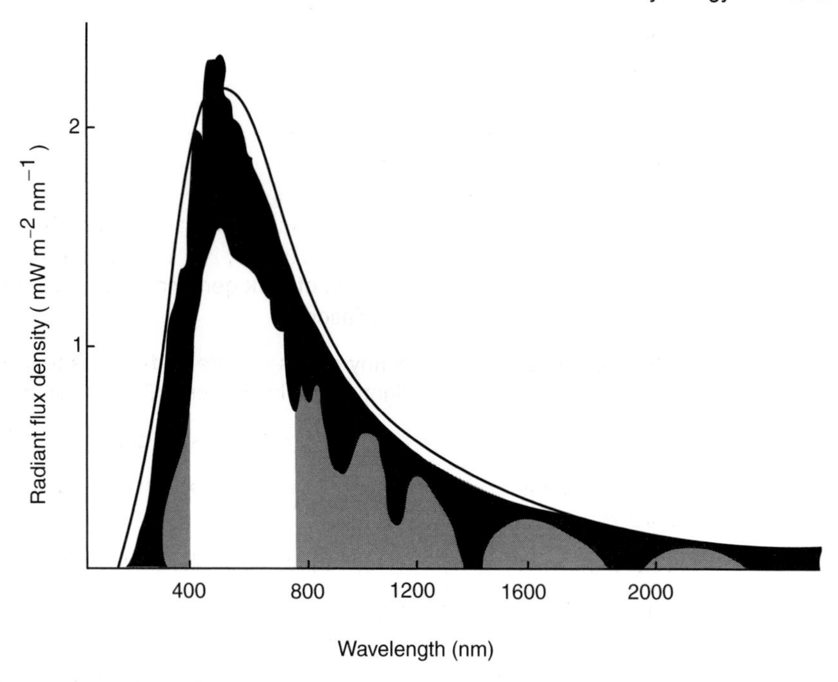

Figure 2.1

Solar radiation flux. The outer solid line represents the ideal output for a 'black body' at 6000 K (the solar surface temperature); the upper rim of the black area is the actual solar flux outside the earth's atmosphere; and the inner open and cross-hatched area the flux at the earth's surface. Only the open part is photosynthetically active radiation (*PAR*).

Physiologically, light has both direct and indirect effects. It affects metabolism directly through photosynthesis, and growth and development indirectly, both as a consequence of the immediate metabolic responses, and more subtly by its control of morphogenesis. Light-controlled developmental processes are found at all stages of growth from seed germination and plumule growth to tropic and nastic responses of stem and leaf orientation, and finally in the induction of flowering (Table 2.1). There may even be remote effects acting on the next generation by maternal carry-over; dark germination of seed of *Arabidopsis thaliana*, a small annual plant, now the standard tool of molecular genetics, is affected by light quality incident on the flower-head. Germination was much greater when the parents had been grown in fluorescent light than in incandescent light, which contains more far-red (Shropshire, 1971), an effect with considerable ecological significance (see below, p. 34).

These responses are mediated by at least four main receptor systems.

(i) Chlorophyll is the key photosynthetic pigment, with several absorption peaks in the red (most importantly at 680 and 700 nm) and also in the blue region of the spectrum;

Table 2.1

Some light-controlled developmental processes

Process	Control
Germination	Light-requiring seeds are inhibited by short exposure to far-red (FR) light; red light usually stimulatory. Seeds capable of dark germination may be inhibited by FR irradiation.
Stem extension	Many plants etiolate in darkness or low light. Red (R) light stops this but brief FR irradiation counteracts R.
	Prolonged FR irradiation can have similar effects to R.
Hypocotyl hook unfolding	Occurs with R or long-term exposure to FR or blue light.
Leaf expansion }	Require prolonged illumination for full expansion.
Chlorophyll synthesis }	Short-term FR inhibitory, long-term may or may not be inhibitory.
Stem movements	Blue light most effective, but UV-A, R and green also effective
Leaf movements	Blue and red light active. R/FR reversible.
Flower induction	In short-day plants, R can break dark period. FR reverses effect.
Bud dormancy	Usually imposed by short-days. Behaves as for flowering.

(ii) Phytochrome, absorbing in two interchangeable forms at 660 and 730 nm, controls many photomorphogenetic responses, and is now known to be a complex family of proteins (~240 kDa) falling into two distinct types (labile molecules found in etiolated tissues and stable molecules in green tissues), encoded, at least in *Arabidopsis*, by five genes (Clack *et al.*, 1994);

(iii) The recently characterized cryptochromes (Lin *et al.*, 1998) are blue light receptors absorbing at around 450 nm, responsible for high-energy photomorphogenesis and for entraining circadian clock phenomena (Somers *et al.*, 1998);

(iv) Tropic responses are controlled by a different blue light receptor, which appears to be a flavoprotein in *Arabidopsis* (Christie *et al.*, 1998).

All plants contain a wider variety of compounds capable of absorbing radiation, and no energy transduction function is known for many; in some the absorption is probably fortuitous. In algae, however, these accessory pigments are known to play an important auxiliary role in photosynthesis.

2. The radiation environment

1. Radiation

Radiant energy is measured in joules (J) and its rate in $J\,s^{-1}$ or watts (W). The rate at which surfaces intercept energy is therefore expressed in $W\,m^{-2}$. When considering the acquisition of energy by plants, however, it is only photosynthetically active radiation (*PAR*, i.e. 400–700 nm) that is of importance and so a measurement that takes this into account is appropriate. This can be achieved in practice by using filters to measure the irradiance within this band.

According to duality theory, radiation can be described either as waves or streams of particles, but for radiometric purposes it is most conveniently treated as if particulate and discretely packaged in photons, whose energy content (quantum) depends on wavelength. The quantum energy (in J) of a photon is $h\nu$, where h is Planck's constant (6.63×10^{-34} J s) and ν (which is the Greek letter *n*, pronounced nu or 'new') is the frequency of the radiation. Since:

$$\text{quantum energy} = h\nu \text{ and} \tag{2.1}$$

$$\nu = c/\lambda \tag{2.2}$$

where c is the speed of light (radiation) (3×10^{8} m s^{-1} = 3×10^{17} nm s^{-1}) and λ is wavelength (in nm), then:

$$\text{quantum energy} \approx \frac{2 \times 10^{-16}}{\lambda}\,\text{J} \tag{2.3}$$

Ecophysiologists distinguish photosynthetic irradiance, which is the total energy falling on a leaf in the waveband 400–700 nm, and measured in $W\,m^{-2}$, and the photosynthetic photon flux density (PPFD), which is the number of photons in the same waveband. The latter can be more usefully related to physiological processes in photosynthesis. The relationship between the two is given by using molar terminology: PPFD is given in moles of photons (a mole of photons is 6.022×10^{23} photons, which is familiar as Avogadro's number). From equation 2.3, therefore, 1 mole of a given wavelength carries $1.2 \times 10^{8}/\lambda$ J. If the wavelength distribution of radiation is known, conversion from $W\,m^{-2}$ to moles $m^{-2}\,s^{-1}$ is therefore possible; for PAR in sunny daylight the appropriate factor is 1 $W\,m^{-2}$ = 4.6 $\mu mol\,m^{-2}\,s^{-1}$ (McCree, 1972). Older papers sometimes quote PPFD in Einsteins (E); 1E equals 1 mole of photons, but the terminology is no longer used.

2. Irradiance

Radiant energy input is greatest on days with a clear, dry atmosphere, and the sun at its zenith. Paradoxically, broken cloud cover locally increases the energy received at ground level, because of reflection from the edges of the clouds. The differences in irradiance between this situation and that on a cloudy winter day, and between that and bright moonlight, encompass several orders of magnitude. Plant responses cover a parallel range (Table 2.2).

Table 2.2

Variation of radiant flux density in the natural environment and of plant response to it (adapted from Salisbury, 1963)

Environment	$W\ m^{-2}$	Plant response
Bright sunshine Sun high in sky	10^3	Photosynthesis saturates, C_3 sun plants
Typical plant growth chamber Daylight. 100% cloud cover	10^2	Photosynthesis saturates, C_3 shade plants
Heavy overcast with rain	10^1	Photosynthetic compensation points, C_3 sun plants
	1	Photosynthetic compensation points, C_3 shade plants
	10^{-1}	
Twilight	10^{-2}	Threshold for incandescent light inhibition of flowering in *Xanthium*
Bright moonlight	10^{-3}	Threshold for red light inhibition of flowering in *Xanthium*
Limit of colour vision	10^{-4}	
	10^{-5}	Threshold for phototropism in *Avena* (blue)
Starlight	10^{-6}	Threshold for unhooking response of bean hypocotyl (red)
Limit of vision	10^{-7}	
	10^{-8}	
	10^{-9}	
	10^{-10}	Threshold for photomorphogenesis *Avena* first internode (red)

The main effects of changes in irradiance occur on the process that uses radiation as an energy source – photosynthesis – rather than on those that use it as an environmental indicator. For most plants, photosynthesis becomes saturated at flux densities well below the maximum they occasionally experience, largely due to problems of CO_2 supply, but in shaded conditions photosynthesis is often limited by the level of radiant energy. Variation in irradiance is a universal feature of habitats colonizable by plants and the complex nature of this variation is well shown in forests where any point under the canopy will experience first, seasonal variation, secondly, a diurnal cycle, thirdly, random 'weather' effects due to cloud cover, and fourthly, canopy shade effects such as sunflecks. In addition to this temporal variation, immediately adjacent points may differ radically in the last two factors (Anderson, 1964). Leaf canopy effects on radiation are discussed later.

Solar radiation reaching vegetation has two components:

(i) irradiance of direct sunlight (I), and
(ii) diffuse irradiance from both clouds and clear sky (D).

Diffuse irradiance increases in importance as the solar beam is attenuated, either by actual obstruction (clouds, leaves, etc.) or by scattering due to particles and molecules in the atmosphere. Scattering is affected by the density of these particles, and also by the path-length of the direct solar beam through the atmosphere, both of which increase the chances of scattering occurring. Particles such as dust and smoke, and molecules such as water vapour, cause scattering in inverse proportion to the wavelength, following a power law relationship; the power function depends on particle size, but the net effect is to reduce the blue content of direct radiation. The scattered blue light contributes to the diffuse radiation, which therefore attains a greater blue content. Thus, although the sunset is red, as a consequence of scattering of blue light along the extended path-length of the beam when the sun is at such a low angle, the overall radiation load is blue-shifted at that time, since diffuse radiation predominates.

In effect the reductions in irradiance caused by occlusion of the direct solar beam are partially offset by the enhanced blue component of diffuse radiation and by the fact that water vapour in particular absorbs in the infra-red region, radiation which is not photosynthetically active. About one-third of direct solar radiation is photosynthetically active (*PAR*, i.e. 400–700 nm) as compared with over two-thirds of diffuse radiation. Under most meteorological conditions, therefore, *PAR* as a fraction of total solar radiation remains virtually constant at 0.5 ± 0.02 (Szeicz, 1974), so that it is convenient to disregard wavelength differences between direct and diffuse irradiance. Direct sunlight will, of course, be more intense than diffuse light, but at least at higher latitudes, diffuse radiation is a major part of the total. Theoretical calculations show that even under cloudless skies diffuse radiation (D) may account for between one-third and three-quarters of the total (T), and in a series of measurements in Cambridge, UK, the ratio D/T was always greater than 0.5 (Szeicz, 1974).

The maximum flux density of bright sunlight (for most purposes) is around 1000 $W\,m^{-2}$ and depends on the solar constant, the radiant flux density at the outer margin of the earth's atmosphere, confirmed by satellite measurements to be 1360 $W m^{-2}$. Typical instantaneous values for *PAR* at vegetation surfaces are 1200–1800 $\mu mol\, s^{-1} m^{-2}$ (*c*. 250–400 $W m^{-2}$) for sunlight and 100–400 $\mu mol\, s^{-1} m^{-2}$ (*c*. 25–100 $W m^{-2}$) for overcast skies (Table 2.2), because only a part of the total radiation flux is photosynthetically active.

Irradiance is reduced not only by cloud, dust, water vapour, and other atmospheric obstructions that increase the D/T ratio, but also through shading by terrestrial objects. Some shadows are caused by selective filters, which let a part of the spectrum through, such as leaves (considered below), water (cutting out long wavelengths) and soil; other objects, such as tree trunks and rocks, are opaque or act as neutral filters. The two are clearly not exclusive; in sites to the north of vegetation (in the northern hemisphere) a mixed shade occurs, having both the enhanced far-red component of leaf-transmitted shade and the enhanced short-wave content (mainly blue) of diffuse radiation.

3. Temporal variation

Temporal variation in irradiance is a universal feature of ecosystems: all places experience day/night fluctuations and the relative duration of these varies systematically with latitude. This is the basis of photoperiodism. Outside the tropics daylength is the most reliable indicator for predicting, and hence avoiding or resisting, unfavourable conditions. Virtually all temperate zone plants exhibit photoperiodic responses, many for flower initiation, but also for seed germination, bud-break, stem elongation, leaf-fall and many other phenomena. Although there is abundant evidence for the photoperiodic preferences of many individual species or cultivars, and although it is known that phytochrome is almost certainly the photoreceptor normally involved, no clear picture has emerged of the mechanisms involved. There is even doubt as to the significance of the 'redness' of end-of-day irradiance, which earlier seemed an obvious signal, given the involvement of the red/far-red reversible pigment phytochrome (Vince-Prue, 1983).

4. Leaf canopies

Many substances absorb radiation preferentially at some wavelengths: for example, water and soil both absorb at long wavelengths, increasing the red/far-red (R/FR) ratio. Leaves, on the other hand, which absorb in blue and red (Fig. 2.2), greatly reduce the ratio. Leaf canopies, therefore, produce a complex radiation climate varying in both irradiance and spectral distribution.

Leaf canopies are not solid sheets, but loosely stacked. Radiation therefore penetrates canopies in four ways, through actual holes as well as by reflection and transmission.

(i) Unintercepted direct irradiation may penetrate either as diffuse or direct radiation, the latter appearing as sunflecks (Plate 2a), which largely have the characteristics of direct irradiation, although gaps in the canopy act as diffusing lenses, spreading the beam as a penumbra (Anderson and Miller, 1974), and reflected radiation (see below) may be focused into the sunfleck, adding some of the high FR component (Holmes and Smith, 1977). In

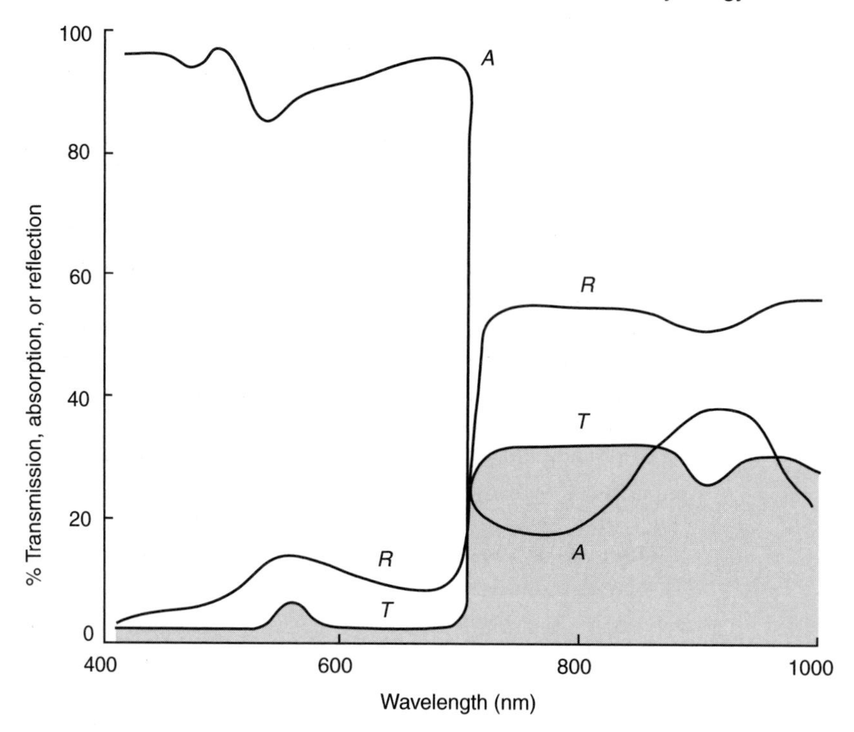

Figure 2.2

Generalized spectral characteristics of plant leaves between 400 and 1000 nm (from various sources). Abbreviations: *A*, absorption; *R*, reflection; *T*, transmission. The stippled area represents transmitted radiation

densely shaded environments, direct sunlight is likely to be of less value to most subcanopy species than might be expected, since their photosynthetic response curves tend to saturate at lower irradiances (see p. 52), though induction may occur. Nevertheless, over half the total daily radiation flux may be accounted for by sunflecks (Pearcy, 1983). In the penumbra, however, which has a larger area than the brightly illuminated sunfleck, the lower flux density will be used in photosynthesis more efficiently. Sunflecks are by nature transitory: a photocell of 1 mm diameter below a crop canopy gave readings fluctuating by as much as 80% several times a second, whereas a larger cell gave an averaged, more or less uniform reading (Anderson, 1970). Flashing light can in fact be as effective for photosynthesis as a continuous source (Emerson and Arnold, 1932).

(ii) Unintercepted diffuse radiation is the diffuse skylight counterpart of the sunfleck. It makes a very large contribution to the total irradiance beneath canopies, as can be envisaged by looking up at a canopy from the forest floor; if the sun is shining, a single sunfleck is likely to penetrate through one canopy gap, whereas all other canopy gaps will allow skylight to penetrate.

(iii) Transmission: the degree of shade clearly depends upon the amount of light absorbed and reflected by the leaves. The radiation flux passing through a leaf is not simply reduced in density, but is also radically altered in terms of spectral quality, due to the action of the various leaf pigments. Typically leaves transmit a few percent of incident *PAR* in the green band at around 550 nm, and are otherwise effectively opaque in the visible range. There is almost invariably a dramatic change from opacity to near transparency above 700 nm, so that transmitted light has a very high FR/R ratio (Fig. 2.3).

(iv) Reflection: leaves not only transmit light, but in common with all other biological surfaces they reflect a proportion. The amount reflected will depend, for example, on leaf shape and the thickness and shininess (chemically determined) of the cuticle. Reflected light is altered spectrally in much the same way as transmitted light (Fig. 2.2).

The relative contribution of these four components, and hence the depth and nature of the shade, depends on the number, thickness, distribution and type of

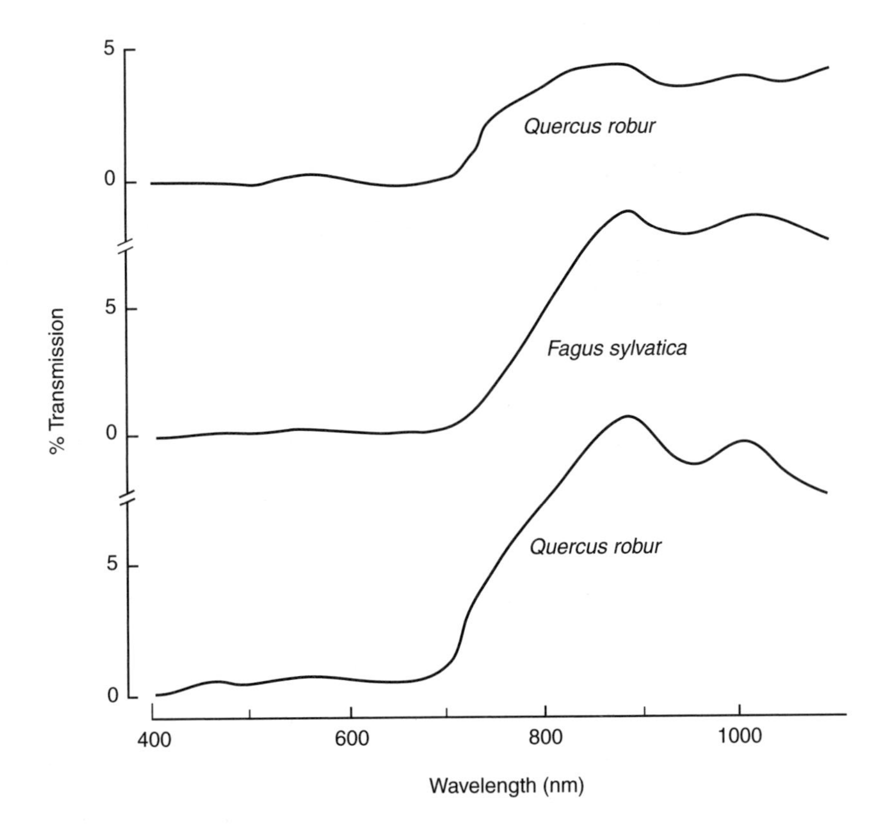

Figure 2.3

Transmission spectra of three woodland canopies. Note the sharp transition at about 700 nm (redrawn from Stoutjesdijk, 1972)

leaves in the canopy. The number is usually expressed as the leaf area index (*LAI* or **L**), a dimensionless parameter representing the area of leaf surface over unit area of ground. It is possible to relate the rate of attenuation of solar radiation through a canopy to **L**; differences between canopies in such a relationship reflect differences in leaf geometry, arrangement, and thickness (Fig. 2.4). This relationship is approximately exponential and can be simply described by the formula:

$$I_L = I_O e^{-kL} \tag{2.4}$$

where I_O and I_L are the irradiance respectively above the canopy and at a point above which are **L** layers of leaf, and k is the extinction coefficient, varying between canopies as just described. A plot of ln I_L/I_O against **L** therefore should reveal a straight line of slope k.

Several more sophisticated models have been advanced to describe this relationship, that take into account such important additional factors as the angle of the leaves (which may itself alter with the angle of the sun in heliotropic species), the solar angle, unequal distribution of radiation coming from different parts of the sky, and the distribution of leaves (Ross, 1981; Baldocchi and Colineau, 1994). These approaches average out a most complex situation, since closely adjacent points within a canopy will vary widely in radiation

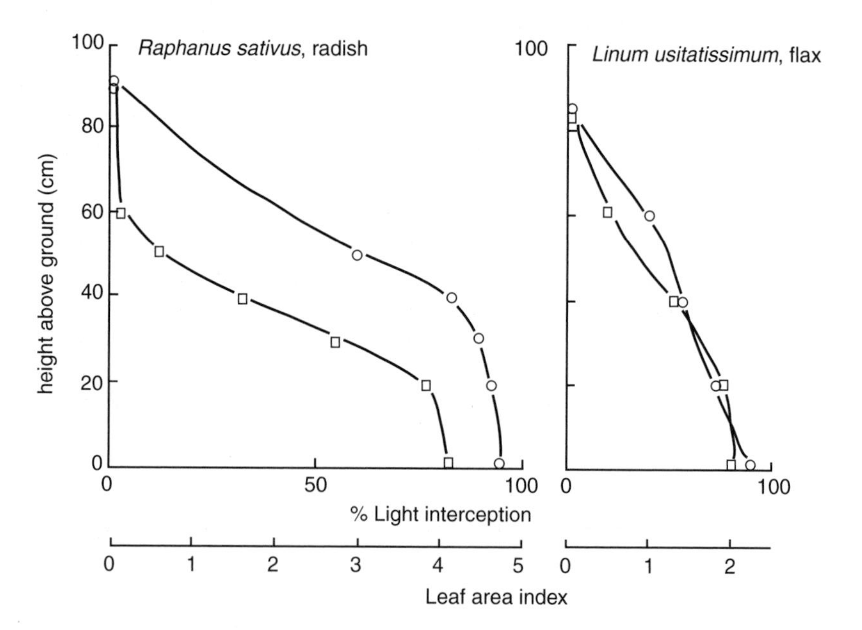

Figure 2.4

Correlation between interception of light and both depth in stand (○) and cumulative leaf area index (□) for two contrasting crops, *Raphanus sativus*, with a high canopy extinction coefficient (spreading leaves) and *Linum usitatissimum*, with a low coefficient (erect leaves) (from Newton and Blackman, 1970)

characteristics, depending on the balance between direct diffuse, transmitted and reflected radiation reaching them. This balance in turn will vary in time as both the sun and the leaves move.

Most work concerned with characterizing the radiation climate within plant canopies has been on agricultural, monospecific stands, with the aim of maximizing photosynthetic production. In a natural situation, however, we must also consider the environment of subordinate layers in a complex community. For instance, a temperate oak wood (*Quercus robur* on mull soil) may have

- a closed canopy of oak leaves with **L** between 2 and 4;
- a lower, usually more broken canopy of shrubs such as hazel *Corylus avellana*;
- a herb layer whose density will depend on that of the upper layers, but which, in the case of *Mercurialis perennis* forming monospecific stands, may itself have **L** = 3.5;
- and beneath all these a layer of mosses such as *Atrichum* and *Mnium* species and possibly soil algae and cyanobacteria as well, which must still achieve net photosynthetic production.

Radiant flux density beneath the tree and shrub canopies can be as low as 1% of levels in the open; beneath the herb canopy it will be even less. In deciduous woods, however, and in most herbaceous plant communities (reedbeds, tall grasslands, etc.) this massive reduction is a seasonal phenomenon (Plate 2b,c). Salisbury (1916) pointed out the importance to the ground flora in oak–hornbeam (*Quercus–Carpinus*) woods in Hertfordshire, UK, of the switch from light to shade conditions that occurred at the time of leaf expansion, resulting in a clear peak of irradiance from March to May, and Blackman and Rutter (1946) showed that the distribution of bluebell *Hyacinthoides non-scripta* was profoundly influenced by it. The effects of this peak on the distribution and phenology of woodland herbs is immense; no analogous situation occurs in evergreen or continuously moist tropical forests.

Under leaf canopies, therefore, all three aspects of the radiation environment, irradiance, spectral distribution and periodicity, are modified to produce a distinct environment requiring photosynthetic and morphogenetic adaptation.

3. Effects of spectral distribution of radiation on plants

1. Perception

Although the dominant pigment in leaves is chlorophyll, plants detect changes in spectral distribution principally through the remarkable family of pigments called phytochromes. These are reversible pigments, absorbing both red light at 660 nm and far-red light at 730 nm. When a phytochrome molecule absorbs red light it is converted from the red-absorbing form Pr into the far-red absorbing form, Pfr, and *vice versa*. It is Pfr that is physiologically active. In practice, plants experience both red and far-red light simultaneously, and the ratio of the two determines the proportion of phytochrome present as Pfr, known as the phytochrome photoequilibrium. As the R : FR ratio declines (as it does under a leaf canopy, for example), so the proportion of Pfr also declines.

These pigments allow plants to distinguish between low radiation fluxes caused by weather and seasonal and diurnal cycles, and genuine shade caused by the presence of other leaves (or in a cell, even other chloroplasts). Much of the early work on phytochromes was performed on etiolated seedlings, where chlorophyll concentrations were very low and the subtle phytochrome responses could be measured. However, we now know that phytochrome is a family of five molecules (denoted A to E), and that the type typically found in etiolated tissue is in many ways atypical, for example in that its active Pfr form spontaneously and rapidly reverts to the Pr form. Different phytochromes have distinct photosensory functions (Quail *et al.*, 1995; Furuya and Schäfer, 1996), with only phyB being active at all stages of the life cycle (Smith, 2000).

2. Germination

The seeds of many weed species are small and easily buried. It is a commonplace that ploughed land in temperate regions quickly comes to support a crop of such plants as chickweed *Stellaria media* and the speedwell *Veronica persica*. These species have a light requirement for germination and grow on soil that is frequently disturbed; they are generally uncompetitive in closed vegetation. Favourable conditions for germination occur when there has been disturbance, that either returns seeds to the surface or even briefly exposes them to light during cultivation. It is not necessary for the seeds to be returned to the surface; the brief exposure to light that a seed experiences as the soil is turned over by ploughing may be sufficient. Scopel *et al.* (1994) found that germination of seeds of weeds such as *Amaranthus retroflexus* and *Solanum* spp. was reduced 4–5-fold by ploughing at night. Quite strong, but very brief illumination is required to trigger germination in this way: vehicle lights at night did not stimulate germination, and covering the plough in the daytime suppressed it effectively. The perception of this signal is mediated by phytochrome and is an example of a very-low-fluence rate (VLF) response. *Datura ferox* seeds respond to the equivalent of less than 0.01 s of full sunlight, supplying only a few μmol of *PAR* per m^2 (Fig. 2.5; Scopel *et al.*, 1991).

Similarly, for those species which produce seeds under canopies (either their own or those of other species), germination is most likely to be successful when the canopy is either temporarily or permanently removed. This is true of such species as tufted hair-grass *Deschampsia cespitosa* and foxglove *Digitalis purpurea*; the latter is a species that can become extremely abundant in clear-cut woodland, though soon disappearing when the vegetation closes. In many of these cases seed is initially insensitive to light, but exposure to FR under a canopy induces a light requirement. Light stimulation of germination is a red/far-red reversible phenomenon, controlled by phytochrome. Typically light-sensitive seed will germinate when exposed to red light, but this stimulation can be erased by subsequent treatment with far-red; in this case it is the final exposure which determines germination.

It has long been known that light-sensitive seed will not germinate under leaf canopies (e.g. Black, 1969). King (1975) showed that several annuals found

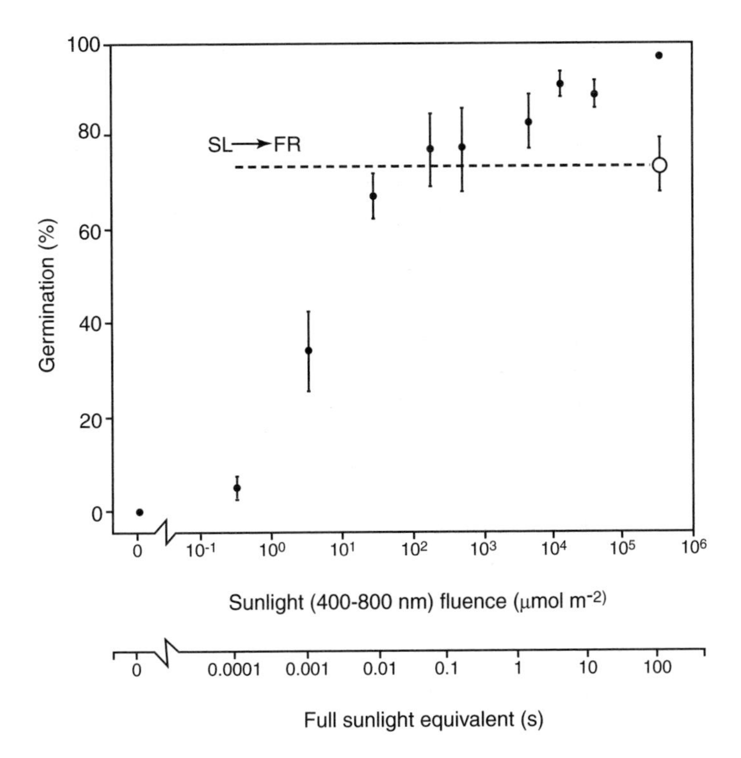

Figure 2.5

The dose response to sunlight of germination of seeds of the arable weed *Datura ferox*. The seeds were dug up, exposed to sunlight for varying periods, and then re-buried at a depth of 7 cm. One-tenth of a second of full sunlight resulted in the same percentage germination as saturating exposure to sunlight, followed by far-red light (SL → FR and open symbol) (from Scopel *et al.*, 1991)

specifically on the relatively bare ground of anthills in otherwise closed chalk grassland (*Arenaria serpyllifolia, Cerastium fontanum* and *Veronica arvensis*) were completely inhibited from germination by leaf-filtered light. There is, however, a problem in interpreting such experiments because of maternal effects on light control of germination (Shropshire, 1971; see also p. 24).

It seems, therefore, that light controlled germination is an adaptation of plants that are intolerant of shade. If the seed of such plants is exposed to high FR levels in nature, implying that it is under a leaf canopy, or if it is in the dark, which would normally mean burial, it does not germinate. Clearly such a mechanism could be used to adjust the timing of germination to take advantage of seasonal 'windows' in the canopy. Seed of shade-resistant plants such as *Rumex sanguineus* on the other hand, will be at less of a disadvantage if it germinates under a closed canopy, as this is the normal habitat for the adult plant. In addition floral structures can alter the germination characteristics of the enclosed seeds. A light requirement may be induced not just by exposure to leaf-filtered light after shedding, but by light filtered through the green bracts or sepals surrounding the seeds in the parent plant. Even the seed capsule may act in this way: seeds of the grass *Arrhenatherum elatius*, which has no chlorophyll in the surrounding structures, germinate fully in the dark, but those of the composite *Tragopogon pratensis*, surrounded by bracts still green when the seeds are mature, have an absolute light requirement (Cresswell and Grime, 1981). The greenness of fruiting heads may therefore be an important focus for selection in relation to germination behaviour.

3. Morphogenesis

The classic plant response to low levels of light is etiolation: placing seedlings of species such as mustard, beans or wheat in dark or near-dark conditions causes them to become elongated. This response is often said to allow the plant to 'reach the light'. However, the species in which this response is typically demonstrated are crop plants and it does not occur in many uncultivated species. Grime (1966) showed that species typical of open, uncompetitive habitats (as well as those found in deep shade) did not etiolate. Species of tall grassland, where elongation could bring competitive advantage by raising leaves to the level of the canopy, were etiolators (Fig. 2.6); importantly, crop plant species are grown in dense stands where again etiolation could be beneficial. It is unsurprising that forest species do not etiolate: the distance to the canopy might be 100 m.

Etiolation can be induced in appropriate species with a simple reduction in the amount of radiation. In nature, shade cast by plants also alters the radiation spectrum, typically radically reducing the R : FR ratio. Plants can, therefore, detect leaf shade by means of the phytochrome receptor. Growing

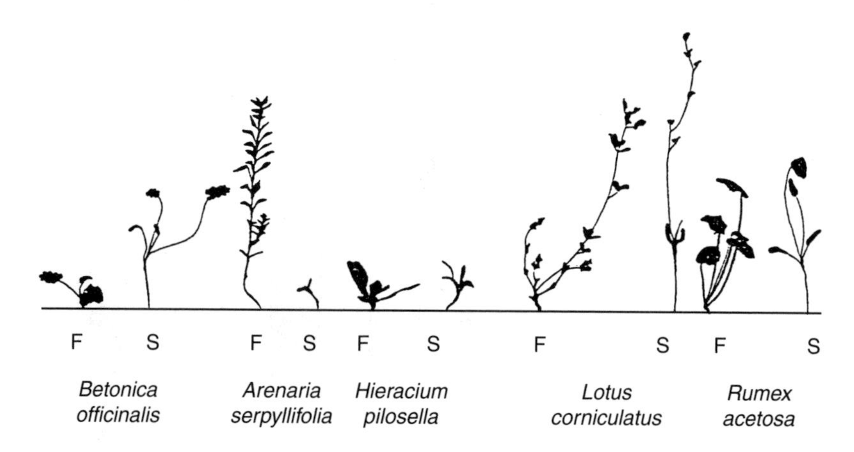

Figure 2.6

Silhouettes of seedlings of five herbaceous species grown either in full light (F) or in shade tubes, 6 cm tall (S). Only some of the species etiolate in shade, notably *Betonica officinalis*, *Lotus corniculatus* and *Rumex acetosa*, all grassland species. In contrast, *Arenaria serpyllifolia*, which stops growing in shade, and *Hieracium pilosella*, which shows little morphological response, are both species of bare ground (from Grime and Jeffrey, 1965)

plants under a light regime that has a very low R:FR ratio, which can be created artificially using filters, normally induces a much greater response than growing them in neutral shade of the same PAR flux (Fig. 2.7). In most cases, stem internodes are longer, leaf areas smaller and plant growth rate lower in low R:FR ratios than in neutral shade. Some species, however, are able to maintain their morphogenetic pattern, even under the low R:FR ratios of leaf-filtered light; invariably these are species characteristic of shaded habitats, such as forest floors (Fig. 2.7). These species experience permanently reduced R:FR ratios and appear to have evolved insensitivity to them. In the annual balsam *Impatiens capensis*, plants from open sites were more responsive to low R:FR ratios than plants from woodlands, suggesting that selection within a set of populations of a single species produces the same differentiation (Dudley and Schmitt, 1995).

In other species, the responses to a low R:FR ratio also appear to be adaptive. When plants are growing surrounded by other individuals, an increase in stem growth and reduced branching, both typical responses, may raise the leaves to a better-illuminated zone of the canopy. Trenbath and Harper (1973) showed that the elongation response of oats *Avena sativa*, a species selected for growth in herbaceous stands ecologically similar to tall grassland, resulted in about 20% extra weight in each seed when grown in mixed culture with *A. ludoviciana*, as compared to the estimate for a shaded plant.

Because the response is to the change in the R:FR ratio, and not to the change in the PAR flux, plants can detect neighbours before they are shaded. In one experiment, mustard *Sinapis alba* plants grown in front of a bank of green grass plants elongated more than those in front of grass seedlings that has been bleached by herbicide treatment, even though both sets of plants received the same amount of radiation (Ballaré *et al.*, 1987). The green plants reflected light with a reduced R:FR ratio and so induced elongation (Fig. 2.8). Similarly the *phyB* mutant of *Arabidopsis*, which lacks the phyB member of the phytochrome gene family, has a slightly reduced elongation response to shade when

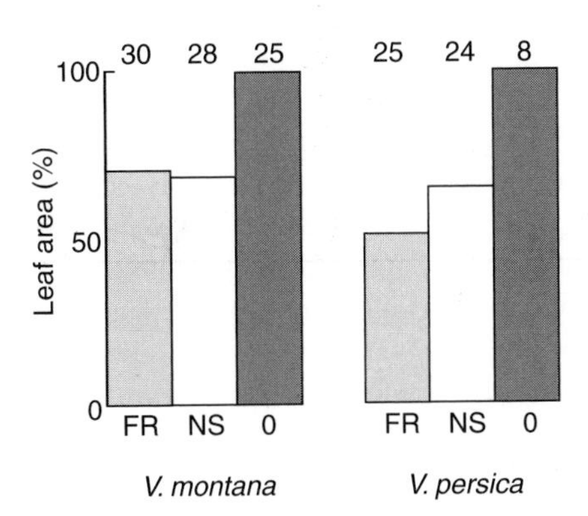

Figure 2.7

Leaf area of *Veronica montana* (a shade plant) and *V. persica* (a sun plant) in far-red supplemented shade (▢), neutral shade (▢) and open (■) conditions, expressed as a percentage of the value in open conditions. The percentage of biomass allocated to stems is given above each column (from Fitter and Ashmore, 1974)

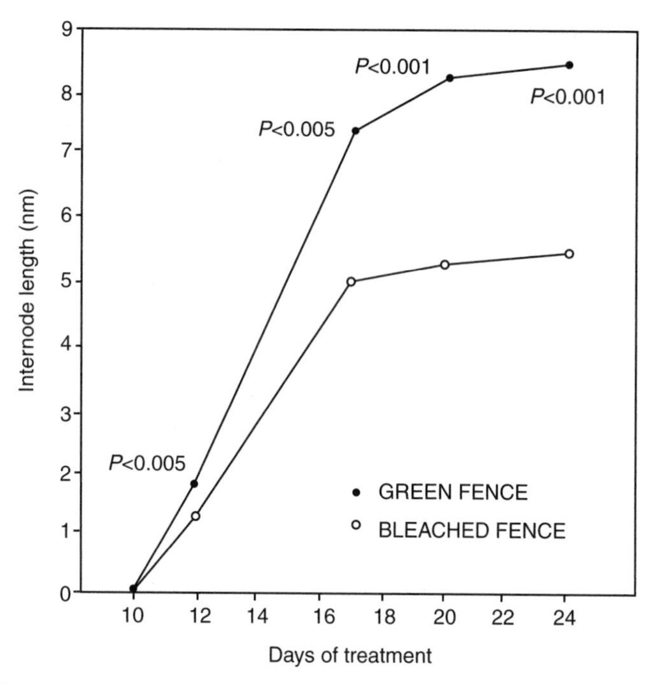

Figure 2.8

The etiolation response of mustard (*Sinapis alba*) plants grown in front of a 'fence' of green grass plants (●) or a similar canopy bleached by treatment with the weedkiller paraquat (○) (from Ballaré *et al.*, 1987). The probabilities are for the difference between the two treatments at each time.

compared with the wildtype, but shows almost no response to neighbours that do not cast shade (Yanovsky *et al.*, 1995). The ability to detect neighbours by this reflection signal is now recognized to be widespread and potentially very significant in plant communities.

4. Placement

Patterns of leaf placement are more complex than a first glance suggests. Horn (1971) suggested two basic architectures for forest trees – monolayer and multilayer. The monolayer is defined as a complete, uniform layer of leaves that lets through little photosynthetically active radiation (*PAR*), but it will have low productivity since it can have fewer layers of leaves (low leaf area index, **L**: theoretically $L = 1$ in a monolayer). The multilayer species has a more dispersed canopy but relies on the facts that:

(i) an individual leaf casts a shadow only for a certain distance (Horn used 70 diameters);
(ii) light-saturation for most species occurs at flux densities well below full sunlight so permitting several layers of leaves (high **L**) all operating at high photosynthetic rates, even if there is a certain amount of mutual shading.

The multilayer can therefore grow faster, but since it lets through more *PAR* to the lower levels of the canopy it is more open to invasion. Multilayers should therefore be characteristic of early successional species and monolayers of species entering later, so-called climax species. The concept of the monolayer implies some mechanism for leaf orientation that minimizes gaps between leaves. It is well known that the leaves on branches of some trees form 'mosaics', interlocking patterns that appear to fill space very effectively (Fig. 2.9). Achieving such patterns requires precise movements during growth, as for example by phototropism.

Tropic responses remain a physiological enigma. Much of the groundwork was laid by Darwin and an apparently adequate explanation was propounded by Went and his co-workers in the 1920s, based on a theory of auxin movement. It is now known that adequate auxin movement does not occur, but as yet no general model of phototropism can be accepted (Iino, 1990; Firn, 1994), although analysis of *Arabidopsis* mutants is revealing the signal induction pathway (von Arnim and Deng, 1996). What is certain is that the action spectrum is complex, with as many as four peaks in the blue region between 370 and 470 nm, that one receptor is likely to be a flavoprotein (Christie *et al.*, 1998) and that the response *is* affected by auxin. Whatever the physiology, the adaptive significance is clear, and is enhanced by the partial reversibility of the process, since although it involves differential growth, this can be reversed by complementary growth opposite the original site. Generally, however, phototropism will be a response to a more or less fixed spatial pattern in the light environment.

Phototropic responses are adaptive; differences between plants adapted to contrasting habitats should therefore occur. Genotypic differences in the patterns of leaf arrangement occur in forage grasses, and the same phenomenon was recognized early for uncultivated plants (Turesson, 1922). Plants with prostrate leaf arrangements have very much higher extinction coefficients for light (k, see equation 2.4, p. 32) within the canopy than those with erect leaves (Cooper and Breeze, 1971), a situation analogous (for a very different growth form) to Horn's (1971) distinction between mono- and multilayer tree strategies. As a result, under agricultural conditions, long-leaved, erect plants of low k can grow faster and be more productive (Table 2.3). Indeed in terms of plant breeding it seems that much more can be achieved by alteration in canopy architecture than in photosynthetic rates, which may be unrelated to growth rate (Sheehy and Peacock, 1975). Crop yields are normally determined by the amount of *PAR* intercepted by the canopy over the growing season (Hay and Walker, 1989).

In natural conditions, however, the prostrate form (high k) will often be favoured, since the advantage of the erect types increases with longer intervals between cutting; under very frequent cutting regimes (intervals of less than 10–14 days), the short-leaved genotypes are favoured (Rhodes, 1969). Under these conditions, which could be produced in the wild by grazing animals, the erect types never produces an adequate canopy. Very severe environments will also enforce the same result, as shown by Callaghan and Lewis (1971) for the grass *Phleum alpinum* in Antarctica: plants from sheltered habitats allocated more biomass to leaf growth than those from exposed sites, but the latter compensated for this by apparently adaptive changes in photosynthetic physiology.

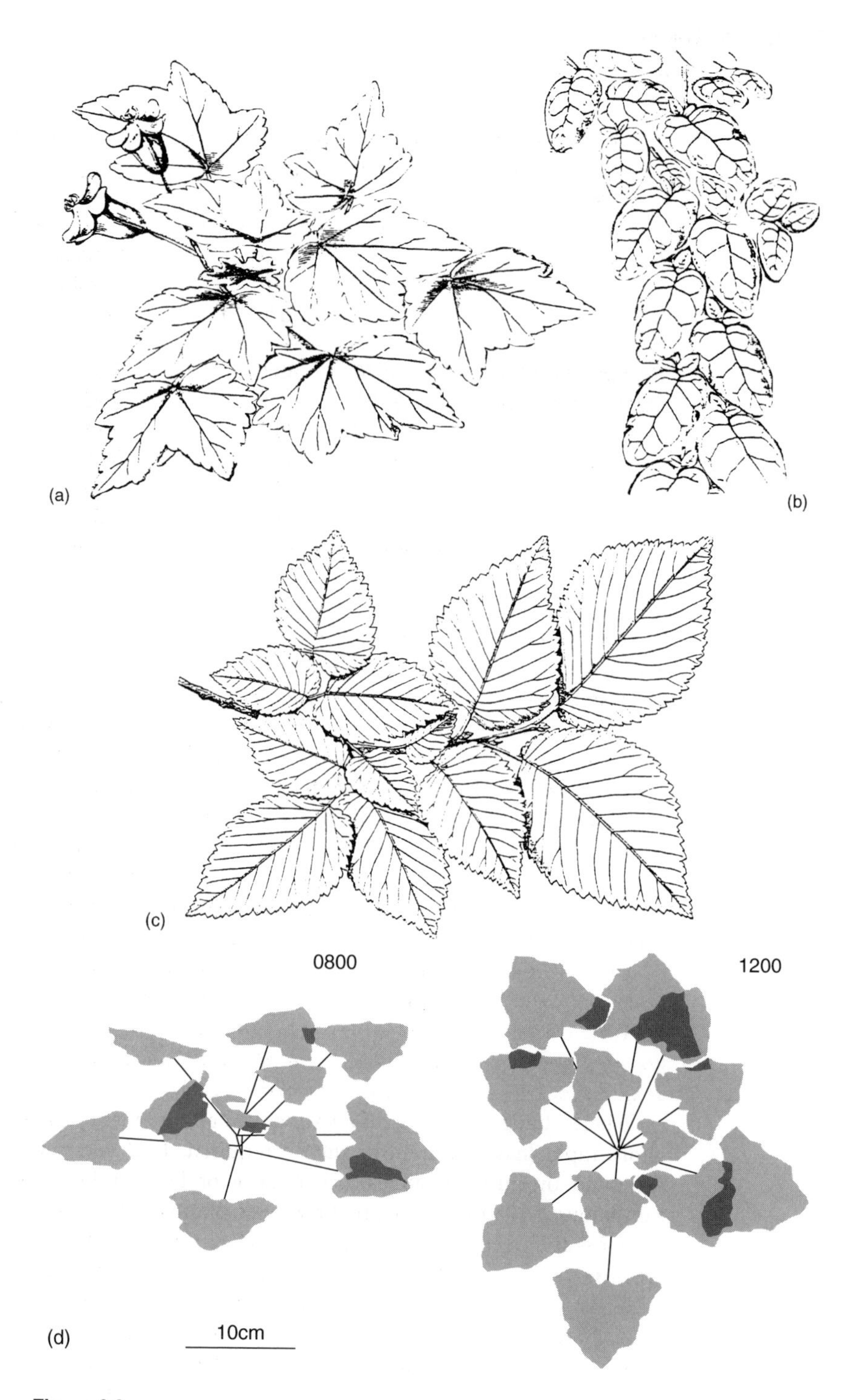

Figure 2.9

Leaf mosaics formed by species with contrasting leaf shapes: (a) *Begonia dregei*; (b) *Ficus scandens*; (c) *Ulmus* sp.; (d) a canopy of *Adenocaulon bicolor* simulated by the model Y-plant at 0800 and 1200 h (from Pearcy and Valladares, 1999)

Table 2.3

Crop growth rates of six common forage grasses in relation to the extinction coefficient of light in the canopy and the leaf area index (Sheehy and Cooper, 1973). Note the negative correlation between growth rate and *k*

Species		Crop growth rate ($g\,m^{-2}\,d^{-1}$)	Extinction coefficient (*k*)	Leaf area index
Festuca arundinacea	S170	43.6	0.34	11.2
Dactylis glomerata	S 37	40.5	0.23	13.7
	S345	25.0	0.91	14.9
Phleum pratense	S 50	36.4	0.30	15.5
	S 48	28.9	0.39	10.4
	S352	21.9	0.55	14.5

The impact of shoot architecture on plant carbon gain can be studied using sophisticated models that link the biomechanical constraints and structural costs of different architectures, the movements of the sun and the physiological responses of photosynthesis to consequent variations in solar radiation. One such model, Y-plant (Pearcy and Yang, 1996), has been used to investigate the optimality of the design of the shoots of *Adenocaulon bicolor*, an understorey herb that grows in the redwood (*Sequoia*) forests of western North America (Pearcy and Valladares, 1999). *A. bicolor* forms rosettes in which the oldest and lowest leaves have longer petioles than the upper, younger leaves (Fig. 2.9d and 2.10a); this is a common pattern for rosette plants and reduces self-shading. The trade-off here is that long petioles carry a construction and maintenance cost that could alternatively be used for photosynthetic tissue. Longer petioles reduce self-shading and increase photosynthesis per unit leaf area at a cost of reduced photosynthetic area. There should therefore be an optimum length for the petioles, and the model suggests that this optimum is almost exactly the length that was measured in the field (Fig. 2.10b).

Where changes in the spatial pattern of light are more short-lived, however, more rapid (nastic) responses will be required. Leaf and petiole movements, operated by turgor changes, occur more or less continuously in controlled conditions, as can be shown by time-lapse photography. The function of these rather slight movements in the field is unclear, as they will often be swamped by air turbulence, though they can be regarded as a mechanism for sensing the environment, particularly in climbing plants. Nevertheless, nastic movements as dramatic as those of the leaves of the sensitive plant, *Mimosa pudica*, whose leaves collapse on physical contact, or the compass plant, *Lactuca serriola*, whose leaves tend to orient themselves north–south, can readily be observed in the field. For plants growing in low flux densities these movements typically follow the sun and ensure maximum illumination; for plants in strong light they are normally avoidance reactions which reduce the heat load on the leaf (as described in Chapter 5) and allow subordinate leaves in the canopy to receive

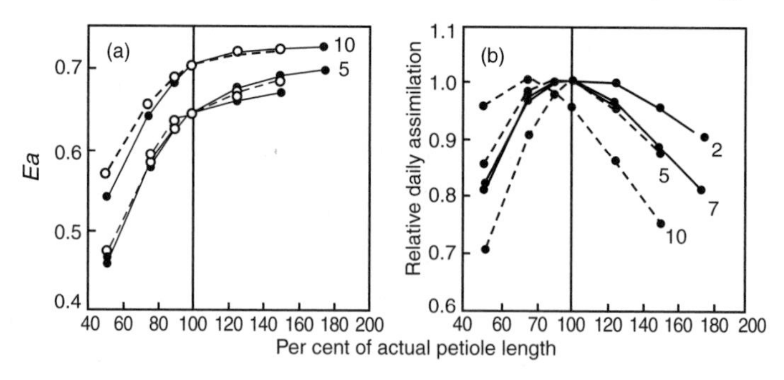

Figure 2.10

Simulations of photosynthetic performance of *Adenocaulon bicolor* show that (a) light absorption efficiency (E_a) increases asymptotically as petiole length increases, because of reduced self-shading, but (b) there is an optimum petiole length caused by the trade-off between E_a and the cost of the petioles, and that the simulated optimum is very close to that measured on field-grown plants. The different lines represent simulations based on actual plants (from Pearcy and Valladares, 1999). In (a) open symbols show simulations in which leaf biomass was adjusted to allow for changes in petiole biomass

more *PAR*. When the sun is far from the zenith such movement can also markedly influence the effective leaf area index. In *Lactuca serriola* both effects are achieved – greater photosynthesis in the morning and lower radiation loads at mid-day (Werk and Ehleringer, 1984).

The magnitude of these effects is seen in an experiment by Jurik and Athey (1994). They tethered some leaves of velvetleaf *Abutilon theophrasti* in a horizontal position, and allowed others to move freely. Total carbon fixation by tethered leaves averaged 318 mmol C m^{-2}, whereas free leaves fixed 457 mmol m^{-2}. Transpiration was also greater in free leaves (152 as opposed to 119 mol H_2O m^{-2}) but even so water use efficiency (carbon fixed per unit water transpired) was still greater at 2.99 mmol mol^{-1} in free than in tethered leaves (2.66 mmol mol^{-1}).

5. Flowering

Reproductive development in many temperate species is determined by photoperiod, detected by phytochrome; in equatorial regions daylength shows little seasonal march and so photoperiodism would be of less value. Keeping a photoperiodic plant under a constant daylength will usually maintain it in one developmental stage, although if the photoperiod promotes flowering, it will normally be stopped eventually by other processes. Thus *Epilobium hirsutum* and *Lythrum salicaria* flower if given 16 h days, but will remain vegetative indefinitely under a 9 h photoperiod (Whitehead, 1971). Not all temperate zone plants are photoperiodic, however: many weedy species, such as groundsel *Senecio vulgaris* will flower in all months of the year if the weather is favourable, and the balsam *Impatiens parviflora* is an example of a large group of plants which flower at a particular stage of development, irrespective of photoperiod.

The photoperiodic response enables the plant to time vegetative and floral growth to fit seasonal changes in environment. If a plant is moved to a different latitude it will then be out of phase and may die from, for example, attempting vegetative growth in the winter or too late in spring. This was what led to the first recognition of the phenomenon (Garner and Allard, 1920) and its importance was quickly seen by silviculturalists, who discovered that a wide-ranging tree species may have well-marked photoperiodic races or ecotypes as for example in birches *Betula* (Vaartaja, 1954; Håbjørg, 1978a). The same is true of grasses (Olmsted, 1944; Håbjørg, 1978b) and of other herbaceous plants such as *Oxyria digyna*, the mountain sorrel, a circumpolar arctic-alpine species that descends as far south as California in the North American Rockies (Fig. 5.7; Plate 14). Mooney and Billings (1961), in a comprehensive study of the different arctic and alpine races of this species, found that the arctic populations required a much longer day to induce flowering; arctic plants could be kept vegetative in growth room photoperiods that stimulated flowering in alpine populations. Despite these important investigations, the ecological role of photoperiodism within communities, for example in determining flowering time phenologies is still unknown.

4. Effects of irradiance on plants

1. Responses to low irradiance

1. Temporary shade

In some habitats the availability of light energy varies seasonally. In temperate deciduous woods there is a distinct peak of irradiance in early spring before the leaves expand (see p. 33), and a well-marked group of plants takes advantage of this. Deciduousness in trees in response to seasonal temperature changes (as opposed to variation in water supply), is, however, an adaptive response that minimizes frost damage to leaves capable of high productivity; for that reason the plants that photosynthesize during this brief seasonal window need to be frost resistant (see Chapter 5). To take full advantage of the radiation peak they must also have fully expanded leaves by April at the latest, so that leaf growth must take place at very low temperatures in February and March, when photosynthetic activity will be low. This growth therefore requires stored reserves, and almost all these species are perennials with underground storage organs (Fig. 2.11), whether bulbs (*Hyacinthoides non-scripta*, *Allium ursinum*), corms (*Cyclamen*), tubers (*Ranunculus ficaria*), or rhizomes (*Anemone*, *Trillium*). These storage organs are re-charged during the radiation peak. This avoidance strategy parallels that of tundra perennials that fund early-season growth from below-ground stores (Chapter 5).

Occupation of this particular niche, therefore, requires modification to all other parts of the life-cycle. Whereas some species, such as *Hyacinthoides*, carry out all their photosynthetic carbon fixation before the canopy closes, others such as *Oxalis acetosella*, remain photosynthetically active for a large part of the shade phase. This activity occurs in very dim light and so requires a change in physiology, which is brought about plastically in two distinct ways. *Oxalis* adjusts both physiologically, by altering photosynthetic behaviour in surviving

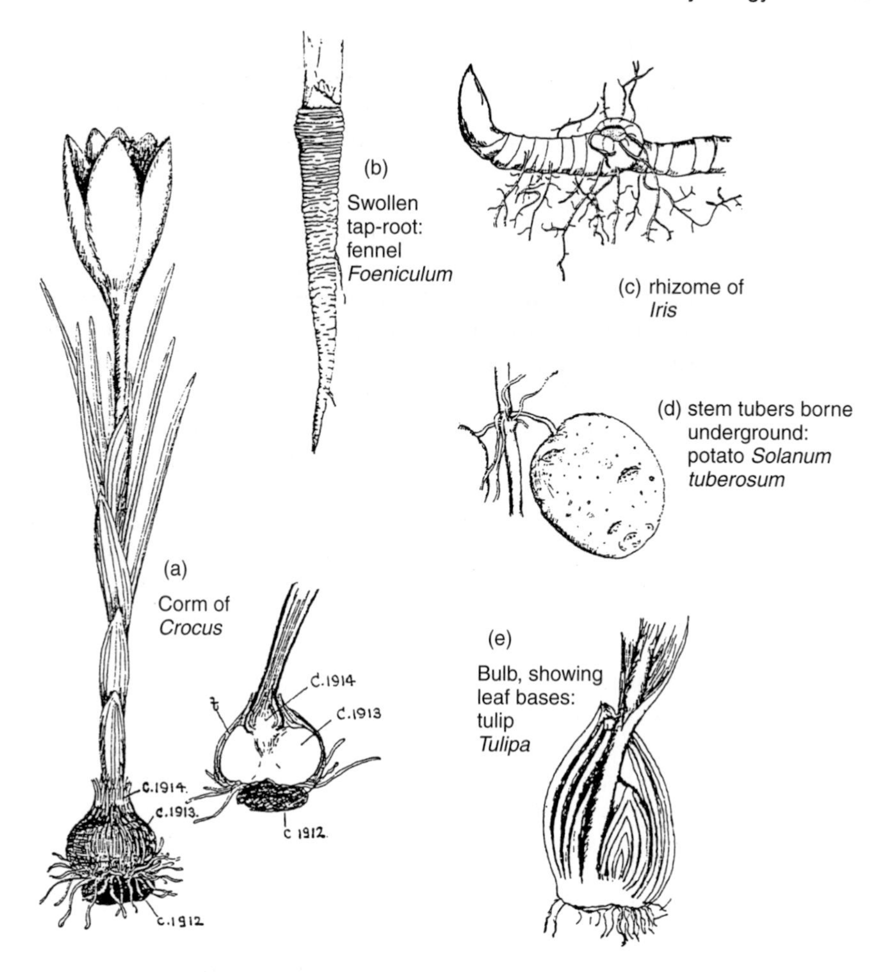

Figure 2.11

Below-ground plant storage organs: (a) corm of *Crocus*, which is a swollen stem, produced annually; (b) swollen tap-root of fennel *Foeniculum vulgare*; (c) rhizome of *Iris*, a typically diageotropic (horizontal) stem, usually below-ground, with continuous growth; (d) stem tuber of potato *Solanum tuberosum*, a swollen stem tip; (e) bulb of tulip *Tulipa* sp., showing the leaf bases – the overwintering stem is at the base of the bulb, and the flowering stem grows from that. From Kerner von Marilaun, A. (1894). *The Natural History of Plants*. Blackie, London.

leaves (Daxer, 1934) and morphologically by producing new leaves (Packham and Willis, 1977). *Aegopodium podagraria*, a garden weed in northern Europe, but a woodland herb in central Europe, also has two leaf types produced by a single plant: thin broad summer leaves, and thicker spring leaves, adapted to higher light levels. These two mechanisms are of general significance in the adaptation of plants to low irradiance and the physiological background to the ecological switch will be examined below.

Of course, any changes in physiology that maintain carbon balance also involve changes in respiration rate. This typically means a lowering of the compensation point, where CO_2 release by respiration and photorespiration exactly balance CO_2 uptake by photosynthesis (Fig. 2.12).

2. Long-term shade

If a plant grows in a shaded environment and is active outside the typically brief periods of high illumination, selection will act more on the photosynthetic process than on the life cycle. Growth in permanent shade is inescapable for the lower leaves of a plant forming a multilayer canopy, and it follows that both plastic (within a genotype) and genetic (between genotypes) differences must exist in the photosynthetic system.

A shaded leaf must at least maintain a positive carbon balance or else it represents a net drain on the plant's overall carbon budget. The flux density at which a positive C balance is reached is called the light compensation point (Fig. 2.12). A leaf below the light compensation point uses more energy for respiration than it can fix and is effectively a non-functional organ. Three variables determine the balance of this equation and the overall shape of the light response curve of photosynthesis:

(i) leaf area per unit mass: an increase provides a greater surface for *PAR* absorption;

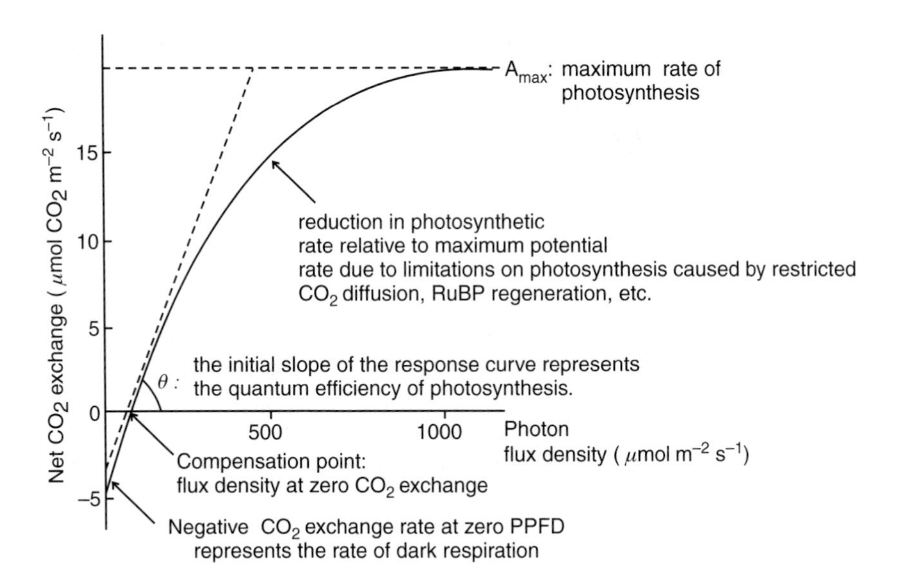

Figure 2.12

Representative photosynthetic response curve showing the quantum efficiency, the initial slope of the response curve; the compensation point, the flux density at which net CO_2 exchange is zero, and therefore the point at which gross photosynthesis equals respiration; and A_{max}, the maximum rate of photosynthesis

(ii) respiration and photorespiration: a reduction lowers the compensation point;
(iii) the capacity of the photosynthetic system, which is determined by the amount of biochemical machinery (principally Rubisco, ribulose bisphosphate carboxylase–oxygenase) available for C fixation.

(a) Increased leaf area. Plant growth can be described by the classic growth equation (see Chapter 1, p. 5):

$$\mathbf{R} = \mathbf{E} \times \mathbf{F} \tag{2.5}$$

where **R** is relative growth rate ($g\,g^{-1}\,d^{-1}$), **E** is net assimilation rate or NAR ($g\,m^{-2}\,d^{-1}$) and **F** is leaf area ratio ($m^2\,g^{-1}$). The immediate, enforced response of a plant removed to a lower flux density will be a reduction in **R** caused by a lowering in **E**, reflecting the effect of *PAR* on photosynthesis. To maintain **R** therefore, assuming no change in the light dependence of photosynthesis, requires an increase in the leaf area ratio (**F** or *LAR*). **F**, as the ratio of leaf area to plant weight, is a complex function without any obvious biological interpretation. It can, however, be thought of as:

$$\begin{gathered} \mathbf{F} = LWR \times SLA \\ A_L/W = W_L/W \times A_L/W_L \\ \text{Leaf area ratio} = \text{leaf weight ratio} \times \text{specific leaf area} \\ (m^2\,g^{-1} = g\,g^{-1} \times m^2\,g^{-1}) \end{gathered} \tag{2.6}$$

LWR is the ratio of leaf weight to plant weight, *SLA* that of leaf area to leaf weight. Leaf area described by *LAR* or **F** can therefore be discussed in terms of two components:

(i) the proportion of plant weight devoted to leaf material; i.e. how much leaf is there?
(ii) the area : weight ratio of the leaf itself, i.e. how thick and how dense is it?

Changes in leaf dry weight. It is well established (cf. Chapter 3) that the ratio of root weight to shoot weight is very plastic; for example, plants grown in infertile soils tend to have very high root : shoot ratios. Since for most herbaceous plants the leaves are a large proportion of the shoot weight, and since the stem acts to place leaves in an appropriate radiation environment, one would expect the *LWR* to be equally variable. Although some evidence is conflicting, it seems that generally *LWR* does increase in shade, although in severe shade, non-adapted species may actually show reduced *LWR*. This shows that *LWR* is a reflection of the plant's ability to maintain its normal developmental pattern; it is regulated over the range of flux densities to which a plant is adapted. Non-adapted plants in shade, however, exhibit etiolation and *LWR* is then reduced.

Specific leaf area (SLA). *SLA* is more variable than *LWR*; in other words, leaf area is more plastic than leaf weight. Immediately subsequent to germination, for example, there is a marked rise in *SLA* caused by expansion of the first leaves, whose dry weight changes only slightly during expansion. *SLA* also responds to environmental changes. Newton (1963) showed that the leaf area of cucumber plants (*Cucumis sativa*), a light-demanding species, was proportional to the total radiation, with a maximum at about 4.2 $MJ\,m^{-2}\,d^{-1}$ (or about 350 $\mu mol\,s^{-1}\,m^{-2}$ for a 16h day). Changes in both daylength and irradiance (as components of total daily radiation) had no effect if the daily total was the same. The reduction in leaf area at the highest irradiances was due to a reduction in cell size, whereas cell number increased to a plateau (Milthorpe and Newton, 1963).

Apparent effects of irradiance may therefore be responses to total radiation load; this is probably the case for *Impatiens parviflora* which shows an almost threefold increase in *SLA* when grown in 7% of full daylight (Evans and Hughes, 1961). When plants are grown under field conditions, responses are even more dramatic, *SLA* changing from 0.05 to 0.13 $m^2\,g^{-1}$ for plants growing under 2.0 as opposed to 0.5 $MJ\,m^{-2}\,d^{-1}$ (Hughes, 1959). The plasticity of the character is emphasized by the rapidity with which *SLA* adjusts when plants are transferred from one light regime to another.

These differences are for whole plants. Similar effects can be found between leaves of a single plant exposed to heterogeneous light, as in a forest canopy. Such differences are the basis of the phenomenon of sun and shade leaves, first recognized in the nineteenth century by Haberlandt. In beech (*Fagus sylvatica*) the phenomenon is very clear-cut (see Table 2.4), but as with *LWR* the degree of plasticity appears to be related to the ecological niche of the species. Some shade-resistant species, such as *Veronica montana* (Fitter and Ashmore, 1974) and *Rhododendron ponticum* (Cross, 1975) show much less striking shade-induced changes in *SLA* than species from open habitats. These species have constitutively low SLA, largely due to a high ratio of non-photosynthetic tissue types (e.g. sclerenchyma) to photosynthetic tissue (mesophyll).

Table 2.4

Characteristics of sun and shade leaves of mature and young beech trees *Fagus sylvatica* (Skene, 1924)

	Mature		Young	
	Sun leaves	Shade leaves	Sun leaves	Shade leaves
Leaf thickness (μm)	210	108	117	90
Number of palisade layers	2	1–2	1–2	1
Thickness of upper palisade layer (μm)	60	28	39	24

Leaf morphology. The relative constancy of *LWR* and plasticity of *SLA* imply that the plant has an optimum developmental pattern in terms of dry weight distribution, achieving adaptation to irradiance by changes in leaf morphology. The increased ratio of leaf area to weight (*SLA*) implies large anatomical changes in the mesophyll and palisade layers, as seen in *Mimulus* (Hiesey *et al.*, 1971) and in a wide range of deciduous trees (Jackson, 1967; see also Table 2.4). In all cases the palisade layer is reduced from 2–3 cells to 1 cell in the shaded or shade-resistant leaves. Such thin leaves have high *SLA*.

These changes affect CO_2 diffusion, but there is little evidence that CO_2 diffusion inside the leaf limits photosynthesis except at very high flux densities (Björkman, 1981). The main effect of a change in leaf thickness is the concomitant change in the quantity of photosynthetic apparatus per unit leaf area, with internal geometry being much less important.

The environmental trigger for most changes in leaf morphology is low irradiance, and a more satisfactory explanation rests on a consideration of the energy balances of different leaf morphologies. Shade leaves tend to be larger and less dissected; as a result they have greater boundary layer resistance, are less susceptible to convective cooling and so may overheat in full sun (see Chapter 5, p. 193; Gates, 1968; Grace, 1983). It is common, for example, to find that lower leaves of a plant have these 'shade' characteristics (Fig. 2.13) and it is notable that cotyledons are rarely dissected.

Leaf cooling is brought about both by evaporation of water as vapour through the stomata and by convective heat loss. The former is controlled by factors such as size and frequency of stomatal apertures and both are affected by the boundary layer conditions of the leaf surface atmosphere (Gates, 1968). If this layer is turbulent (which is likely except in very calm conditions)

Figure 2.13

Silhouettes of successive leaves from base to top of a stem of musk mallow *Malva moschata*, showing the progressive increase in leaf dissection

evaporation will be faster; a stable boundary layer will provide a large resistance to the movement of water vapour. Boundary layer resistance can be reduced by leaf dissection, as long as each individual leaf segment maintains a discrete boundary layer. Where the distance between segments is too small, a common boundary layer will form over the whole complex and boundary layer resistance will be considerably increased. A fuller discussion of leaf cooling is given in Chapter 5.

Apart from water vapour, the other important gas moving through the stomata is CO_2, and the same constraints will apply. It becomes of particular importance in water, however, where CO_2 diffusion is much slower and where its concentration is very sensitive to pH as a result of bicarbonate ion formation; the characteristic feathery leaf of water plants (*Myriophyllum*, *Ranunculus* subgenus *Batrachium*, *Hottonia palustris*, see Fig. 2.14) effectively increases the surface area and allows greater CO_2 uptake. Significantly, the response is usually light-triggered, by perception of photoperiod (Cook, 1972).

Leaf morphology, therefore, affects photosynthesis in four main ways:

(i) interception of *PAR*;
(ii) temperature regulation;
(iii) water balance;
(iv) CO_2 diffusion.

Points (ii) and (iii) are probably the most important and are explored in detail in Chapters 4 and 5 (cf. equations 5.1–5.4, p. 194). In all cases the environmental stimulus to which the plant responds is radiation, whether as irradiance, duration, or their product.

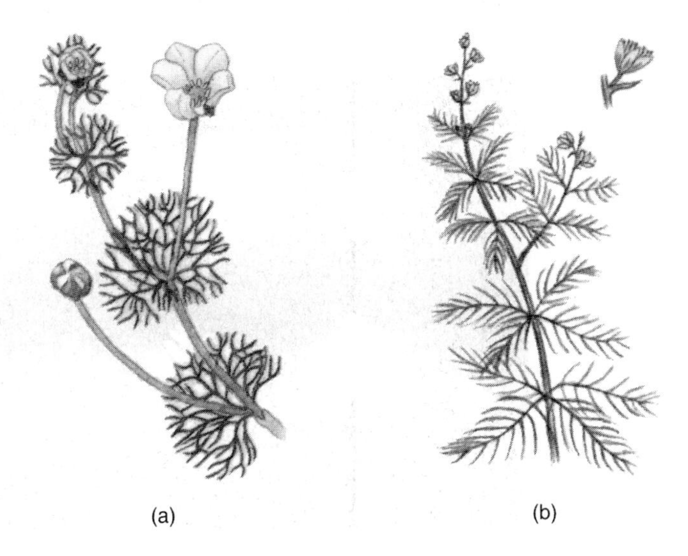

Figure 2.14

Leaves of (a) *Ranunculus circinatus* and (b) *Myriophyllum alterniflorum*, showing dissection of underwater leaves

(b) Respiratory rate. At the compensation point, net photosynthetic carbon fixation (i.e. carbon fixation less photorespiratory carbon loss) equals respiratory loss. A reduction in respiration will therefore lower the compensation point, but a reduction in respiration rate is likely to reduce growth rate, which could lower the competitive ability of the plant in relation to faster-growing species. As a response to low irradiance it is therefore likely to be advantageous only in severe shade where growth rates are so much reduced that competition is no longer a significant force. Plants normally found in deep shade may have inherently low relative growth rates: Mahmoud and Grime (1974) found that at extremely low illumination (down to $0.07\ \mathrm{W\,m^{-2}}$) *Deschampsia flexuosa* showed zero growth rate, whereas *Festuca ovina* and *Agrostis capillaris (A. tenuis)* had large negative growth rates, when maintained below their compensation points, which were at about $0.7\ \mathrm{W\,m^{-2}}$ (Fig. 2.15). *D. flexuosa* survived four weeks in deep shade; the two latter species senesced. Although respiratory rates were not measured, the clear implication is that *D. flexuosa* owed its resistance to a virtual cessation of respiratory activity. Certainly McCree and Troughton (1966) found that white clover, a species unable to survive in deep shade, when grown at $65\ \mathrm{W\,m^{-2}}$ and transferred to a range of flux densities from 88 to $3.7\ \mathrm{W\,m^{-2}}$, had respiration rates varying from 7.0 to $0.7\ \mathrm{mg\ CO_2\,dm^{-2}\,h^{-1}}$. In other words, respiration rate adjusts in response to light availability.

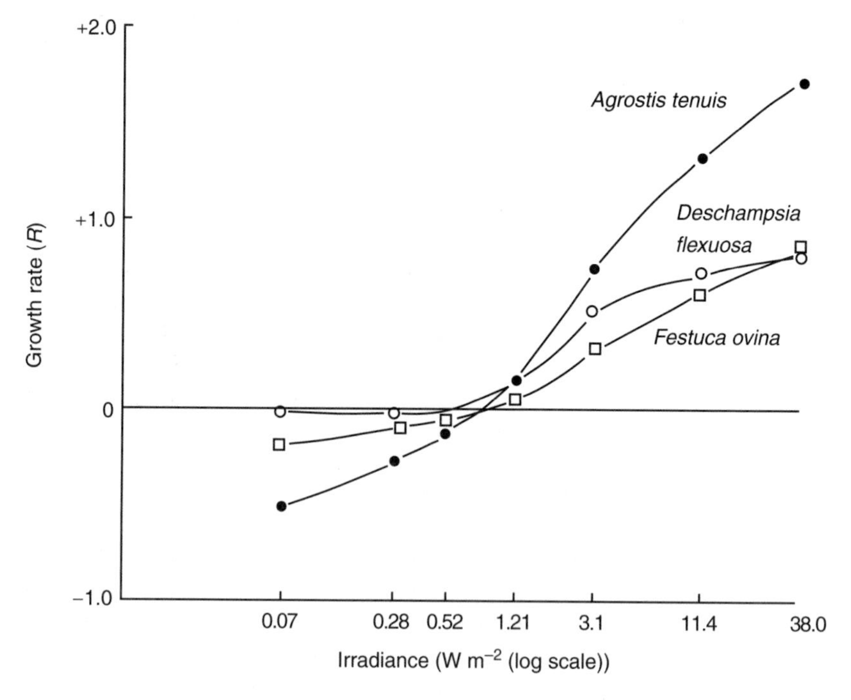

Figure 2.15

Effect of very low irradiance on growth rate in three grass species. Note the very small negative **R** of *Deschampsia flexuosa* (from Mahmoud and Grime, 1974)

A small reduction in respiratory rate is then a fairly general response to reduced irradiance, but as a major adaptive response it is only of value to severely shaded plants, and particularly as a survival mechanism for long-lived plants to persist during periods of temporary stress (cf. above). This is clearly the value of the near-dormant behaviour of *D. flexuosa* as has previously been recognized, for example by Chippindale (1932), who observed that seedlings of *Festuca pratensis* were able to survive long periods without growing when under severe competitive stress from older plants of *F. pratensis* and *Lolium italicum*, but could resume normal growth when the stress was removed. He termed the phenomenon inanition. Similarly, Hutchinson (1967) showed that woodland plants, such as *Digitalis purpurea* and *Bromus ramosus* could survive for 5–6 months in complete darkness, and very strikingly that *Deschampsia flexuosa* could persist for over 7 months. In contrast, small-seeded pioneer plants, such as *Betula pubescens* and *Erophila verna*, were generally least resistant to this stress. Equally, plants grown on nutrient-deficient soils, which reduced growth and probably also respiration rates, could survive for longer in complete darkness than those grown on fertile soil. In these cases, it is whole-plant carbon balance that is important: respiratory activity in all parts of the plant is reduced.

Not all respiratory CO_2 loss results in ATP production. Plants also operate a non-phosphorylating ('alternative') pathway (Lambers, 1997) whose function is unclear but which can account for a significant part of CO_2 production. It may act as an overflow, maintaining sink activity in parts of the plant undergoing little metabolism or growth, and minimizing potential down-regulation of photosynthesis caused by build-up of photosynthetic products. Another potential function is in the generation of heat in flowers of species as diverse as lotus (*Nelumbo nucifera*: Seymour and Schultze-Motel, 1996) and aroids (*Lysichiton* and *Arum:* Knutson 1974). Switching between these pathways can alter plant growth and metabolism without any change in overall respiration rate and carbon balance.

(c) Photosynthetic capacity. Since shading tends to cause an increase in *SLA* (and hence, at constant *LWR* in **F** also), relative growth rate (**R**) can theoretically be maintained, even with falling **E**. *SLA* and **F** are generally inversely correlated with irradiance, whereas **E** is directly proportional to it (Blackman and Wilson, 1951; Newton, 1963; Coombe, 1966), because **E** is principally a manifestation of the efficiency of the photosynthetic system. One possible mechanism of shade resistance, therefore, is an increased responsiveness of the photosynthetic system to irradiance. Sun and shade plants do indeed have very different light response curves (Fig. 2.16).

Generally a leaf with a sun physiology has high A_{max} (the saturating irradiance above which no further increase in photosynthesis occurs) and high saturating irradiance, and high compensation point; a shade physiology is the converse. *Cordyline rubra*, an extreme shade plant found in Australian rainforests has A_{max} of 2 μmol $CO_2\,m^{-2}\,s^{-1}$ and saturates at around 80 μmol photons $m^{-2}\,s^{-1}$. Corresponding values for the sun species *Atriplex triangularis*, are over 30 and 1000, respectively. There is therefore a clear distinction between sun and shade species on physiological grounds, specifically in compensation point and A_{max}. Both of these responses can be attributed mainly to changes in leaf

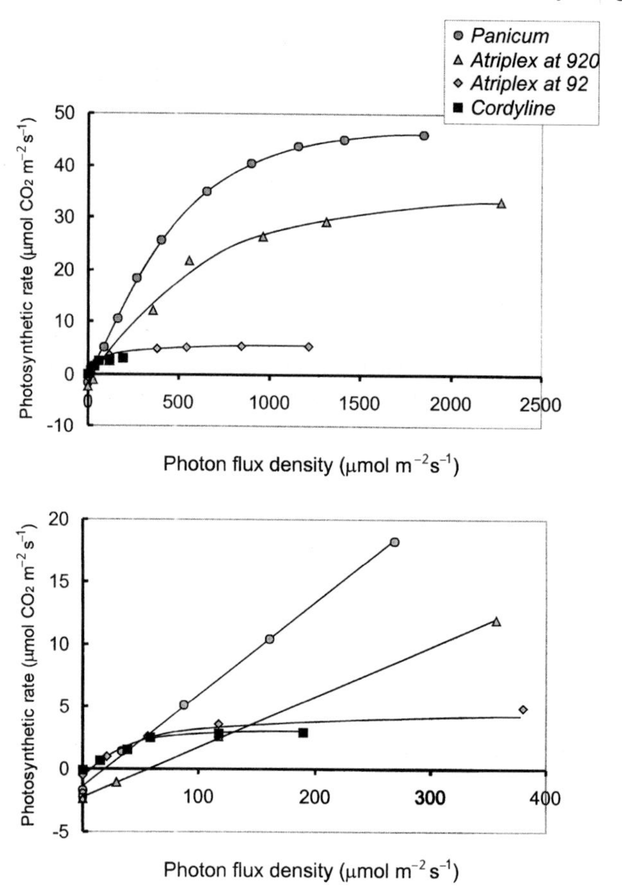

Figure 2.16

(a) Photosynthetic response curves to photon flux density for the C_4 plant *Panicum maximum* grown at 42 mol m^{-2} d^{-1}, the C_3 sun plant *Atriplex triangularis* grown at high (920 µmol m^{-2} s^{-1}) and low (92 µmol m^{-2} s^{-1}) irradiance regimes, and of the extreme shade plant *Cordyline rubra* grown under a rainforest canopy at 0.3 mol m^{-2} d^{-1}. (b) An expansion of the bottom left-hand corner of (a) to show the greater quantum efficiency of the shade-grown plants than the C_3 plant grown in bright light (redrawn from Björkman, 1981)

morphology; the thinner leaves of shade plants have less biochemical machinery and so respire less and can fix less carbon. When these variables are expressed on a unit mass basis, much less difference is seen between sun and shade leaves. Importantly, there is no difference between sun and shade leaves in terms of quantum efficiency, the initial slope of the photosynthetic response curve. Quantum efficiency is determined by the efficiency of the light-harvesting equipment, and there is no trade-off which ensures selection for less efficient light harvesting in sun leaves and species.

Sun and shade leaves on a single plant can be as distinct morphologically as those on sun and shade plants (cf. Table 2.5) and they show parallel differences in response to irradiation (Boysen-Jensen and Muller, 1929). This is a widespread, plastic response, not confined to plants characteristic of shaded habitats. Leaves of the forage grass *Festuca arundinacea*, grown previously at around 180 μmol $m^{-2} s^{-1}$, had faster rates of photosynthesis at high illumination (higher A_{max}) than those from plants previously grown in dim light (Woledge, 1971). The same is true for *Atriplex triangularis* in Fig. 2.16; when grown at 92 as compared to 920 μmol $m^{-2} s^{-1}$, A_{max} fell to 5 μmol $CO_2 m^{-2} s^{-1}$ and the saturating irradiance to around 400 μmol $m^{-2} s^{-1}$. Surprisingly, truly shade-resistant plants lack this plasticity and suffer damage which leads to reduced CO_2 fixation rates if exposed to high irradiance.

When a leaf is kept in the dark or in dim light and then brightly illuminated, there is a lag period of several minutes to a few hours before it can adjust its maximum photosynthetic rate to the new high radiant flux, this is known as induction. Two species native to deeply shaded rainforests in Australia, *Alocasia macrorrhiza* and *Toona australis*, illustrate this well. *Alocasia* is an understorey herb, which never grows in full sun, whereas *Toona* is a tree that develops in gaps caused by tree-fall. Their maximum photosynthetic rate declines by around 80% if they are maintained in low irradiance conditions (10 μmol $m^{-2} s^{-1}$), but exposure to 500 (*Alocasia*) or 1200 (*Toona*) μmol $m^{-2} s^{-1}$ for 30–40 min restores full activity (Chazdon and Pearcy, 1986a). Brief periods of high irradiance, simulating sunflecks and lasting 30–60 s, are just as effective (Fig. 2.17); the more shade-tolerant *Alocasia* benefits more from very brief illumination – the first 30 s fleck results in twice as much carbon fixation as in *Toona* – but a series of short flecks or as few as two longer (60 s) ones benefit *Toona* more (Chazdon

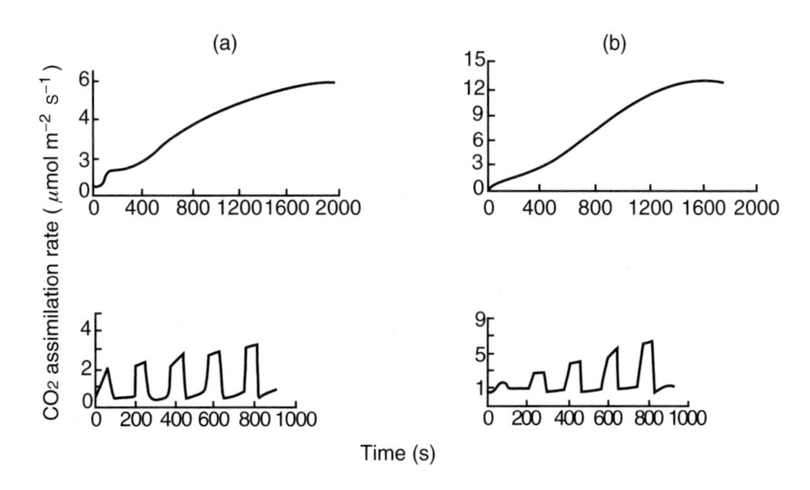

Figure 2.17

A comparison of the photosynthetic induction of (a) *Alocasia macrorrhiza* and (b) *Toona australis* in response to continuous illumination or a series of 60 s lightflecks. In each case the upper figure represents the time-course of induction when irradiance was raised from 10 to 400 (*Alocasia*) or from 20 to 1200 μmol $m^{-2} s^{-1}$ (*Toona*) permanently; the lower figure represents the induction under 60 s flecks (from Chazdon and Pearcy, 1986a)

and Pearcy, 1986b). For such plants to utilize sunflecks effectively, they must either be of long duration, or occur in rapid succession. In fact, flecks do often come in groups and may be of major importance in the carbon economy of shaded leaves.

The induction process by which plants react to fluctuating light has three components (Pearcy *et al.*, 1996). The most rapid response is due to an increase in the capacity to regenerate ribulose bisphosphate, which takes place over 1–2 min. The ecological role of this response is unclear, since it does not differ systematically among sun and shade species; it is important in both sun plants such as soybean and shade plants such as *Adenocaulon bicolor*, but not in the extreme shade plant *Alocasia macrorrhiza*. The next most rapid element in induction is the activation of Rubisco. In the dark, Rubisco reverts to an inactive form, that presumably minimizes resource or energy demands within the chloroplast (Salvucci, 1989). Sun and shade plants do appear to differ in the degree to which they inactivate Rubisco when under low light conditions. Neither the shade plant *Paris quadrifolia* nor the sun plant *Hordeum vulgare* (barley) showed inactivation after 2 min at 25 $\mu mol\,m^{-2}\,s^{-1}$. However, after 20 min barley photosynthesis was strongly repressed on return to high light (800 $\mu mol\,m^{-2}\,s^{-1}$) whereas *Paris* showed a much smaller effect (Fig. 2.18: Ögren and Sundin, 1996). Finally, induction can be affected by stomatal responses that determine the flux of CO_2 into the leaf.

There is no doubt that the ability to utilize sunflecks is of great importance to the survival of plants in deeply shaded environments, and probably also to the carbon balance of plants in complex stands, where lower leaves are growing in shade. However, the quantitative data to support this assertion do not agree on the fraction of carbon acquisition that is achieved during sunflecks. One species that has been well studied is *Adenocaulon bicolor* in redwood forests; depending on location and the precise local light climate, between 30 and 65% of daily carbon gain on sunny days (Pfitsch and Pearcy, 1989), and between 9 and 44% of annual carbon gain (Pearcy and Pfitsch, 1991) was accounted for by sunflecks.

3. Mechanisms

Populations of plants from habitats differently illuminated, and individuals or parts of individuals similarly distinguished, have differing photosynthetic responses to irradiance. Since these differences are expressed on a leaf area basis, they could have three causes:

(i) Improved access to substrate, i.e. higher CO_2 diffusion rates;
(ii) Increased capture of photons, i.e. more chlorophyll in the light-harvesting apparatus;
(iii) Increased activity of the biochemical apparatus of photosynthesis.

Changes in morphology and anatomy found in sun and shade leaves can certainly affect internal CO_2 diffusion (Holmgren, 1968; see also p. 48), but this probably rarely limits photosynthesis. Changes in resistance to CO_2 movement between leaf and air, however, may be important. *Atriplex triangularis* showed a threefold increase in stomatal conductance for CO_2 when grown at 920 as compared to 92 $\mu mol\,m^{-2}\,s^{-1}$, but at any given conductance, the high irradiance

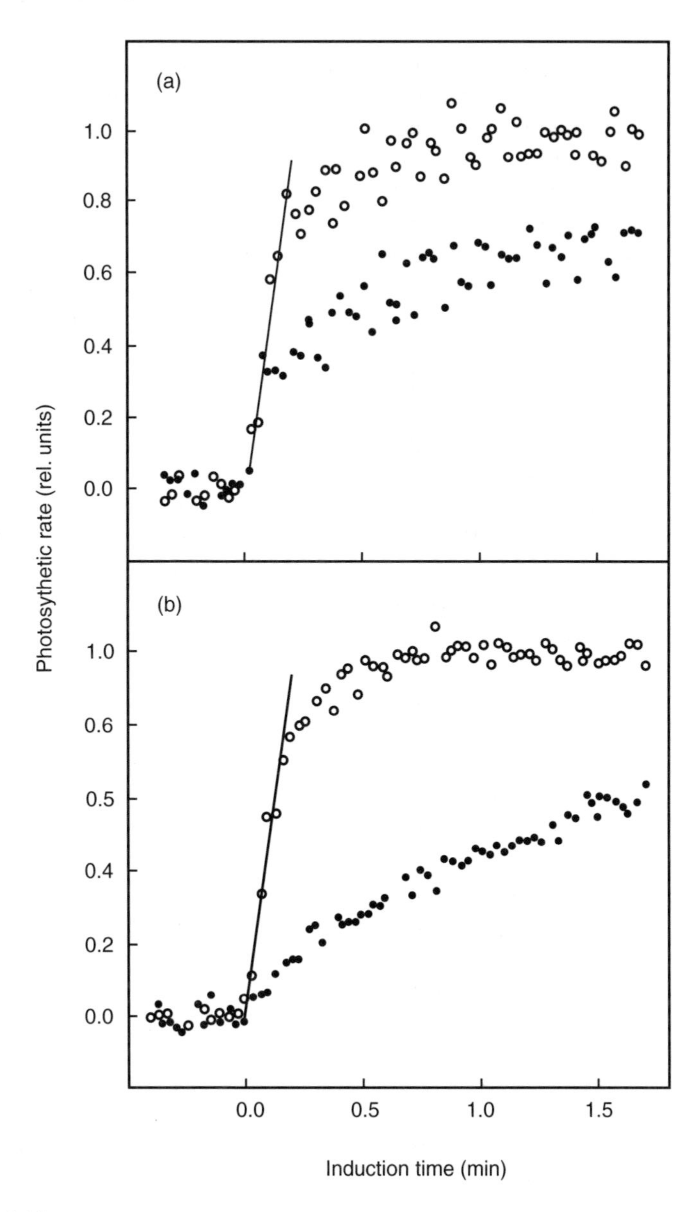

Figure 2.18

Exposing leaves of (a) the shade plant *Paris quadrifolia* and (b) the sun plant *Hordeum vulgare* (barley) to a high PPFD (800 μmol m^{-2} s^{-1}) after either 2 minutes (○) or 20 min (•) in dim light (25 μmol m^{-2} s^{-1}) produced contrasting results. *Paris* maintained a high Rubisco activity even after 20 min in dim light, whereas barley took much longer to recover full photosynthetic activity after such a long period. After 2 min in dim light, the species behaved similarly (from Ögren and Sundin, 1996).

plants photosynthesized much faster (Björkman *et al.*, 1972), a difference that must have been due to changes in the photosynthetic system.

Shaded leaves generally have enhanced chlorophyll concentrations per unit weight, though often not per unit area (Shirley, 1929), largely because of differences in the allocation of nitrogen within the leaf. The increase in chlorophyll concentration is especially in the so-called 'antennal' chlorophyll molecules that are excited by photons, as opposed to those that comprise the reaction centres in each photosystem (Evans, 1989), and shaded leaves conversely have lower concentrations of the primary CO_2-fixing enzyme, Rubisco; Rubisco and chlorophyll are effectively competing sinks for limiting N supplies. The lower Rubisco concentration reduces maximum photosynthetic capacity, whereas the increase in antennal chlorophyll maximizes light capture rates, increasing photosynthetic rates in dim light.

Björkman and Holmgren (1963) grew plants of *Solidago virgaurea* from shaded and exposed habitats in low ($30\ W\,m^{-2} = 6.6\ mol\,m^{-2}\,d^{-1}$) and high ($150\ W\,m^{-2} = 33\ mol\,m^{-2}\,d^{-1}$) irradiances and found that A_{max} was generally higher in the plants from exposed habitats, an indication of genetic differentiation (Table 2.5). In these plants A_{max} was reduced if the plants were grown in low light, showing physiological plasticity. The plants from shaded habitats, by contrast, had the same maximum photosynthetic rate irrespective of the irradiance at which they were grown; but the quantum efficiency was increased by growth in low irradiance in these shade plants, while remaining unchanged in the exposed plants. The two sets of populations are therefore physiologically distinct, but both are capable of plastic modification, which occurs by raising A_{max} in the sun plants and by raising the quantum efficiency in the shade plants. These adaptations are appropriate to the situations likely to be encountered by the plants, although plants with high leaf turn-over may be able to adjust by producing new leaves with altered physiology, rather than by altering the physiology of extant leaves.

Table 2.5

Differences in photosynthetic characteristics and plasticity in clones of goldenrod *Solidago virgaurea* from shaded and exposed habitats (Björkman and Holmgren, 1963)

	Habitat of plants	
	Shaded	Exposed
1. Light-saturated photosynthetic rate (A_{max} μmol $CO_2\,m^{-2}\,s^{-1}$); plants grown in well-lit conditions ($33\ mol\,m^{-2}\,d^{-1}$)	11.3	16.1
2. Ratios of A_{max} for plants grown at $33\ mol\,m^{-2}\,d^{-1}$ to that of plants grown in dimly lit conditions ($6.6\ mol\,m^{-2}\,d^{-1}$)	0.97	1.84
3. Ratios of quantum efficiency for plants grown at $33\ mol\,m^{-2}\,d^{-1}$ to that of those grown at $6.6\ mol\,m^{-2}\,d^{-1}$	0.64	1.04

In *Solidago virgaurea* (Björkman, 1968) Rubisco activity was 0.25 μmol $CO_2\,mg^{-1}$ protein min^{-1} for plants from exposed habitats grown in strong light, and only 0.21 for those grown in weak light. Plants from shaded habitats gave values of 0.15 and 0.12 for strong and weak light, respectively. Here, both genetic differentiation and plasticity are apparent. On the other hand, although Hiesey *et al.* (1971), found a strong correlation between Rubisco activity and radiant flux density in two species of *Mimulus*, when expressed on a fresh weight basis (4.4 μmol $CO_2\,g^{-1}\,min^{-1}$ at 18 $W\,m^{-2}$ to 13.5 at 106 $W\,m^{-2}$), this was associated with a parallel increase in total protein; all activities were around 0.5 when expressed per unit protein. The increase in Rubisco activity may not, therefore, always be specific.

Generally, therefore, strict shade plants (as opposed to shade leaves on sun plants) do not respond to increases in irradiance by increasing enzyme activity. Since Rubisco is the largest single component of leaf protein, this is a major saving in nitrogen allocation, that becomes available for antennal chlorophyll, maximizing light capture. Not all plants, however, show the same forms of adaptation. In *Solidago* the sun and shade races are clearly differentiated in their response to irradiance, but in *Solanum dulcamara* at least one shade race had a 'sun' physiology and was more sensitive to water status (Gauhl, 1969, 1979), and two species of *Mimulus*, the coastal *M. cardinalis* and the subalpine *M. lewisii*, were able to grow equally well in both low (25 $W\,m^{-2}$) and high (120 $W\,m^{-2}$) irradiance (Hiesey *et al.*, 1971).

2. Photosynthesis at high irradiance

1. Photoprotection and photoinactivation

High irradiance is a relative term. Shade plants suffer reversible damage when grown in full sunshine (Björkman and Holmgren, 1963): shade-adapted plants of *Solidago virgaurea*, grown for one week at a high irradiance, had reduced quantum efficiency compared to control plants (Table 2.5), but after a week at low irradiance this damage was repaired. Similar effects have been shown for seedlings of *Quercus petraea* (Jarvis, 1964): in sun leaves, photosynthesis at 40 $W\,m^{-2}$ was reduced by 12% by prior exposure for 2 h to 400 $W\,m^{-2}$, whereas the corresponding reduction for shade leaves was 45%. This effect is known as photoinhibition, but it is a complex phenomenon and the more precise terms photoprotection and photoinactivation are preferred (Osmond *et al.*, 1999).

The problem of coping with high radiation flux is not unique to shade plants. All plants experience wide variations in radiation flux, and the adaptive mechanisms discussed above, that effectively economize on the investment of N in the leaf, may result in the photosynthetic reaction centres in the chloroplast being unable to utilize the absorbed quanta. The response of photosynthesis to radiant flux is always hyperbolic (Fig. 2.12), whereas the relationship between quanta absorbed and the flux is linear. There will always be the potential, therefore, for the chloroplast to find itself with excess energy to be dissipated. Most of this excess is lost as heat, with a small proportion also by fluorescence (Krause and Weis, 1991); this mechanism is photoprotection and it is rapidly reversible. It depends on the conversion of xanthophyll molecules within the chloroplast; violaxanthin plays a key role in the electron transport chain. If metabolic activity is unable to utilize the ATP generated by charge separation in photosynthesis, a proton gradient builds up, causing a rise in pH in the thylakoids and the conversion of violaxanthin to zeaxanthin. The zeaxanthin is responsible for dissipation of energy as heat. This xanthophyll cycle seems to be universal in photosynthetic organisms (Demmig-Adams and Adams, 1996).

If excess energy is not dissipated, it can cause photoinactivation of the photosystem II reaction centre. Although photoinactivation is reversible, the process is slower, and there is an immediate loss of photosynthetic competence and a direct cost of the repair. Both photoprotection and photoinactivation can be quantified by detecting the fluorescence emitted by photosystem II, where water is split generating protons and electric charge.

The hyperbolic light response curve for photosynthesis shown in Fig. 2.12 demonstrates that most plants cope poorly with very high PAR fluxes. Even when other factors, such as water, are not limiting, the efficiency of photosynthesis for most plants at PFD values above around 400 $\mu mol\,m^{-2}\,s^{-1}$ is far below that which would be obtained by extrapolating the initial slope; in other words the quantum efficiency progressively declines, because of processes such as Rubisco oxygenation and limitations to electron transport. From biochemical first principles, the efficiency of photosynthesis should be 1 mol CO_2 assimilated per 8 photons absorbed. That would give a quantum efficiency of 0.125. In practice, values of around 0.1 are typically found, because some of the electron flow generated by the photosystems goes to other electron acceptors. When plants are exposed to excess light, the quantum efficiency typically declines markedly (e.g. Fig. 2.19), in the case illustrated because of a great increase in dissipation of light energy, in other words photoprotection (Osmond *et al.*, 1999).

Photoinhibition becomes particularly acute when other factors, notably low temperature, reduce photosynthesis, for example at the alpine treeline (see

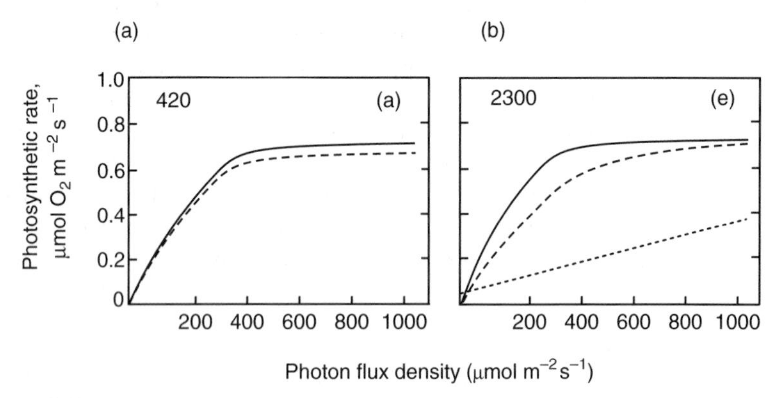

Figure 2.19

Photosynthetic rate of *Arabidopsis thaliana* grown at 230 $\mu mol\,m^{-2}\,s^{-1}$ and then exposed to either (a) 420 $\mu mol\,m^{-2}\,s^{-1}$ or (b) 2300 $\mu mol\,m^{-2}\,s^{-1}$. The solid line in each case represents the response curve at 230 $\mu mol\,m^{-2}\,s^{-1}$; in (a) the dashed line represents the rate after 6 h, and in (b) after 20 min (dotted line) or 1 hour (dashed line). Note the much reduced quantum efficiency of plants exposed to very bright light for a long period in (b) (from Osmond *et al.*, (1999, after Russell *et al.*, 1995)

Chapter 5). Evergreen trees appear to be unable to recolonize clear-felled areas of old forests in Australia largely because the combination of high irradiation and low temperatures in the exposed cleared areas causes severe damage to photosynthetic systems of species such as *Eucalyptus polyanthemos* and *E. pauciflora* (Ball, 1994). A similar effect was found with the evergreen beech *Nothofagus solanderi* in New Zealand. Seedlings may be able to establish only underneath the canopy of the parent tree, where they are protected from early morning sunlight at a time when leaves may be frosted.

There is genetic variation within species for susceptibility to photoinhibition, suggesting that selection acts on the trait; plants of the sedge *Cyperus longus* taken from the northern edge of its range in Europe had a greater capacity to repair photoinhibitory damage than did populations from southern Europe (Gravett and Long, 1990). The ability to cope with high irradiance during cold weather is an essential part of the adaptive strategy of all non-tropical species, and it appears that there is a range of mechanisms involved, which may be appropriate to different conditions. Evergreen species, alpine species and winter annuals (such as many cereals) will experience different patterns of this stress and they show different responses (Huner *et al.*, 1998). Conifers, for example, accumulate large concentrations of xanthophylls, which provide a high level of photoprotection, whereas alpine species such as *Oxyria digyna* have an increased capacity to repair damaged reaction centres in PSII, and winter cereals have increased photosynthetic capacity.

2. Alternative photosynthetic pathways

In 1965 Hatch and Slack demonstrated the existence of a new photosynthetic pathway in sugar cane, *Saccharum officinarum*. The C_4 pathway uses the enzyme phosphoenolphosphate (PEP) carboxylase for the primary fixation of CO_2, and the fixed carbon travels initially through a sequence of 4-carbon dicarboxylic acids – oxaloacetate and either malate or aspartate, but the CO_2 is eventually released and re-fixed by the universal carboxylating enzyme Rubisco. It is best thought of therefore as a CO_2 concentrating mechanism. Plants possessing this pathway are usually anatomically distinct as well, having 'Kranz' anatomy, with a well-marked bundle sheath surrounding the vascular bundles and the chloroplasts concentrated in a ring of mesophyll cells radiating out from the sheath (Fig. 2.20), and in the sheath itself. It is in these bundle sheath cells that the malate or aspartate is decarboxylated, liberating CO_2 to be re-fixed by Rubisco (Fig. 2.21a). There are also ultrastructural differences in the chloroplasts. The major distinctions between C_3 and C_4 plants are listed in Table 2.6.

Carbon dioxide fixation by PEP carboxylase is not in itself novel; it is the initial event of another photosynthetic syndrome known as crassulacean acid metabolism (CAM), characteristic of desert succulents but also of a surprising range of other plant types (Ting, 1985) and it also occurs in non-photosynthetic tissues. In CAM, however, fixation occurs at night and CO_2 is released again in the daytime for photosynthesis; the malic acid produced by fixation is stored in the vacuole at night and transported back into the cytoplasm and decarboxylated in daylight (Lüttge and Smith, 1982). The quantity of acid produced can be very high – up to 1.4M H^+ in *Clusia minor* (Borland *et al.*, 1992)

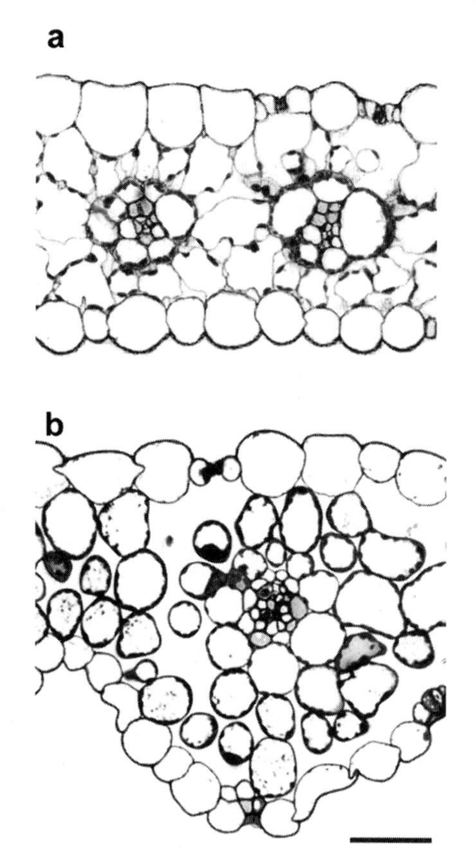

Figure 2.20

Contrasting leaf anatomies in (a) C_4 maize and (b) C_3 wheat primary leaves, stained with toluidine blue. Note the prominent sheath of photosynthetic cells, characteristic of Kranz anatomy, around the vascular bundles in the maize leaf. Scale bar = 50 μm. Photographs courtesy of J Marrison.

– and the cytoplasm could not function at the pH values that this would produce. In C_4 plants, the PEP carboxylase is active at the same time as Rubisco, resulting in very high saturating PPFD, high temperature optima, and the ability to reduce mesophyll space CO_2 concentrations to very low levels (around 100 $\mu l\, l^{-1}$ or 10 Pa). This very low internal CO_2 concentration increases the concentration gradient across the stomata and hence also the rate of CO_2 diffusion, and alleviates one of the limiting factors in photosynthesis at high photon flux densities. C_3 plants cannot lower internal CO_2 concentrations below about 250 $\mu l\, l^{-1}$ (25 Pa) because of the oxygenase capacity of Rubisco, which results in the phenomenon known as photorespiration – effectively reverse photosynthesis. In C_4 plants PEP carboxylase acts as a CO_2 scavenging system, and as a result photosynthetic rate is independent of O_2 concentration, in contrast to C_3 plants in which the rate increases markedly with declining O_2 concentrations.

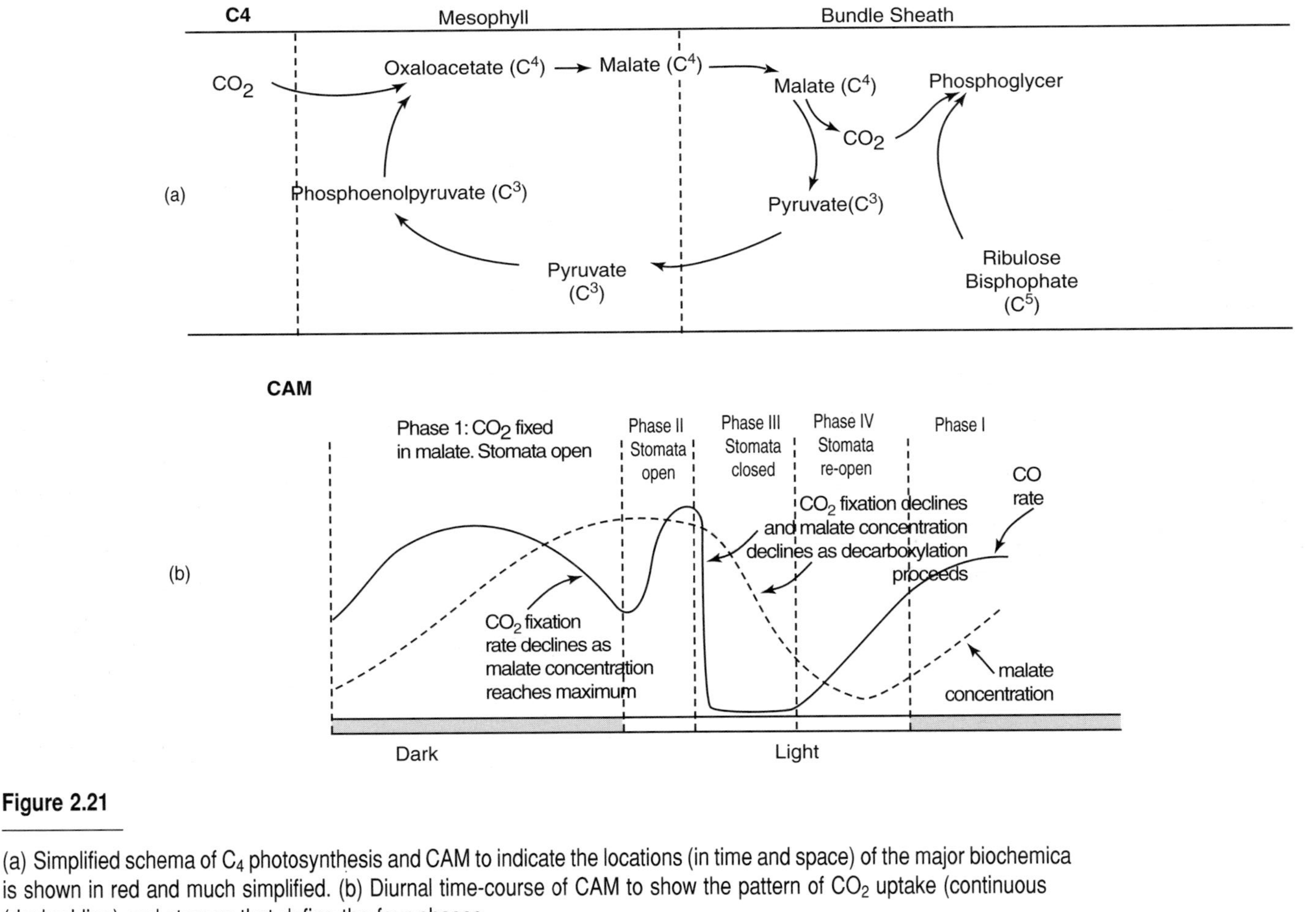

Figure 2.21

(a) Simplified schema of C_4 photosynthesis and CAM to indicate the locations (in time and space) of the major biochemica is shown in red and much simplified. (b) Diurnal time-course of CAM to show the pattern of CO_2 uptake (continuous (dashed line) and storage that define the four phases

Table 2.6

Characteristics of C_3 and C_4 photosynthesis

Characteristic	C_3	C_4	CAM
Initial CO_2-fixing enzyme	RuBP carboxylase	PEP carboxylase	PEP carboxylase
Time of initial C fixation	Day	Day	Night
Operating internal CO_2 concentration ($\mu l\ l^{-1}$)	220–260	100–150	Variable (to >1000)
Effect of O_2	Inhibitory (photorespiration)	None in range 2–21 kPa	Slight
Temperature response 20–40 °C at 330 $\mu l\ CO_2\ l^{-1}$	Usually slight	Strong	Strong
Water use efficiency	Low	High	Very high
Typical $\delta^{13}C$ (see p. 179)	−27‰	−14‰	Variable

From an ecological standpoint, however, the most obvious feature of C_4 photosynthesis is the greater water use efficiency it permits (cf. Chapter 4, p. 179). Because the internal CO_2 concentration in the leaf is lower than in C_3 plants, and the gradient between air and mesophyll across the stomata is steeper, the same rate of CO_2 diffusion can be achieved with stomata less open, giving a lower stomatal conductance. This low conductance reduces water loss and therefore unit carbon gain can be achieved for reduced water loss. Unsurprisingly, C_4 plants are most frequent in hot, dry sunny environments and are largely confined to low latitudes, but even there they are rarely more abundant than C_3 plants (see Fig. 4.16). There are a few temperate C_4 species, for example *Spartina anglica* (Mallot *et al.*, 1975) and *Euphorbia peplis* (Webster *et al.*, 1975); both are coastal species where irradiance is high and salinity may affect water relations.

Other temperate species, such as *Jovibarba sobolifera*, *Sedum acre* and several *Sempervivum* species have CAM (Osmond *et al.*, 1975); all are succulent members of the family Crassulaceae, from which the syndrome takes its name. CAM differs from C_4 photosynthesis principally in the temporal separation of PEP carboxylase and Rubisco fixation steps in CAM. There are four distinct phases in a 24 hour cycle of CAM. In Phase I, at night, the stomata are open, CO_2 is combined with pyruvate by PEP carboxylase (PEPC) to form malate and the malate is transported into the vacuole. As *PAR* increases after sunrise, there is a brief Phase II, in which the stomata are still open and both modes of CO_2 fixation operate; then the stomata close, PEPC becomes inactive, malate is transported from vacuole to cytosol, decarboxylated, and the liberated CO_2 fixed by Rubisco (Phase III). Phase IV is the transition back to Phase I which is triggered by the exhaustion of the malate store (Fig. 2.21b).

One morphological consequence of CAM physiology is succulence, the accumulation of large volumes of water in the cells, since a large vacuolar volume is required to store the malic acid produced at night. In strict CAM plants, stomata are closed in daytime and open at night, so minimizing water loss; again succulence is a necessary feature since evaporative cooling cannot occur and the plant relies on the high specific heat of the stored water acting as a heat sink (cf. Chapter 4, p. 131). In extreme situations, where water supply is so limited that the stomata are permanently closed and all gas exchange ceases, some CAM plants continue to display acidity fluctuations apparently because they re-utilize CO_2 produced by respiration. This behaviour has been termed CAM-idling (Rayder and Ting, 1983) and is distinct from CAM-cycling (Ting, 1985), where organic acid levels fluctuate, although normal daytime C_3 fixation continues. This happens in many bromeliads and may be a 'tick-over' mechanism, that allows facultative CAM plants to respond rapidly to environmental changes.

Since C_4 photosynthesis was initially discovered in and is largely confined to tropical plants, it was at first thought to be a distinct evolutionary line, but since both C_3 and C_4 species occur in single genera such as *Atriplex* (Björkman *et al.*, 1971) and *Euphorbia* (Webster *et al.*, 1975) this is obviously not so. Both C_4 photosynthesis and CAM have evolved frequently, principally in response to the selective pressure of hot, dry environments. For example CAM is currently known from 33 families (Smith and Winter, 1996), including monocots, dicots, ferns and the primitive seed plant *Welwitschia* (Gnetopsida). There is little novel biochemistry involved in the C_4 pathway, and the classification of a plant as C_3, C_4 or CAM therefore depends on the degree of morphological specialization (Kranz anatomy, succulence), metabolic behaviour (e.g. acidity fluctuation) and the relative activity of the two carboxylating enzymes, and so, possibly, on environmental conditions. Epiphytes, for example, even in wet forests, may suffer water deficits because they have limited access to water. Of 27 epiphytic bromeliads examined for CAM by Medina and Troughton (1974), 13 showed no dark CO_2 fixation and discriminated strongly against the stable isotope ^{13}C (a characteristic of initial Rubisco fixation and not shown where initial fixation is by PEP carboxylase; see p. 179); these were considered to be C_3 plants. A further 13, which exhibited dark fixation and had low discrimination, were classified as C_4 or CAM plants. One species, *Guzmania monostachya*, had pronounced dark fixation (accumulating 27.5 μmol malate g^{-1} in 12 h in the dark) but strong discrimination, implying that both enzyme systems were operating. CAM plants are often found in dry habitats where dark fixation is an advantage, as stomata can be closed in daytime and so conserve water. *Guzmania monostachya* is an epiphyte in sunny, humid habitats, where possibly both pathways can be valuable, but bromeliads in general have very flexible photosynthetic systems, many having CAM-cycling.

Carbon isotope discrimination analyses show that many species can switch between pathways, and in *Frerea indica*, the perennial succulent stems have CAM whereas the seasonal leaves, only produced in the wet season, have C_3 photosynthesis (Lange and Zuber, 1977). Even more striking evidence of the adaptability of these various CO_2 fixation pathways is the inducibility of CAM metabolism in the succulent *Mesembryanthemum crystallinum* (Winter, 1974; and see Fig. 1.9, p. 18). Induction can be signalled by a reduction in water availability to the roots, from which a signal moves to the shoots (Eastmond and Ross, 1997). Within 7–14 days of the onset of water stress the activity of PEP carboxylase rises and dark fixation starts, which can be shown to be due to *de novo* synthesis of the enzyme (Von Willert *et al.*, 1976; Queiroz, 1977), and there is a general trend from C_3 to CAM as the season progresses and water becomes scarcer (Fig. 2.22; Winter *et al.*, 1978). This switch has been shown to increase reproductive success, clearly demonstrating the adaptive significance of CAM (Winter and Ziegler, 1992). In fact many CAM species, particularly those in the Crassulaceae (e.g. stonecrops), and including the major commercially significant CAM plant, pineapple, which is a bromeliad, show a mixture of C_3 photosynthesis and CAM, largely controlled by their water status. It would be wrong, however, to see CAM solely in terms of water stress. The important feature is the flexibility of time of CO_2 fixation it confers. The aquatic pteridophyte *Isoetes* and the three unrelated aquatic angiosperm genera *Crassula*, *Littorella* and *Alisma* all exhibit CAM (Keeley, 1996) and clearly they do not suffer water deficits. The postulated explanation is that they are unable to compete for CO_2 (as dissolved HCO_3^-) with other aquatic plants in the daytime, but can utilize HCO_3^- at night by a CAM mechanism.

3. Photosynthetic pathways compared

As soon as the intrinsically greater potential photosynthesis of C_4 plants was recognized, much interest centred on the implications for plant productivity, for

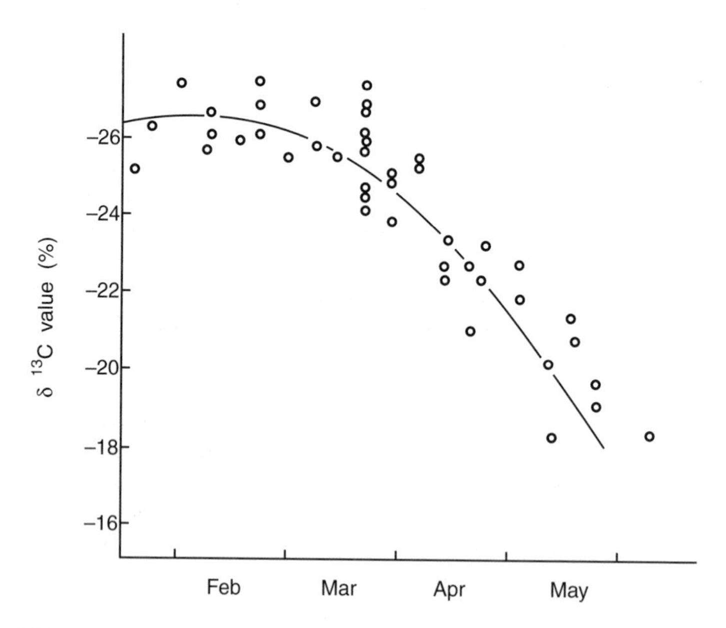

Figure 2.22

Induction of CAM in *Mesembryanthemum crystallinum* growing in the wild on rocky cliffs by the Mediterranean in Israel. The declining $\delta^{13}C$ value shows decreasing discrimination against the heavy carbon isotope in CO_2, which indicates that RuBP carboxylase was no longer the primary CO_2-fixing enzyme (redrawn from Winter *et al.*, 1978)

example by breeding C_4 characters such as lack of photorespiration into C_3 crop plants, such as wheat. The ecological implications of an apparent division of plants into classes of high and low productivity also excited interest (Black, 1971). The basis for these comparisons was, however, often inappropriate and many C_3 plants have a high photosynthetic capacity and, under favourable conditions, can achieve photosynthetic rates comparable with those of C_4 plants (Pearcy and Ehleringer, 1984).

The C_4 pathway represents a distinctive adaptation. Under conditions of high irradiance, high temperature, and long growing seasons, and particularly where water deficits play a greater role in determining plant distribution, PEP carboxylase is commonly the primary CO_2-fixing enzyme in grasses but less so in dicots. Even in environments of this type, however, C_3 plants are usually more frequent, and at high latitudes C_4 plants are rare (Fig. 4.16).

The large differences in photosynthetic rates often reported have as much to do with the environmental conditions as with photosynthetic capacity. Hot deserts are one of the classic C_4 environments, and yet they also contain large numbers of C_3 species. In the flora of short-lived, ephemeral species that develop following sporadic rainfall in the Sonoran desert, there are both C_3 and C_4 species. C_3 ephemerals predominate in the flora that develops following winter rainfall when daily maximum air temperatures are 15–30 °C. In summer these temperatures increase to 35–45 °C, and the corresponding flora is almost exclusively C_4. Nevertheless, the two groups have similar A_{max} at appropriate temperatures (Fig. 2.23). The perennials *Atriplex hymenelytra* (C_4) and *Larrea tridentata* (C_3), which grow in the same habitat, also give no evidence to support the contention that C_4 photosynthesis is inherently more productive, even in this extreme habitat, for they have almost identical A_{max} values (Pearcy and Ehleringer, 1984). Since, in less extreme environments, photosynthesis is probably as much limited by the rate of utilization of fixed carbon in the rest of the plant (i.e. it is sink-limited) as by environmental factors, it is probably futile to attempt to breed C_3 crop plants for more efficient photosynthesis. It may be significant that leaves of C_3 plants are more palatable than those of C_4 plants to grasshoppers (Caswell *et al.*, 1973) and to sucking insects (Kroh and Beaver, 1978), since very productive plants are normally considered more palatable to insects (Harper, 1969). Possibly herbivore pressure in the tropics has been so great that extra photosynthate in C_4 plants is diverted into chemical defence (see Chapter 7).

The critical variables, however, are probably water use efficiency and temperature. The balance between the carboxylation and oxygenation activities of Rubisco is temperature sensitive, the latter being progressively more important as temperature rises. As a result the quantum yield of C_3 photosynthesis (that is the number of moles of CO_2 fixed per mole of photons absorbed) declines by about 35% over the range 15–40 °C (Ehleringer and Werk, 1986). In contrast the quantum yield of C_4 photosynthesis is more or less constant, because the transfer of CO_2 from PEP carboxylase means that Rubisco is operating at a high enough CO_2 concentration to inhibit the oxygenase reaction. The quantum yield of C_3 photosynthesis is greater than that of C_4 photosynthesis at temperatures below about 30 °C and less above. As much as anything else, this may explain the general distribution of C_3 and C_4 plants, although sensitivity to low night-time temperatures is also

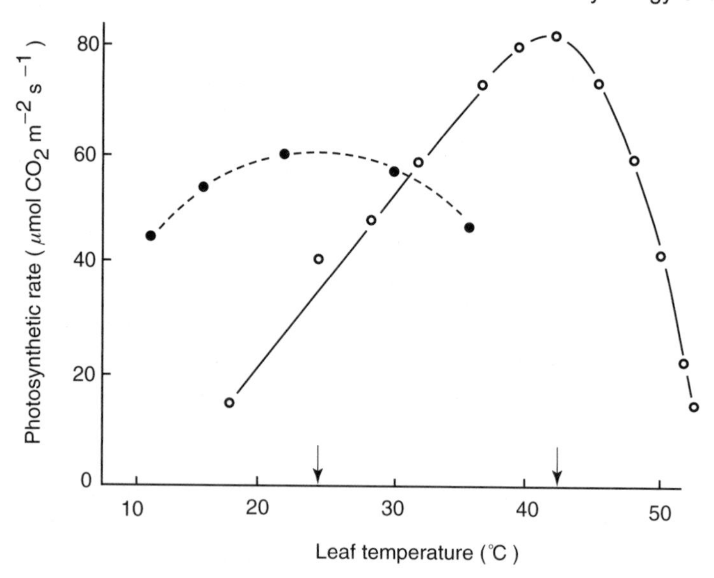

Figure 2.23

The relationship of photosynthetic rate to leaf temperature in two desert ephemerals, the cool-season C_3 *Camissonia claviformis* (- - - - -) and the hot-season C_4 *Amaranthus palmeri* (——). The temperature optimum for *A. palmeri* was almost 20 °C higher, but the maximum photosynthetic rates were similar, and at 25 °C, which is roughly the temperature optimum for *C. claviformis* (A_{max} about 60 μmol CO_2 m^{-2} s^{-1}), the photosynthetic rate of *A. palmeri* was only about 30 μmol CO_2 m^{-2} s^{-1}. From Pearcy and Ehleringer (1984).

considered important. Several studies (Ehleringer, 1978; Vogel *et al.*, 1986) have shown distribution patterns of C_4 plants which correspond to temperature or rainfall gradients (see Fig. 4.16), but it is not possible to draw firm conclusions about causation from such correlative data. The distribution of CAM plants, both geographically and taxonomically, is more complex than that of C_4 plants. Although the very high water use efficiency of CAM may be the dominant selective force in many current situations, the existence of aquatic photosynthesis in the most ancestral of all CAM taxa, *Isoetes*, suggests that other features of CO_2 fixation of CAM may have been more significant originally. Because CAM photosynthesis (Phase III) occurs behind closed stomata, the $CO_2 : O_2$ ratio is very high and Rubisco function consequently more efficient.

5. Responses to elevated carbon dioxide concentrations

1. Photosynthetic responses

One of the few secure generalizations about global change is that atmospheric CO_2 concentrations have increased as a result of anthropogenic emissions (principally burning fossil fuels and clearing and burning forests) from a preindustrial level of around 270 $\mu l\,l^{-1}$ to a current concentration of about 365 $\mu l\,l^{-1}$, and that this concentration will continue to increase for the foreseeable future. What eventual concentrations are reached will depend on the abilities of governments and individuals to regulate the two main causes. Concentrations of 500 $\mu l\,l^{-1}$ are very likely by the year 2100 (Houghton *et al.*, 1990), and some predict that 700 or greater is possible. These changes will have a profound influence on photosynthetic carbon fixation.

When plants are grown at elevated CO_2 concentrations, the rate of photosynthesis is typically increased. The principal effect is on A_{max}, the maximum carbon fixation rate. At 350 $\mu l\,l^{-1}$ CO_2, the photosynthetic response to *PAR* levels off at about 20 $\mu mol\ CO_2\,m^{-2}\,s^{-1}$ in C_3 plants such as wheat, largely because of the limitation caused by CO_2 availability (Cf. Fig. 1.5, p. 9). At 700 $\mu l\ CO_2\,l^{-1}$, A_{max} typically rises to 25–30 $\mu mol\ CO_2\,m^{-2}\,s^{-1}$, although the stimulation in photosynthetic rate may be inversely proportional to the rate (Fig. 2.24; Greer *et al.*, 1995). Species with high photosynthetic rates are probably limited by their ability to regenerate ribulose bisphosphate, and so are less able to respond to increased availability of CO_2. Nevertheless, there is potential for elevated atmospheric CO_2 concentrations to result in increased productivity, with obvious benefits to agriculture, though less certain consequences for natural communities. In practice, other factors discussed below complicate the story considerably.

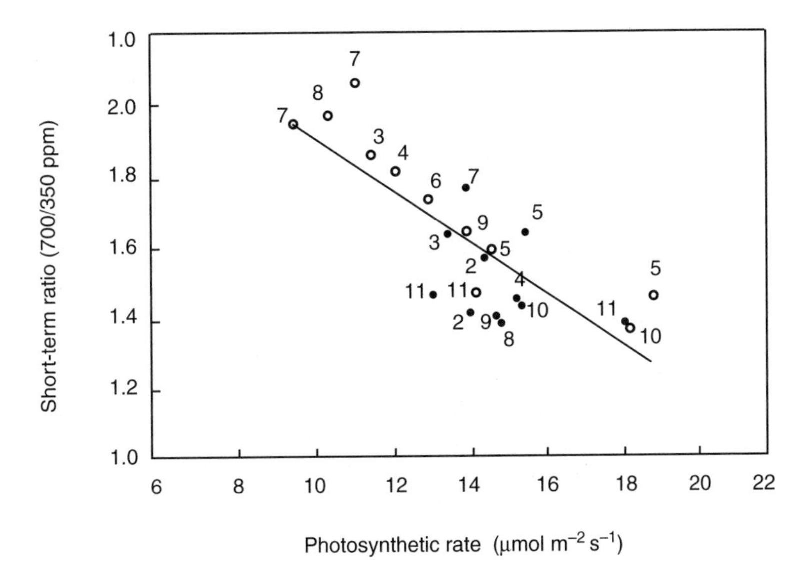

Figure 2.24

Photosynthetic rate at 700 $\mu mol\ CO_2\,mol^{-1}$ relative to that at 350 $\mu mol\,mol^{-1}$ was inversely proportional to the rate at 350 $\mu mol\,mol^{-1}$, for a range of species tested by Greer *et al.* (1995). Solid circles represent plants grown at 350 and open circles plants grown at 700 $\mu mol\,mol^{-1}$; numerals refer to plant species. All plants were grown and measured at 18 °C

Most of the studies on the effect of atmospheric CO_2 concentration on photosynthesis are carried out over short time periods. When a leaf is moved from ambient (~350 μl $CO_2 l^{-1}$) to an elevated concentration, it will normally respond by an increase in photosynthetic rate, depending on the extent to which CO_2 concentration was limiting the rate. If the plant is maintained at the higher CO_2 concentration for a long period, however, the photosynthetic rate may decline; in other words photosynthesis is down-regulated. There are two possible reasons for this. First, it may be that the leaf nitrogen concentration falls; lower leaf N concentrations are one of the most consistent observations in plants grown in elevated CO_2. In that case, the plant may simply have thicker leaves (lower *SLA*, see p. 47) and a reduced photosynthetic rate on a unit mass basis but not per unit area, and consequently whole-plant carbon gain will not be affected. An alternative, however, is that accumulation of soluble carbohydrates in the leaf will result in direct down-regulation of genes for photosynthetic enzymes (van Oosten *et al.*, 1994), with real reductions in photosynthetic activity and of carbon gain on a whole-plant basis. Evidence on this is unclear: Baxter *et al.* (1995) found a strong negative relationship between photosynthetic capacity and total soluble carbohydrate in leaves for one of three grasses they studied (*Poa alpina*; Fig. 2.25) but not for two others, for which leaf N concentration was a better predictor of photosynthetic capacity. This probably indicates that comparisons between plants grown under standardized growth-room conditions are hard to make, since, although the conditions are the same

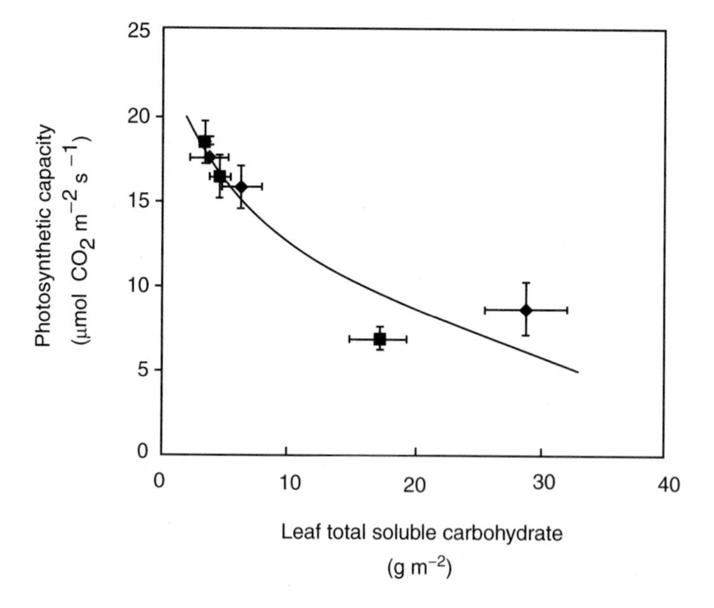

Figure 2.25

Photosynthetic capacity of *Poa alpina*, an alpine grass, was a negative function of leaf carbohydrate concentration, suggesting down-regulation of photosynthesis caused by low sink activity (from Baxter *et al.*, 1995)

for all species, each species responds differently to those standard conditions and hence the comparisons are not like-with-like.

Raised CO_2 concentrations are not unprecedented. The relative frequency of the stable carbon isotopes ^{13}C and ^{12}C in ancient carbonate deposits reveals that during the Devonian period (~400 million years ago) atmospheric CO_2 concentration was probably 20 times greater than now (Berner, 1993). It is probably no coincidence that this was the time when plants first colonized the land; the high CO_2 concentrations would have reduced the problem of controlling water loss in plants that had not yet evolved root systems to acquire water, or stomata to regulate its loss. Just as the low internal CO_2 concentration achieved by C_4 plants steepens the CO_2 concentration gradient across the stomata and so enables C_4 plants to fix more carbon than C_3 plants at a given leaf diffusive resistance, so increasing the external CO_2 concentration similarly increases water use efficiency. Given time, many species exposed to elevated atmospheric CO_2 concentrations evolve to have a lower frequency of stomata on their leaves. This has been shown by examining plants growing around thermal vents that emit high concentrations of CO_2 and even by measuring stomatal density of leaves of herbarium specimens; leaves from plants collected in preindustrial times when atmospheric CO_2 concentration was $\sim 280 \mu l\, l^{-1}$ have higher stomatal densities than modern leaves (Woodward, 1987). In the same way, leaves from fossil deposits ranging from 280 to 310 million years ago (Carboniferous to Permian, when atmospheric CO_2 concentration was low) to over 400 million years ago (Devonian, high CO_2) show variation in stomatal density that parallels the CO_2 concentration.

The last 30 000 years have seen dramatic changes in the earth's climate, accompanied by closely correlated changes in atmospheric CO_2 concentration, from very low values during the glacial maximum to the rapidly increasing concentrations resulting from industrialization. Van de Water *et al.* (1994) obtained a long series of leaves of the pine *Pinus flexilis* from packrat middens which covered the whole of this period, and showed that not only the stomatal density but also the stable carbon isotope composition of the needles behaved as predicted from the known atmospheric CO_2 concentrations. Stable carbon isotope composition is typically quoted as a $\delta^{13}C$ value, which represents the ratio of ^{13}C to ^{12}C in the sample, relative to a standard (see also Chapter 4, p. 179). The $\delta^{13}C$ value is quoted in ‰ (not %) and is usually negative; the greater the discrimination against the heavy isotope ^{13}C, which occurs in photosynthesis because it diffuses more slowly, the more negative the $\delta^{13}C$ value becomes. C_4 plants, which are less limited by CO_2 diffusion because of their lower intercellular CO_2 concentration, discriminate less than C_3 plants and have less negative $\delta^{13}C$ values, typically around −14‰, as compared with −27‰ for C_3 plants. Beerling (1994) used the *Pinus flexilis* data to test a photosynthesis model that could predict the $\delta^{13}C$ value of the needles, and the fit was good (Fig. 2.26)

These climatic fluctuations have had a powerful impact on human history. There is even a suggestion that atmospheric CO_2 concentration may have had a direct impact, rather than one mediated *via* climatic change. Sage (1995) proposed that the rise in CO_2 concentration from around 200 $\mu mol\, mol^{-1}$ in glacial times (up to 15 000 years ago) to 270 $\mu mol\, mol^{-1}$ in the late glacial period (up to 12 000 years BP) may have made agriculture feasible, by

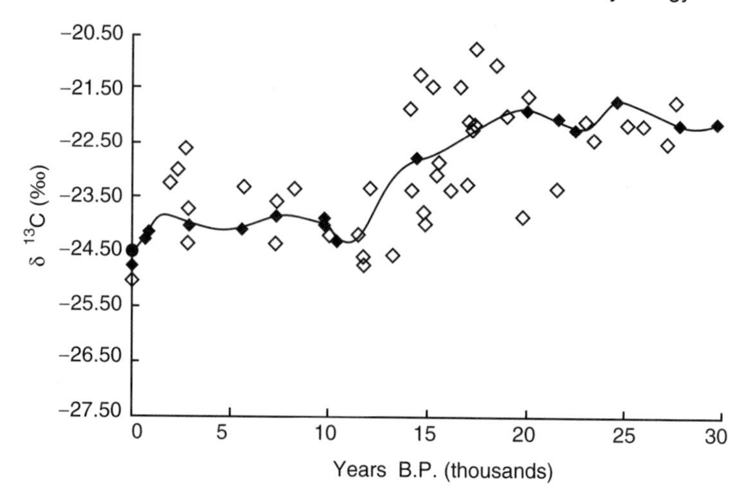

Figure 2.26

Predictions of a photosynthesis model (solid line and symbols) which estimates the $\delta^{13}C$ of plant carbon as a function of environmental conditions, tested by using the data of van de Water *et al.* (1994) of $\delta^{13}C$ of *Pinus flexilis* leaves from packrat middens (open symbols). The model predicts $\delta^{13}C$ accurately, as environmental conditions vary (from Beerling, 1994)

increasing productivity of C_3 crops such as the ancestors of wheat. Such a hypothesis is hard to test, but it emphasizes the profound effects of atmospheric CO_2 concentration on human society.

2. Whole-plant responses

When Hunt *et al.* (1991) grew 27 herbaceous species in controlled conditions at four atmospheric CO_2 concentrations ranging from 350 (current ambient) to 800 $\mu l\,l^{-1}$, they found a wide range of responses (Fig. 2.27). Some species (7 out of 27) were unaffected by the concentration (e.g. *Brachypodium pinnatum*, Fig. 2.27a), whereas others showed the expected increase (e.g. *Chamerion angustifolium*, Fig. 2.27b). An increase in growth is not, therefore, a universal response. The range of responsiveness was very great, the most responsive species was *C. angustifolium*, which increased its biomass by 3.66-fold as CO_2 concentration was doubled from 350 to 700 $\mu l\,l^{-1}$. No other species even doubled its biomass over this range, sunflower *Helianthus annuus* coming closest at 1.97-fold. All other species showed increases of less than 60% and 14 out of the 20 that responded increased by less than 40%.

There is some pattern to this huge variation. Fast-growing species tend to respond more than slow-growing ones (Poorter *et al.*, 1996). When Atkin *et al.* (1999) grew four slow-growing *Acacia* species from semiarid habitats and six fast-growing *Acacias* from mesic environments in ambient and elevated CO_2, they found that the fractional increase in **R** (and in its components, **E** (or *NAR*) and *SLA*) due to CO_2 was the same for all species, at around 10%. The absolute increase in growth rate was therefore much greater for the fast-growing species (Fig. 2.28).

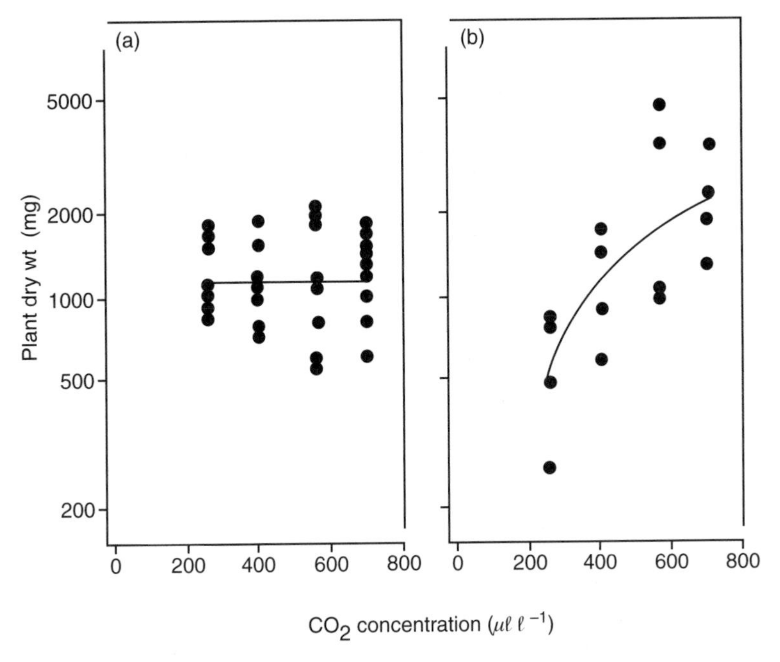

Figure 2.27

(a) *Brachypodium pinnatum* and (b) *Chamerion angustifolium* show quite different responses to elevated CO_2 (from Hunt *et al.*, 1991)

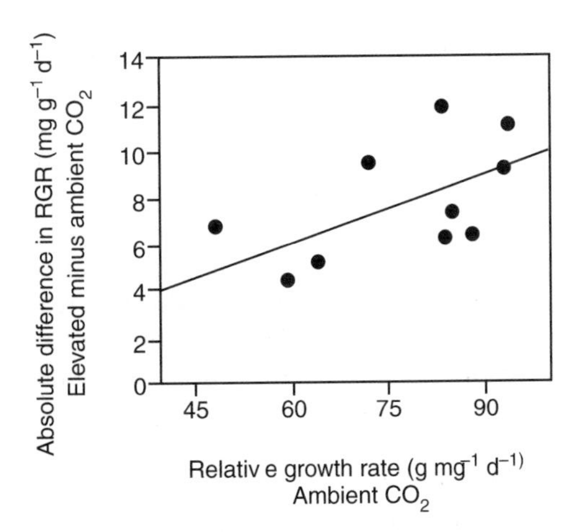

Figure 2.28

The absolute increase in relative growth rate (**R**) caused by growing ten *Acacia* species in elevated CO_2 (700 vs 350 μmol mol^{-1}) was a function of **R**: fast-growing species responded absolutely more, but proportionally the same (from Atkin *et al.*, 1999)

When data from the enormous number of experiments that have looked at responses to CO_2 in controlled conditions are summarized, the picture seen by Hunt *et al.* (1991) is confirmed: on average, doubling CO_2 leads to an increase in yield of around 30% (Rogers *et al.*, 1994). However, most of these experiments have been performed under very unrealistic conditions, typically plants were grown in pots in controlled environment chambers or glasshouses. Does the same response occur in the field? This can be tested only by experiments that allow the atmosphere around field-grown plants to be manipulated, and there are two main ways of doing this: open-top chambers (OTC) and free-air carbon dioxide enrichment (FACE). In an OTC, a translucent box is built around the plants, open at the base and the top, and CO_2 released within it, with a monitoring device that controls the concentration. Conditions inside the chamber are not identical to those outside and so two controls are needed: chambers maintained at ambient CO_2 concentration, and plants outside the chamber. FACE experiments more nearly mimic natural conditions: a ring of CO_2-releasing jets is built and a sensor placed in the centre that records CO_2 concentration and windspeed and direction (Plate 3). When the CO_2 concentration falls below a pre-set figure, the jets on the windward side release CO_2 which is blown across the ring. In this design there are no 'chamber effects', and the scale of the plots can be larger, although not big enough to measure real community level effects (e.g. those involving mobile insects). FACE experiments use a lot of CO_2 and are expensive to run.

Do field-grown plants respond to elevated CO_2? The answer appears to be yes for crop plants growing in irrigated, fertilized soil. Orange trees *Citrus aurantium* grown in open-top chambers in Phoenix, Arizona, grew 2.8 times more in 655 than in 355 $\mu l\ CO_2\ l^{-1}$ (Idso and Kimball, 1993), but that is an exceptional result. In a FACE experiment, also in Arizona, wheat grain yield was 8% higher in 550 $\mu l\ l^{-1}$ than in ambient ($\sim 370\ \mu l\ l^{-1}$) under irrigated conditions, and 12% higher in dry conditions (Kimball *et al.*, 1995). Most data on crop plants suggest that these lower figures are more likely.

Plants in natural communities may behave differently, however, possibly because they are mostly limited by factors other than the availability of CO_2, typically water or nutrients or even the action of grazing animals. In the longest-running OTC experiment, which is in a salt-marsh in Chesapeake Bay, there was a small increase in net ecosystem productivity, but less than predicted because of down-regulation of photosynthesis (Drake *et al.*, 1997). In an arctic tundra experiment, there was no increase in net ecosystem productivity (NEP) because the acclimation of the dominant species, the cotton-grass *Eriophorum vaginatum*, was so complete (Oechel *et al.*, 1994). Several FACE experiments have found the same result; generally there seems to be a small increase in NEP but little or, more usually, no change in above-ground biomass. However, below-ground more happens; when blocks of soil from two upland grasslands were maintained at 700 $\mu l\ l^{-1}$ in solardomes there were no changes in above-ground biomass after 2 years (Wolfenden and Diggle, 1995), but a 40% increase in below-ground biomass and an even larger increase in root production (Fig. 2.29), the difference being due to high levels of root turnover (Fitter *et al.*, 1997). The greater response of root growth than of shoot growth is

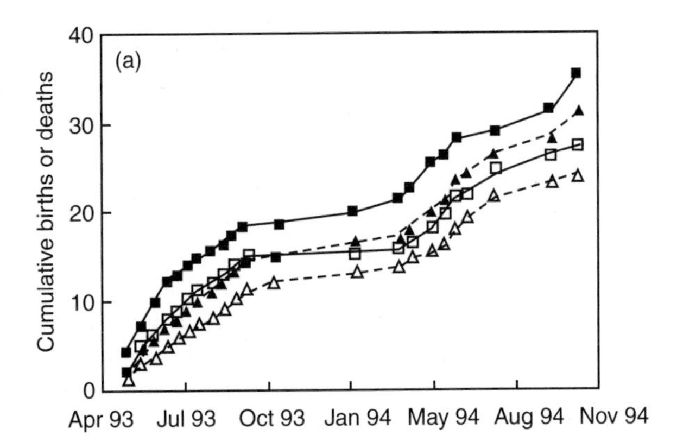

Figure 2.29

Cumulative root production measured as births (squares) and deaths (triangles), over a 2-year period under two grassland types maintained in a solardome at ambient (350 μl l^{-1}) (□, △) and elevated CO_2 (700 μl l^{-1}) (■, ▲). The data were obtained from minirhizotrons (from Fitter *et al.*, 1997; and see Plate 5)

exactly what would be expected from plants growing in water- or nutrient-limited conditions, and is often found in such experiments.

If the rising concentration of atmospheric CO_2 results in increased root growth in natural ecosystems, then soils may come to store progressively more carbon. For example, Ineson *et al.* (1996) measured the carbon entering the soil by growing birch *Betula pendula* on a soil that had previously had C_4 grasses growing on it. Since C_4 plants discriminate much less strongly against the heavy carbon isotope ^{13}C, it is easy to distinguish the 'new' carbon added by growth of the birch trees (a C_3 plant) from the 'old' carbon derived from the C_4 grasses. They found that growing the plants under elevated CO_2 increased root mass by 69%, but carbon added to soil by 150%. It is already believed that terrestrial ecosystems act as a sink for about 2 of the 7 gigatonnes of carbon that human activity releases into the atmosphere each year. Whether this 'missing sink' (the rest of the emissions can be accounted for either by the accumulation of CO_2 in the atmosphere or by its absorption by the oceans) is in soil or vegetation is unknown, but currently twice as much carbon is held in soils as in vegetation, and soil is a likely location. The continued operation of this terrestrial sink is vital: if it ceases to act, the accumulation of CO_2 in the atmosphere will accelerate, leading to a runaway greenhouse effect.

3

Mineral nutrients

1. Introduction

The fundamentals of plant biochemistry vary little between species: all green plants require the same set of essential mineral nutrients, which are used by different plants for the same functions (Table 3.1). An essential element is one without which the plant cannot complete its life cycle; most of those in Table 3.1 fall into this category, but some are required in such minute concentrations that demonstrating essentiality is difficult because of contamination, seed reserves and other problems. Essentiality has only relatively recently been demonstrated for nickel (Ni; Brown *et al.*, 1987), an element that is only found in one enzyme in most organisms (urease) but is more widely used by Archaea. Others may yet be added to the list. In addition, there is a class of elements sometimes called beneficial, which promote survival and therefore are likely to be ecologically, if not physiologically, essential; silicon (Si), which accumulates in grasses and forms crystals in the stems and leaves, is a good example, since plants can be grown to maturity in water culture containing no Si with no ill-effects. Si-free plants under natural conditions, however, would probably be eaten or be unable to compete for light (Epstein, 1999).

A few species have specialized mineral requirements, such as that for cobalt by legumes co-existing symbiotically with nitrogen-fixing *Rhizobium* bacteria, or for sodium by plants utilizing the C_4 photosynthetic pathway (Brownell and Crosland, 1972) and by some salt-marsh and salt-desert plants. There have been reports that some rare earth elements, such as lanthanum and cerium, are micronutrients, but careful experimentation has not revealed any growth promotion by them (Diatloff *et al.*, 1999). Species differ, however, in both their absolute ion concentrations and the relative proportions of different ions they contain. Plants growing on salt-affected soils, for example, may have very high sodium concentrations, whereas those on soils derived from limestone rocks (calcium carbonate) have internal calcium concentrations 10 to 100 times greater than plants growing on acid soils. Some species are typically found on nitrogen-rich soils, for example red goosefoot *Chenopodium rubrum* which is abundant on manure heaps, and others, such as stinging nettle *Urtica dioica*, on phosphate-rich soils. The same is true for other nutrients.

All plants possess uptake mechanisms capable of moving ions of essential elements across their cell membranes. In addition an enormous range of other elements, ranging from the abundant (aluminium, sodium) to the obscure

Table 3.1

The elements required for healthy plant growth, excluding carbon, hydrogen and oxygen, which are incorporated in photosynthesis

Element	Symbol	Taken up as	Typical concentration in plant tissue ($\mu g\ g^{-1}$)	Function
Nitrogen	N	NO_3^-, NH_4^+, amino-acids	20 000	Constituent of amino acids and hence proteins
Potassium	K	K^+	10 000	Balances electric charge in cells; osmoregulation
Phosphorus	P	$H_2PO_4^-$	2000	Nucleic acids, ATP, membranes
Calcium	Ca	Ca^{2+}	500–5000	Component of many molecules; cell wall and membrane integrity and function; second messenger. In base-rich soils, both Ca and Mg are present in plant tissue at very high concentrations
Magnesium	Mg	Mg^{2+}	200–2000	Part of chlorophyll molecule; required for many enzyme reactions; control of cation–anion balance
Sulfur	S	SO_4^{2-}	1000	Component of many primary and secondary plant metabolites, including some amino acids
Iron	Fe	Fe^{3+}	100	Operation of redox systems
Manganese	Mn	Mn^{2+}	50	Redox processes; enzyme activation
Zinc	Zn	Zn^{2+}	20	Enzyme functioning; membrane structure
Boron	B	$B(OH)_3$	20	Uncertain: cell walls, lignification?
Copper	Cu	Cu^{2+}	5	Redox reactions
Molybdenum	Mo	MoO_4^{2-}	0.2	Enzymes of N metabolism
Nickel	Ni	Ni^{2+}	0.5	Enzymes of N metabolism
Chlorine	Cl	Cl^-	1000–10 000	Various but poorly defined; deficiency rarely observed as requirement is ~200 μg^{-1}
Sodium	Na	Na^+	500–50 000	Only in halophytes and plants with C4 photosynthesis (see Chapter 2); can accumulate to very high levels in plants on saline soils
Cobalt	Co	Co^{2+}	100	Only in legumes (required for N-fixation)
Silicon	Si	$Si(OH)_4$	1000–10 0000	Not physiologically essential except possibly in a few species, but ecologically essential for grasses in providing support and deterrence to grazers

(zirconium, titanium, and similar elements) are accumulated by plants. Sometimes this accumulation is characteristic of a particular group: selenium is accumulated by some *Astragalus* species and silicon may be a major component of grass stems. These elements have no known metabolic function in higher plants (although as with nickel, Archaea have selenocystein residues in their enzymes), but they may be ecologically important, for example, in providing support (silicon) or protection from grazing animals (both silicon and selenium).

Concentrations of the macronutrients N, P and K in mature leaves of a wide range of plant species typically vary by a factor of about 5, suggesting that plants regulate their internal concentrations within rather narrow limits, although there are species that maintain exceptionally low or high tissue concentrations (Table 3.2). Some dicotyledonous species have high K contents, and legumes (Fabaceae), which have symbiotic N-fixing bacteria in their roots, have high N concentrations. For Ca the range is greater, reflecting the enormous differences in soil calcium between soils formed over calcareous (e.g. chalk, limestone) and non-calcareous (e.g. sandstones, granite) substrata and the largely non-nutritional role played by very high concentrations of Ca.

Soil calcium content is one of the major determinants of soil pH, since Ca^{2+} ions occupy the exchange sites on the soil minerals and act as a buffer system, and pH is intimately involved in the availability of many nutrients. At low pH, a range of different phenomena affects nutrient availability, including:

- microbial (especially bacterial) activity is inhibited, so that decomposition and the release of ions (especially of nitrogen) from organic matter, and the conversion of ammonium into nitrate are slowed;
- phosphate ions react with aluminium hydroxides, which are highly active below pH 4–5, and may become unavailable;
- iron is soluble as ferric ions;
- molybdenum is unavailable.

Many of the effects of low pH, particularly those involving increased solubility of Al^{3+} and Mn^{2+}, are of toxicity rather than nutrient availability and are considered in Chapter 6, but even from its effects on the major nutrients it can be seen why pH has such a powerful influence on community composition in the field. It has long been known that limestone added to acid soils improves crop yield. Where limestone is added to natural vegetation on acid soils, large changes in species composition may ensue; in one case, after about 50 years and with a pH change from under 5 to over 7, 15 of the original species had been replaced by 41 new invaders (Hope-Simpson, 1938).

Adding nutrients to soil can also transform vegetation. The finest example of this is on the Park Grass plots at Rothamsted Experimental Station near Harpenden (UK), a field that was divided up in 1856 into plots which have received, in most cases, the same controlled nutrient application ever since. An unfertilized plot has had a hay crop removed each year, representing a substantial loss of nutrients; that plot now has about 40 species, none dominant, and produces less than 1500 kg of hay per hectare per year (Plate 4a). By contrast a plot receiving a complete fertilizer and limestone

Table 3.2

Concentrations (mg g^{-1} dry weight) of four macronutrient elements in leaf tissue of plants from contrasting habitats. Data from various sources

Species	Family	Habitat	N	P	N : P	K	Ca
Juncus squarrosus	Juncaceae	Acid heath	7.9	0.9	7.8	15.0	0.6
Nardus stricta	Poaceae	Acid heath	10.2	1.0	10.2	7.9	1.3
Festuca ovina	Poaceae	Acid heath	14.0	1.5	9.3	14.3	2.0
		Calcareous grassland	13.6	1.2	11.3	13.6	3.3
Dactylis glomerata	Poaceae	Neutral grassland	14.7	1.8	8.2	19.1	3.7
Vicia sepium	Fabaceae	Neutral grassland	36.6	2.6	14.1	17.5	8.8
Urtica dioica	Urticaceae	Disturbed ground	39.0	4.4	8.9	27.0	71.0
Actinodaphne ambigua	Lauraceae	Montane tropical forest	11.4	0.7	17.0	7.6	4.7
Allophylus varians	Sapindaceae	Montane tropical forest	22.7	1.5	14.9	12.5	17.3
Acronychia pedunculata	Rutaceae	Montane tropical forest	21.0	0.9	22.3	9.7	10.0
Rhodamnia cinerea	Myrtaceae	Dipterocarp forest	12.6	0.9	14.0	10.6	4.7
Macaranga triloba	Euphorbiaceae	Dipterocarp forest	27.8	1.3	21.7	12.3	9.5
Caryocar glabrum	Caryocaraceae	Amazonian tierra firme	17.1	0.9	19.2	5.8	3.2
Eperua leucantha	Fabaceae	Amazonian caatinga	11.5	1.4	8.4	7.5	5.8
Remijia involucrata	Rubiaceae	Amazonian low bana	5.5	0.4	15.7	5.1	2.7
Range (highest/lowest)			7.1	11.0	2.9	5.3	>100

dressing annually has a yield about four times as great, but with only 10–15 species, two or three of which are clearly dominant. Other treatments produce even more extreme results: the grass *Anthoxanthum odoratum*, for example, comprises almost 100% of the biomass on some unlimed plots that have received heavy doses of ammonium, producing a nutrient-rich but very acid (pH 4.0) soil (Plate 4b). A similar treatment omitting P, K and Mg, however, permits *Agrostis capillaris* to dominate, and if the soil is limed, raising the pH above 4, *Alopecurus pratensis* is the major species (Table 3.3; Plate 4c). Such experiments provide valuable ecological information on the nutrient requirements of individual species; for example, dandelions *Taraxacum officinale* thrive best on the Park Grass plots that have been limed (pH >6.5) and fertilized with potassium but not other nutrients (Tilman *et al.*, 1999). This distribution in the field matches the sensitivity of dandelion to potassium deficiency in laboratory experiments. Similarly, high P and low N favours legumes, notably red clover *Trifolium pratense* (plate 4d).

Field experiments such as that at Park Grass show that species respond in different ways to nutrient availability. Because they are performed in natural vegetation, they measure a real response, but they lack precision and cannot be used to understand the mechanisms of the response. When phosphate is added to soil, for example, it might benefit a species in several ways:

(i) stimulation of growth to a greater extent than that of a competitor;
(ii) depression of the growth of a competitor if the competitor is more prone to P toxicity (Green and Warder, 1973);
(iii) precipitation at the root surface of toxic Al^{3+} ions or of Fe^{3+}, which is a nutrient (Brown, 1972);
(iv) stimulation of N-fixing activity of associated legumes, resulting in increased availability of N in the soil (Israel 1987);
(v) alteration of the soil microbial populations (especially fungi and bacteria) which might indirectly affect plant growth (cf. p. 115).

Table 3.3

The abundance (as % by weight of the total grass biomass) of the commonest grasses (those representing >20%) in a range of plots on the Park Grass Experiment during 1947–1949. N1, N2 and N3 represent increasing amounts of N fertiliser application. Data from Johnston, A.E. (1994). The Rothamsted classical experiments. In 'Long-term Experiments in Agricultural and Ecological Sciences' Eds Leigh, R.A. and Johnston, A.E., pp 9–38. CABI, Wallingford

Fertiliser	pH 3.7–4.1	%	pH 4.2–6.0	%	pH 6.0–7.5	%
N1	*Agrostis capillaris*	79	*Dactylis glomerata*	36	*Dactylis glomerata*	27
					Festuca rubra	26
					Avenula pubescens	22
N2P	*Agrostis capillaris*	44	*Festuca rubra*	59	–	
	Festuca rubra	23	*Alopecurus pratensis*	28		
N2PK	*Holcus lanatus*	91	*Alopecurus pratensis*	38	*Arrhenatherum elatius*	48
			Arrhenatherum elatius	28		
N3PK	*Holcus lanatus*	96	*Alopecurus pratensis*	72	–	

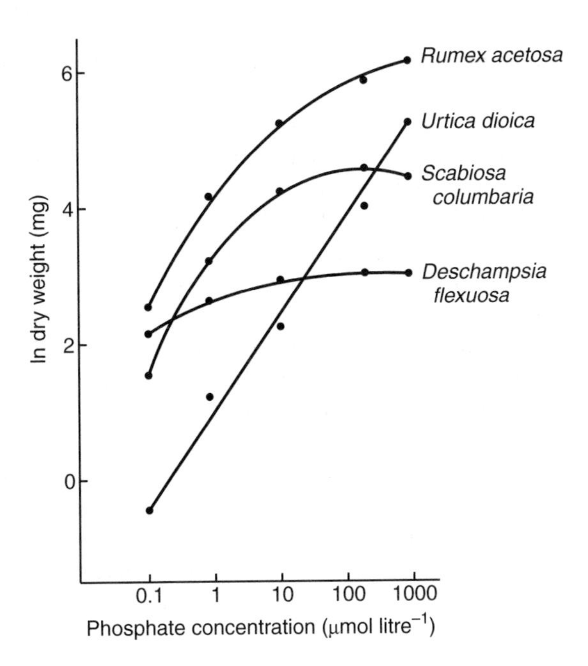

Figure 3.1

Growth response of four ecologically contrasted species to phosphate concentration in solution culture, after 6 weeks (from Rorison, 1968). Note that *Urtica dioica* (stinging nettle) showed a linear response up to the highest concentration supplied, whereas *Deschampsia flexuosa*, a grass of nutrient-poor soils, ceased to respond at a concentration at least 100 times lower

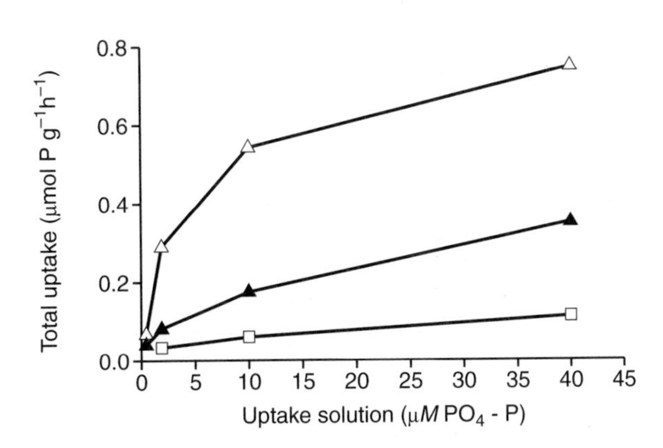

Figure 3.2

Uptake rate of phosphorus by seedlings of three tree species from eastern North America, as a function of phosphate concentration (from Lajtha, 1994). Tulip tree *Liriodendron tulipifera* (□) and yellow birch *Betula alleghaniensis* (△) showed markedly different responses; hickory *Carya ovata* (▲) was intermediate

Although most species do increase their growth when nutrients are added, the nature of the response varies greatly. *Urtica dioica* is a common plant of disturbed ground, especially where there has been eutrophication, often a consequence of human activity. It responds linearly to increasing P supply over 4 orders of magnitude (Fig. 3.1), whereas the grass *Deschampsia flexuosa*, a common species on very infertile soils, barely responds at all. There are two possible reasons for such differences in behaviour: either the less responsive species do not take up more of the nutrient when the supply increases, or they do take it up but fail to utilize it.

Instances of differences in uptake rates at varying P supply are easy to find. If almost any two species are compared, such differences can be found, although interpreting them can be more difficult. For example, yellow birch *Betula allegheniensis*, shagbark hickory *Carya ovata* and tulip tree *Liriodendron tulipifera*, are all common trees of deciduous forests in eastern North America. When grown at a range of P supply rates, they had markedly different uptake responses, with birch being the most and tulip tree the least responsive (Lajtha, 1994; Fig. 3.2). However, these rates are calculated per unit mass of root; the species have

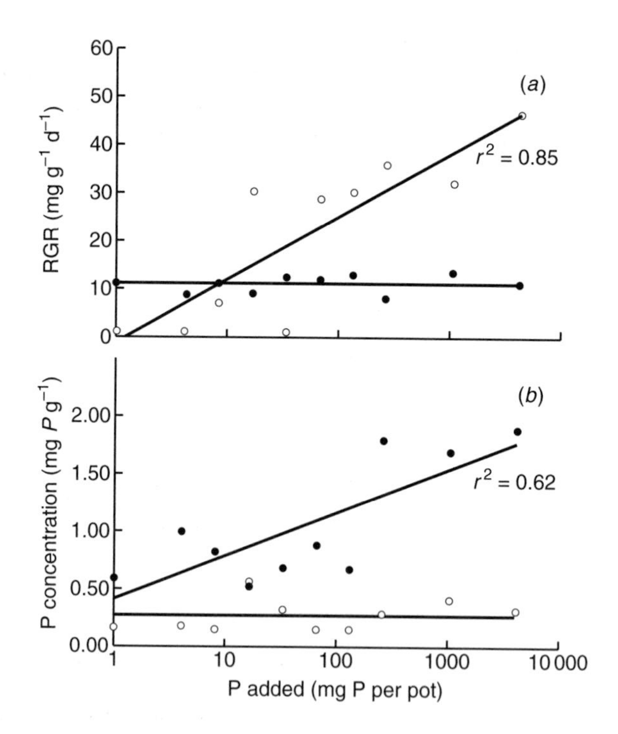

Figure 3.3

(a) Relative growth rate (RGR) and (b) P concentration of the leaves of two contrasting tropical trees as a function of P supply (from Raaimakers and Lambers, 1996). *Tapirira obtusa* (O) is a fast-growing pioneer species and *Lecythis corrugata* (●) is a species of mature forest. *T. obtusa* regulated P concentration and altered growth rate, whereas *L. corrugata* did the opposite

very different root morphologies and the patterns might look different if the results were calculated per unit root length. Equally, the plants were grown beforehand for 4 weeks under standard conditions, so that they may have been in different physiological states before the experiment started. Nevertheless, the data show that the species do behave differently in terms of P uptake.

Failure to utilize nutrients is apparently responsible for the difference in response to P between two tropical tree species. *Tapirira obtusa* is a pioneer species, capable of exploiting canopy gaps by rapid growth, whereas *Lecythis corrugata* is a slow-growing canopy species. When the two species were grown with P supplies ranging from 1 to 10 000 mg pot^{-1}, the pioneer *T. obtusa* responded to increasing P supply by matching growth rate to P uptake rate, so that its internal P concentration was constant (Raaimakers and Lambers, 1996; Fig. 3.3). In contrast, the canopy species, *L. corrugata*, did not vary its growth rate at all, with the result that its internal P concentration rose steadily. Paradoxically, the plant with the highest internal P concentration grew slowest, simply because it failed to convert the extra P into growth. Such responses may be common; in many habitats, nutrients may be supplied intermittently and species that store nutrients in this way may be able to maintain growth over periods of inadequate supply. The grasses *Festuca ovina* and *Arrhenatherum elatius* are found in very different habitats, *F. ovina* growing in nutrient-poor sites where storage is likely to be important, and *A. elatius* in nutrient-rich sites. When they were both grown in solutions where the same amount of nutrient was made available but in pulses varying in duration from

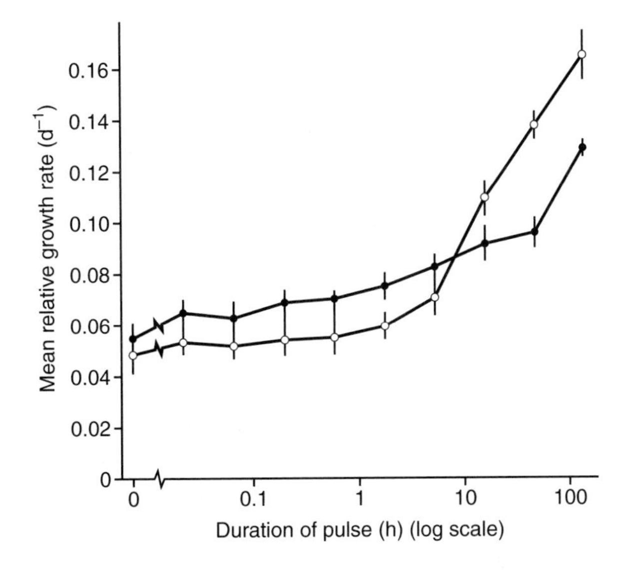

Figure 3.4

Relative growth rate of two contrasting grass species when exposed to pulses of nutrients of different duration, ranging from 80 s (0.02 h) to 6 d (144 h). *Arrhenatherum elatius* (○) is a fast-growing and *Festuca ovina* (●) a very slow-growing species (from Campbell and Grime, 1989)

80 s to 6 days, *F. ovina* was able to maintain its growth better than *A. elatius* in the treatments with the most intermittent supply (Fig. 3.4; Campbell and Grime, 1989).

Similar differences also occur between populations of many species. Plants of *Anthoxanthum odoratum* from various parts of the Park Grass plots, in some cases separated by only a few metres, show close correlation of response to environment (Davies, 1974). These differences have arisen in some cases in less than 50 years, and have a genetic basis (Crossley and Bradshaw, 1968; Ferrari and Renosto, 1972); they are not just examples of plasticity in response to environment. That such differences in response to nutrients are hereditary should not be a surprise to anyone who considers the achievements of plant breeders in moulding the fertilizer responses of crop plants.

2. Nutrients in the soil system

1. Soil diversity

Soils vary in their ability to supply nutrients, and the factors that give rise to this variation are those which determine the development of soils, namely, parent material, climate, topography, age, and organisms. *Parent material* normally sets the upper limits on nutrient content of soil (except for N which can enter by fixation), so that soils formed on granite are base-poor, for example. Soils formed from limestones, on the other hand, initially are very base-rich, since pure chalk is 100% $CaCO_3$, and they typically lose Ca^{2+} ions through the action of rainfall (*climate*), which leaches out carbonates with dissolved carbon dioxide (carbonic acid), a process dependent on *topography*. Such effects are clearly time-dependent, hence the *age* factor. Topography may reverse the usual situation by allowing accumulation of base-rich water: fen peats are wholly organic, and typically have no native inorganic calcium, but are base-rich from continual flushing by ground-water. *Climate*, of course, has more general weathering effects and in some areas nutrient inputs in rain can be important, particularly of Na and K in coastal areas.

The effect of *organisms* is more complex, involving ecological succession. Soil formation does not take place without vegetation so that these two time-sequences may be considered as essentially interlinked, but at any stage in either process the vegetation both reflects existing soil conditions and influences the future course of soil formation, thereby secondarily determining succession. A classic instance of this is the acidification of soil by alders (*Alnus crispa*) in the vegetation succession on newly exposed glacial moraines in Alaska (Crocker and Major, 1955): the pH of soil under alder declined by 3.0 units over 35 years, a thousand-fold change in H^+ ion concentration, whereas soil under other species showed almost no change.

Any of the factors producing large-scale variation in soil nutrients can also be active on a small scale. Leaching can cause local acidification of hummocks in raised bog (Pearsall, 1938) and in fen (Godwin *et al.*, 1974), outcropping of calcareous parent material in otherwise acid turf allows calcicole species to invade. The details of the vegetation mosaic of most communities can be related to variation on the appropriate scale. The mechanisms are various but the following occur.

(i) Litter fall: the litter of plant species varies greatly in nutrient content, with species of infertile soils generally returning least to the soil, simply because their leaves have the lowest nutrient concentrations and live longer, not because they are better able to recycle nutrients internally from senescing leaves (Aerts, 1996). Most species withdraw about half the N, P and K from their leaves before abscission, but Ca concentrations normally build up (Fig. 3.5). Nutrients in organic matter that are cycled through herbivores may be returned to the soil, for example as urea, to give even more concentrated pockets of high fertility, as a glance at a cattle field will show.
(ii) Acidification: many species produce base-poor litter which can have marked effects on gross soil pH (cf. Chapter 6). More subtle influences arise from the secretion by roots of protons, especially when cation exceeds anion uptake. This occurs, for example, when NH_4^+ rather than NO_3^- is the N source (Riley and Barber, 1971), and because plant species differ in their tendency to absorb these two ions, rhizosphere pH can be up to 2 pH units

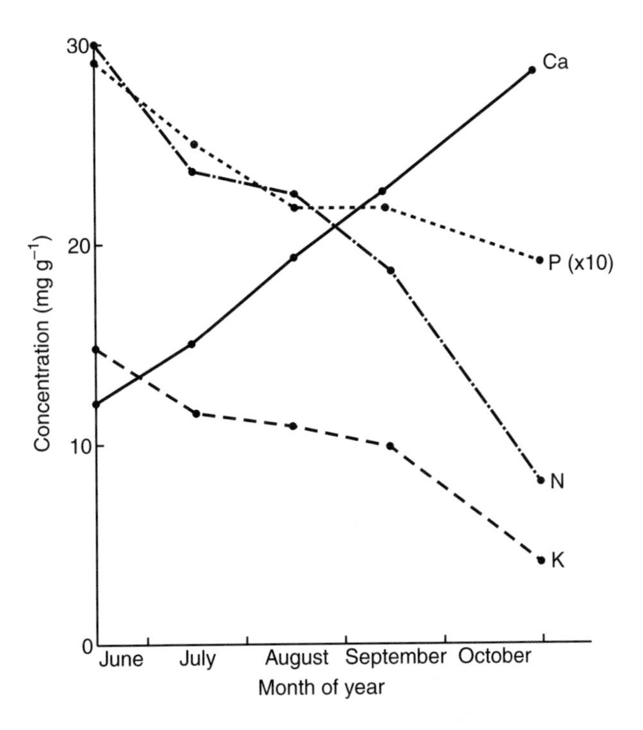

Figure 3.5

Mean mineral concentration of leaves of nine broad-leaved deciduous trees as a function of time of year. Note the large increase in Ca concentration. The scale for P has been expanded by a factor of 10. Data from Alway, F.J., Malli, T.E. and Methley, W.J. (1934) *Soil Survey Assoc. Bull.* **15**, 81

higher or lower than the bulk soil (Marschner and Römheld, 1983). Equally the oxidation of NH_4^+ to NO_3^- in soil causes acidification (Plate 7).

(iii) Depletion: plant uptake causes depletion of the soil, yet many plant roots are relatively short lived. There will therefore be a mosaic of depleted, recharging, and unexploited soil zones, at least for immobile nutrients.

(iv) Fixation: many bacteria can fix atmospheric nitrogen to NH_4^+ either independently or symbiotically with fungi (lichens) or higher plants. Where these organisms are active, local areas of high NH_4^+ or NO_3^- concentration may occur. Bacteria such as *Azotobacter* and, in tropical soils, *Azospirillum* tend to be specifically associated with roots (Brown, 1975). In agricultural soils an analogous effect is produced by the addition of granular fertilizers.

These mechanisms will tend to produce a patchwork of low and high nutrient concentrations, revealed by intensive sampling. The scale of the variation can be fine, with samples only a few centimetres apart having very different properties, especially where microbial processes generating nutrient variation are examined. Jackson and Caldwell (1993) showed that soil around individual plants of sagebrush *Artemisia tridentata* and wheatgrass *Pseudoroegneria spicata* had a pronounced pattern of nutrient availability, at a scale of less than 1 m, effectively imposed by the pattern of the plants (Fig. 3.6). Nitrogen mineralization rate, however, exhibited no spatial pattern at sampling scales down to 12.5 cm, suggesting that if there was pattern, it was at a very fine scale. In other words, adjacent samples only a few centimetres apart may have quite different N mineralization rates, and that is well within the range encountered by the root systems of all but the smallest plants.

The effects of vegetation and leaching normally interact to give a more ordered, and very widespread variation pattern–soil zonation. Except in soils where intense earthworm activity causes continuous mixing (for example the brown earths typical of temperate deciduous forests) or in ploughed or otherwise disturbed soils, plant roots draw up minerals from deep layers which are then deposited on the surface as litter. Rainfall moves ions down the profile again, but since the soil is not chemically inert, it reacts with some ions more than others, giving characteristic horizons. These may vary widely in pH, nutrient content, and organic matter, and provide distinct soil environments within a single soil profile, allowing species with different habitat requirements to co-exist.

Soil formation is a time-dependent process and so long-term changes may be brought about by any of the factors discussed above, but of more importance to most plants are effects on a seasonal or shorter cycle. In natural, as opposed to agricultural, soils there may be seasonal nutrient variation related to the amounts passing through the organic cycle. The most obvious input is leaf-fall, in the autumn in temperate climates; in moist tropical and subarctic evergreen forests this seasonality is less clear, and in the former may be related to rainfall. In the typical temperate cycle, however, the input coincides with the slowing down of microbial activity caused by falling temperatures, and so though there may be a small autumn peak, particularly of elements such as potassium easily washed out of leaves, the bulk of the release would be expected to occur the

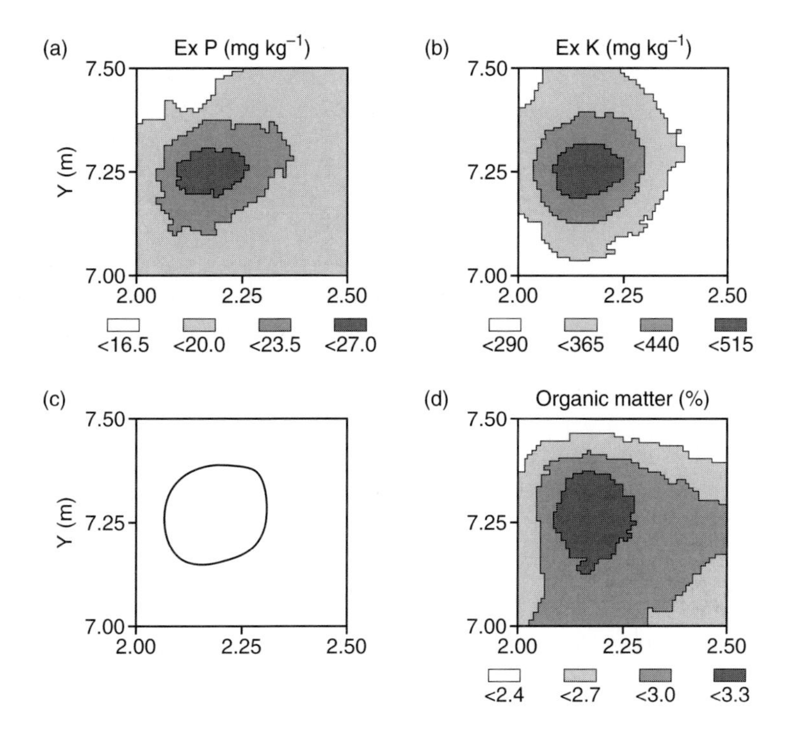

Figure 3.6

Contours of (a) extractable P, (b) extractable K and (d) organic matter concentration in soil around a single tussock of the grass *Pseudoroegneria spicata* (from Jackson and Caldwell, 1993). The axes represent horizontal distances (0.5 m); each diagram displays the same 0.25 m^2 area. The circle in (c) represents the outline of the tussock

following spring. This spring peak is followed by a summer fall, the result of root and microbial activity, although further differences may arise from the differential uptake activity of the roots of various species. However, when Farley and Fitter (1999a) measured nutrient availability over two seasons in a natural woodland soil, strict seasonal patterns were not apparent, although there were very large temporal fluctuations (Fig. 3.7). In most natural ecosystems roots account for over half of production and root death must represent the major input of organic matter to soil. The timing of root mortality is not well understood and it may be that this is the reason why the expected seasonal patterns are not always seen.

2. Concentrations

The concentrations of nutrients in solutions expressed from natural soils are typically very low (Table 3.4). Even in fertilized soils the values may be well below those that would saturate ion uptake systems. Nevertheless, plants will grow well in solution culture at these or sometimes much lower concentrations. Asher and Loneragan (1967) grew several species in water culture and found that some, such as *Erodium botrys*, *Bromus rigidus*, and *Trifolium subterraneum*,

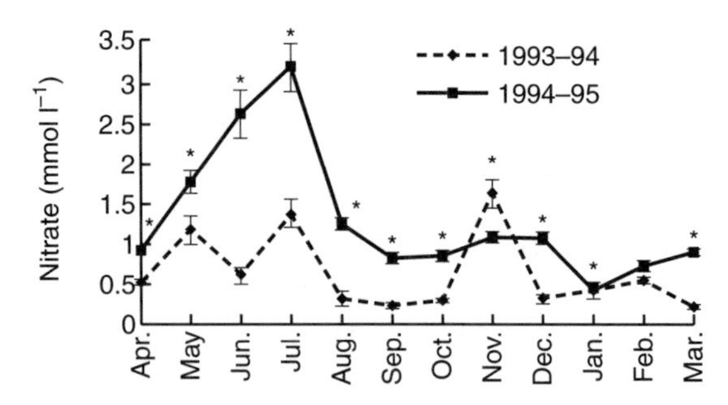

Figure 3.7

Long-term fluctuations in nitrate concentration in the soil solution of a temperate woodland soil reveal that there were large changes in nutrient supply between sampling times, but that these changes did not follow a precise seasonal pattern (from Farley and Fitter, 1999a)

Table 3.4

Mean values and range of soil solution concentrations for seven ions (mM)

	Ca^{2+}	Mg^{2+}	K^{+}	NO_3^{-}	NH_4^{+}	$H_2PO_4^{-}$	SO_4^{2-}
n	9	7	9	4	3	6	5
Mean	8.9	3.7	1.7	9.1	5.7	0.02	1.6
Min	1.7	0.3	0.1	1.0	5.0	0.00	0.3
Max	19.6	10.3	6.8	27.6	6.1	0.09	3.5

Most available data are from agricultural soils. Uncultivated soils will have lower values but few data are available. These data are from Adams, F. (1974). The soil solution. In 'The Plant Root and Its Environment' (Ed. E. W. Carson). University Press of Virginia, Charlottesville.

showed growth responses up to 5 μM P, whereas one, *Vulpia myuros*, reached a peak at 1 μM. For four species good growth was made at less than 1 μM P. It might, therefore, be a cause for surprise that conventional nutrient solution recipes recommend P levels around 1 mM, apparently three orders of magnitude more than required. The difference is that these are for traditional static culture, whereas Asher and Loneragan used a flowing culture system, so that the supply of 1 μM was maintained. This concept of supply is fundamental and is covered below briefly, and in more detail by Tinker and Nye (2000).

Ion concentrations vary greatly among soils and spatially within them, depending on factors involved in pedogenesis (parent material, climate, topography, organisms, and the age of soil), but of more importance to the individual plant is variation in time. The process of uptake in static solution culture reduces concentration (a fact, incidentally, that often complicates the interpretation of physiological experiments), and in soil the same will occur, its extent depending on how well the soil is buffered against depletion.

Soil buffering powers vary greatly for different ions, depending on the extent to which they are adsorbed by soil. Strongly adsorbed ions such as phosphate are powerfully buffered, since the bulk of the ion is in an insoluble state and acts as a reservoir. By contrast, nitrates are very soluble and so are not stored in soil to any extent; the reservoir for nitrate in soil comprises N in organic matter and adsorbed NH_4^+, both of which must be converted by microbial activity into NO_3^-, and indirectly atmospheric N_2 which is fixed by various bacteria and cyanobacteria, to give NH_4^+.

3. Ion exchange

The bulk of most ions in soil occurs in the solid phase, either as a component of soil minerals, or as part of the soil organic matter. Because of their reactivity, the most important inorganic components of typical soils are clay minerals, which are predominantly aluminosilicates. They consist of aluminosilicate sheets, bound together by numerous cations such as Al^{3+}, Mg^{2+}, K^+ and Fe^{3+} (Fig. 3.8). Where they fracture they expose surfaces which have a fixed negative charge. This charge is balanced by cations that are therefore held exchangeably on such surfaces, giving rise to the cation exchange capacity of the soil. In practice, diffusive forces cause cations held adjacent to the fixed negative charges to migrate away from them, so that the cation layer around such a surface is more diffuse than would otherwise be the case, giving a distribution known as a Gouy double layer.

In soils of pH greater than 5, the principal cations found in this layer are Ca^{2+}, Mg^{2+}, K^+ and Na^+, and the concentrations of these ions in solution are largely controlled by equilibria set up between surfaces and solution. If these cations are lost from soil, for example by leaching, the unsatisfied fixed negative charge is increasingly counteracted by protons or by Al^{3+} ions derived from the weathering of the clay minerals. Such soils suffer progressive acidification (cf. Chapter 6). One of the main problems caused by acid deposition from power stations and other fuel combustion sources, is this leaching of basic cations, and their consequent replacement on the exchange sites by Al^{3+}. This leads to a rapid decline in soil fertility.

The supply of most cations and anions depends largely on inorganic equilibria, best illustrated by phosphorus. A typical soil may contain anything from 0.05 to 1.0 g P per kg of soil, but the concentration in soil solution is likely to be between 0.1 and 10 μM, equivalent to less than 60 μg P per kg of a moist soil, or about 0.005% of the total amount. The remaining 99.995% is in the solid phase, in a bewildering variety of compounds, both inorganic and organic, most of which are not simply characterizable chemically. Solid phase phosphate is best classified according to its relationship with the solution phase, producing a labile pool in equilibrium with the solution, and a non-labile pool.

The non-labile pool is usually much the largest, but the distinction is empirical. The labile pool is that part that will exchange with phosphate in solution, determined experimentally with $H_2{}^{32}PO_4^-$. Its size depends on the time allowed for equilibriation, as over long periods ions that are less labile will become exchangeable. Solid phase phosphate does not therefore exist as two discrete phases, but in a continuous spectrum of lability, depending on the nature of the binding to calcium, aluminium, ferric, and also organic compounds in the soil.

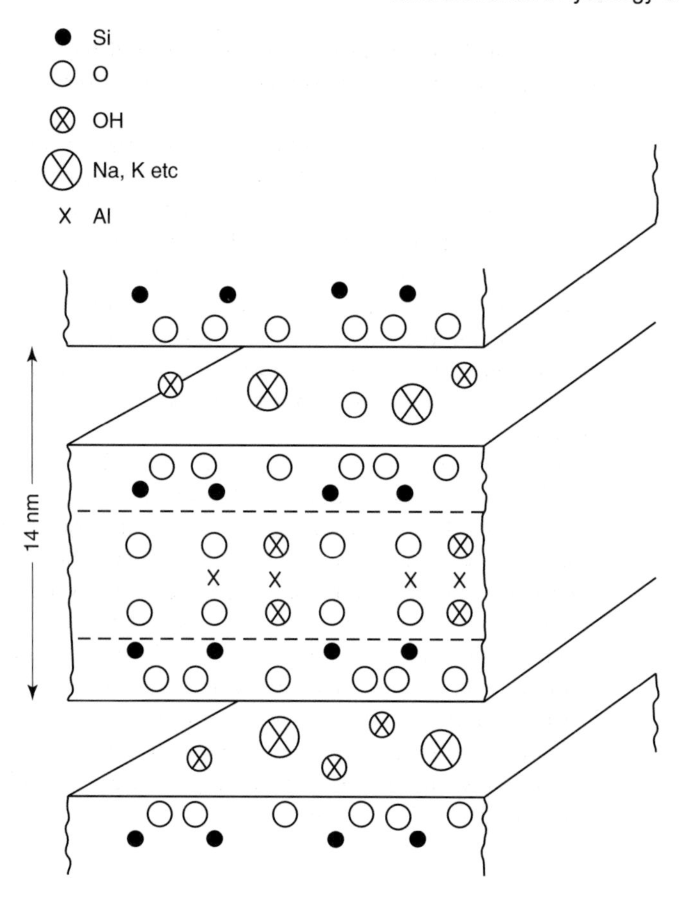

Figure 3.8

Diagrammatic representation of the structure of a typical clay mineral, montmorillonite, to show the layered structure and the position of the exchangeable ions

The value of this model is that it enables rates of supply to be quantified. A soil solution at 10^{-6} M phosphate represents about 10 g phosphate in solution per hectare; assuming crop uptake of 20 kg ha^{-1} and a growing season of 2000 hours, this solution would be wholly depleted in 1 hour, and long before that uptake would have been greatly reduced. Necessarily, labile phosphate must come into solution to maintain the supply, and so the labile pool-solution equilibrium is of crucial importance. This equilibrium can be measured by taking advantage of the fact that if phosphate is added to soil in solution, the bulk is immediately adsorbed. By varying the amount added at constant temperature, an adsorption isotherm can be constructed, relating the amount adsorbed to the concentration in solution after a given time, and indicating the equilibrium between labile and solution P. Equally, as phosphate is removed from solution, by plant uptake or artificially, ions come into solution from the labile phase by desorption, and an analogous desorption isotherm exists. In practice such isotherms exhibit hysteresis and the desorption isotherm, which controls the supply of phosphate to absorbing roots, cannot be predicted from the adsorption isotherm.

Figure 3.9 shows two adsorption isotherms for contrasting soils, one steep, one shallow. For the soil with a steep isotherm (Hathaway), large changes in the quantity of adsorbed phosphate (C) are needed to bring about a small change in solution concentration (C_l); for the other soil (Quast) the reverse is true. For a given amount of sorbed phosphate the first soil will therefore show a lower concentration in solution, but will maintain this concentration better against depletion; the first soil is therefore more strongly buffered than the soil with the flatter isotherm. The buffer power of a soil for an ion represents its ability to maintain solution concentrations in response to the removal of the ion from solution.

Buffer power measurements can be used to predict fertilizer additions required to maintain plant growth (Ozanne and Shaw, 1967), but increasing the labile pool by fertilization will also alter the equilibrium between labile and non-labile pools (Fig. 3.10), and some labile will be immobilized as non-labile. This description for phosphate applies equally to other adsorbed ions such as potassium, molybdenum, manganese, sulphate and cadmium.

The supply of adsorbed ions in soil should therefore be less variable seasonally than that of ions for which organic matter is a primary source, in particular nitrate. The dynamic balance of the inorganic adsorption system provides a buffer against fluctuations such as those exhibited by nitrate in soil. Nevertheless, phosphate is also supplied from organic matter as well, and marked temporal variation can occur (Gupta and Rorison, 1975).

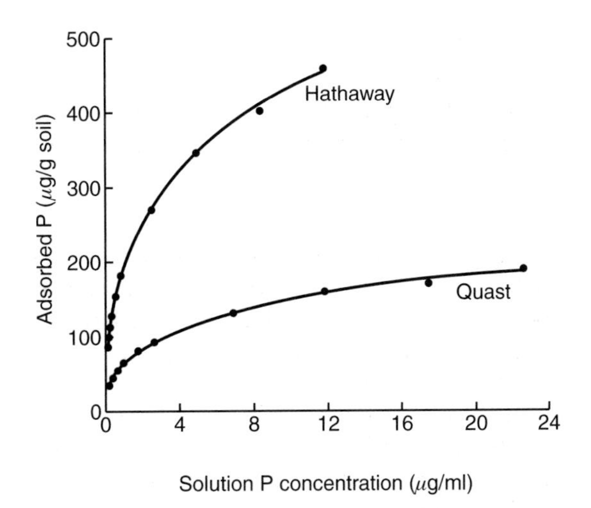

Figure 3.9

Adsorption isotherms for phosphate of two contrasting Australian soils. The steeper slope of the isotherm for Hathaway soil means that it is more strongly buffered against changes in soil solution concentration (from Holford *et al.*, 1974)

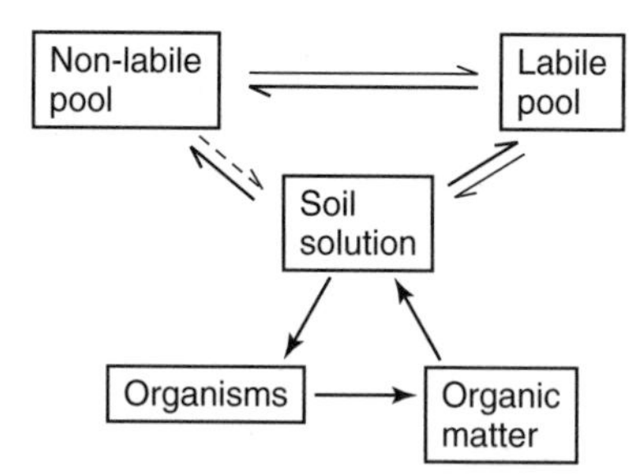

Figure 3.10

Model of compartmentation of adsorbed ions in soil. The labile pool is the proportion of non-dissolved ion that is readily exchangeable, in the case of phosphate for ^{32}P. The thickness of the arrows indicates the balance of the equilibrium

4. Cycles

The supply of ions from soil can be viewed in terms of mineral cycles, with inputs to and losses from the ecosystem and rates of transfer between components of the system. For some nutrients, especially nitrogen, for which inputs (due to rainfall and fixation) and losses (due to leaching and denitrification) represent a large proportion of the total amount cycling through the system, the cycle is open. In contrast, in the cycles of strongly adsorbed ions such as phosphate, these inputs are usually trivial in comparison with the total amounts present, and the cycle is more nearly closed.

The nitrogen cycle is complex (Fig. 3.11) and the organic matter compartment is dominant. Most plants can take up both ammonium and nitrate. Several bacteria are involved in the conversion of ammonium into nitrate, both ammonia oxidizers, such as *Nitrosomonas*, that convert NH_4^+ into NO_2^-, and nitrite oxidizers converting NO_2^- into NO_3^-, including *Nitrobacter*, *Nitrospira*, *Nitrospina* and *Nitrococcus*. The rate-limiting step is often NH_4^+ production from organic matter, brought about by a wide range of fungi and bacteria. These decomposers also require N, however, and if the material is low in N, it will be incorporated into their biomass and not liberated until carbon supply is reduced. There appears to be a critical value of between 1.2 and 1.8% N in litter (corresponding to C : N ratios of between 30 : 1 and 20 : 1), below which little or no NH_4^+ is released. Since most plants conserve N and P by recovering about 50% of them before leaf-fall (Aerts, 1996; Fig. 3.5), freshly deposited leaves tend to have a high C : N ratio (up to or exceeding 100 : 1), which slows down the rate of decomposition and N mineralization. The more slowly N is made available to plants, the lower will be the N uptake rates and the higher the C : N ratio of the litter, which accentuates the infertility. One of the best-characterized effects of growing plants in elevated CO_2 concentrations is a reduction in leaf N concentration; the higher C : N ratio of leaf litter under such conditions also slows decomposition rates (Cotrufo and Ineson, 1996).

There are two important short-cuts to the N cycle, both of which allow plants to acquire N in organic form and thus avoid the need to wait for microbes to release it as inorganic N and then to compete with them for it. First, it is

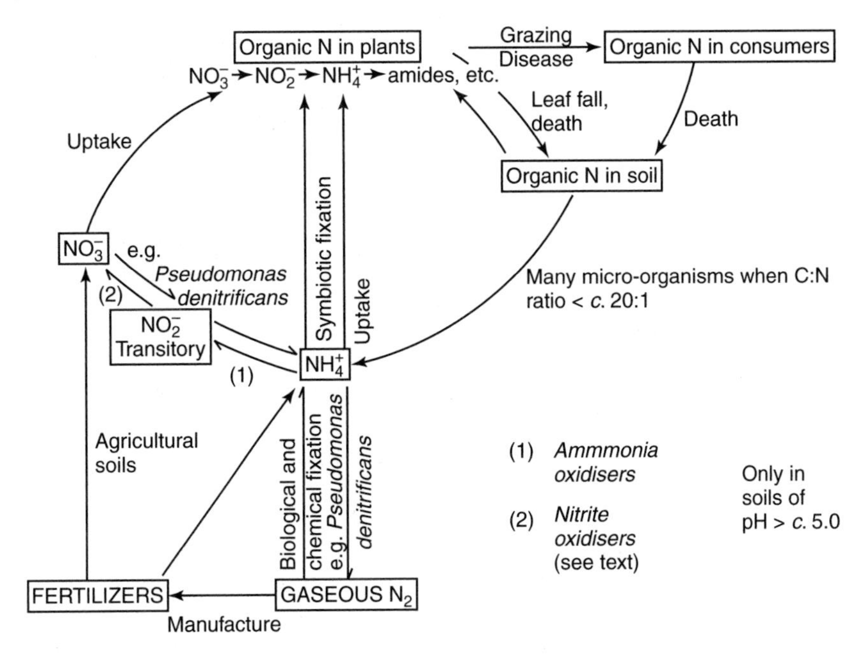

Figure 3.11

Simplified scheme of nitrogen cycle, showing major sources and pools, and primary processes involved

increasingly clear that some (but not all) plants can absorb organic N compounds through their roots. This has been shown to be of ecological importance in the field: for example, 12 out of 13 sedges (Cyperaceae) examined by Raab *et al.* (1999) could absorb the amino acid glycine, and cotton grass *Eriophorum vaginatum* (also in the Cyperaceae) can apparently acquire 60% of its N in the field by absorbing amino acids (Chapin *et al.*, 1993). In the latter study, barley *Hordeum vulgare* was unable to acquire significant amounts of N in that way and Hodge *et al.* (1999a) could not detect organic N uptake when *Lolium perenne* was supplied with patches of the amino acid lysine in soil. However, since Jones and Darrah (1993a) have shown that maize *Zea mays* can readily take up organic N compounds in this way in solution culture, the distinction is not between plants of nutrient-rich and nutrient-poor environments. Second, several types of mycorrhizal fungus can not only take up organic N compounds and transfer the N to the plant partner in the symbiosis, but in some cases can apparently degrade organic matter to acquire the N directly; the functioning of mycorrhizas is discussed later (see p. 120).

5. Transport

The supply parameters considered above adequately describe nutrient availability at a fixed point in soil, but absorption by a root necessarily involves movement of the ion from the soil to the root surface. Since roots are scattered in soil and even in a densely rooted horizon are unlikely to occupy more than

10% of the space, with 1% being a more usual maximum (Dittmer, 1940), such movement must occur over considerable distances if the plant is to exploit more than a fraction of the soil volume. The mobility of ions in soil is comprehensively discussed by Tinker and Nye (2000). In practice there are two ways in which an ion can move towards a plant root:

(i) by mass flow or convective flow down the water potential gradient caused by transpiration. Leaching is a form of mass flow in which the water flow is gravitational, down the soil profile. It is of great importance in agriculture and in the development of soils over long periods of time, but is little influenced by roots.
(ii) by diffusion down concentration gradients created by the uptake of ions at the root surface.

The speed at which ions can move by mass flow is vastly greater than by the very slow process of diffusion, but in practice it is only of importance for some ions. The movement of ions by mass flow can be represented as follows (Tinker, 1969):

$$F = V \times C_l$$

where F is the flux across the root surface in moles $m^{-2}\,s^{-1}$, V is the water flux in $m^3\,m^{-2}\,s^{-1}$ or $m\,s^{-1}$, C is the concentration of ions in moles m^{-3}, and the suffix l represents liquid phase. When a single root grows into a volume of soil, the entire soil volume can in theory be exploited by mass flow; the adequacy of the supply, however, depends entirely on whether the product of water flux and bulk soil solution concentration satisfies the plants requirements.

Uptake at the root surface lowers the concentration. If the uptake rate is greater than the rate of supply by mass flow, either because V is inadequate, as for example at night or under daytime water stress when transpiration ceases, or because C_l is too low, a concentration gradient will develop from the bulk soil towards the root. Diffusion will then occur down this concentration gradient. Conversely ions may arrive faster than they are taken up, so that concentrations build up at the root surface and diffusion will then operate in the opposite direction. This may happen for calcium in calcareous soils (Barber, 1974). The full equation for the flow of nutrient ions through soil to a plant root must therefore include both mass flow and diffusive components:

$$F = VC_l + (-D\,\mathrm{d}C/\mathrm{d}x)$$

where D is the diffusion coefficient in soil ($m^2\,s^{-1}$) and $\mathrm{d}C/\mathrm{d}x$ the concentration gradient; the negative sign allows for the fact that diffusion occurs down a concentration gradient.

Several predictive models have been constructed to examine the effects of mass flow and diffusion in the supply of ions to plant roots (e.g. Passioura, 1963; Baldwin, 1975; see Nye and Tinker (1977) for a full discussion). They show that as the water flux increases the zone of depletion around the root becomes a zone of accumulation (Fig. 3.12). Where the line (A) lies above the dotted line in Fig. 3.12, mass flow would be the sole source of supply, and this can

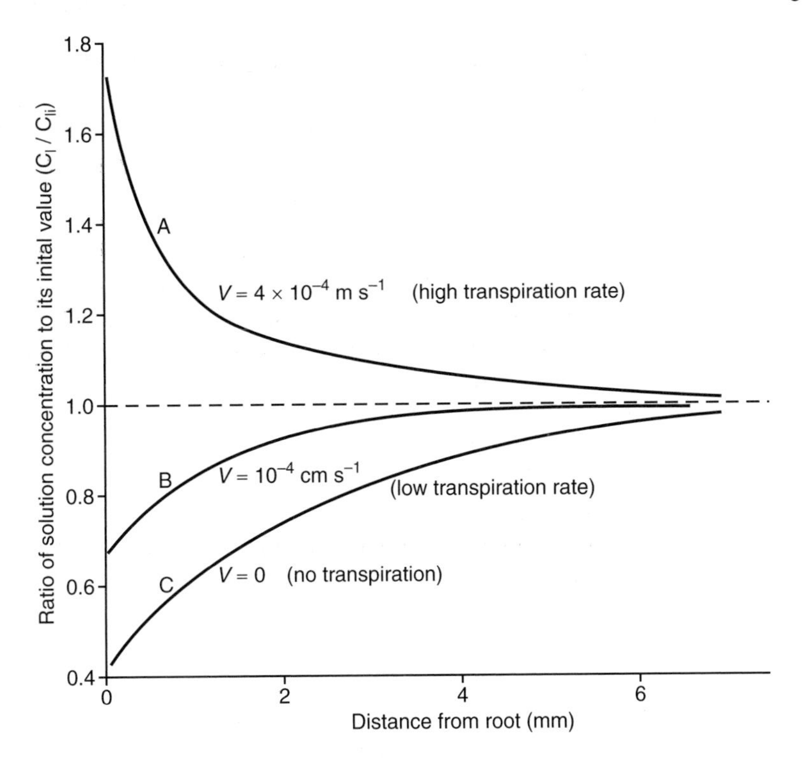

Figure 3.12

Changes in ion concentrations around a root after about 12 days (10^6 s) at different mass flow rates, with a diffusion coefficient of 2×10^{-11} m^{2}s^{-1}, typical for K^+. (After Nye and Marriott, 1969)

therefore be viewed, hypothetically, as a relationship between ion flux and water flux, as in Fig. 3.13. Here the line through the origin shows what would occur with mass flow only (i.e. $F = VC_l$); the dotted line represents the actual situation for magnesium – at values of V less than about 2×10^{-10} m s^{-1} ion flux across the root surface is greater than mass flow would allow, so that a concentration gradient must exist towards the root and the diffusive contribution must be positive. Above this critical value of V, mass flow supplies ions faster than they are taken up and diffusion operates away from the root, lowering the flux into the root below the figure that could theoretically be supplied by mass flow (cf. A in Fig. 3.12).

Using these theoretical considerations and an increasing number of experimental confirmations, the relative importance of mass flow and diffusion for all important nutrient ions can be approximately stated. From data of available P, K, Ca and Mg and known values of water use and nutrient uptake by a maize crop, Barber (1974) estimated the relative contributions of mass flow, and (by difference) diffusion (Table 3.5). For potassium this suggests that 11% was supplied by mass flow, which agrees well with data obtained

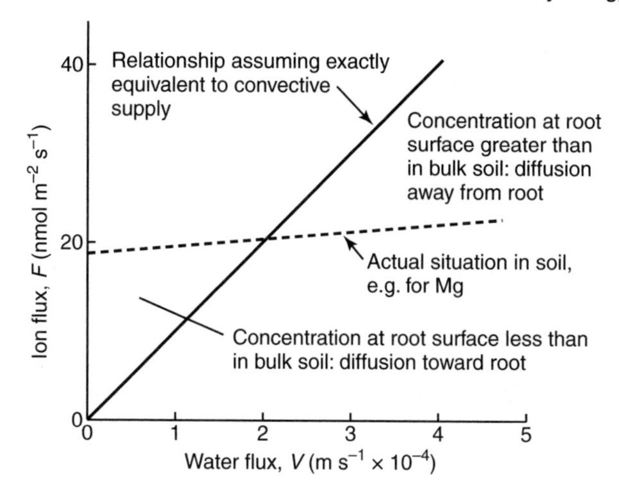

Figure 3.13

Relations between ion flux and water flux (from Tinker, 1969)

Table 3.5

Estimated contributions to crop nutrient uptake of mass flow, and diffusion during one season's ion uptake by a maize crop (Barber, 1974)

Ion	Uptake by crop (kg ha^{-1})	kg ha^{-1} supplied by	
		Mass flow	Diffusion
Ca^{2+}	45	90 (200)	negative (—)
Mg^{2+}	35	75 (214)	negative (—)
K^{+}	110	12 (11)	95 (89)
$H_2PO_4^-$	30	0.12 (0.4)	29 (>99)

Figures in parentheses are percentages of uptake for each ion.

experimentally by Tinker (1969) by growing leek seedlings in soil and measuring water and ion uptake, ion status of the soil, and the relevant plant parameters. He found that VC_l (the mass flow contribution to flux) represented between 4 and 13% of the total flux, F, depending on harvest interval.

Mass flow is a rapid process if water flux and soil solution concentrations are great enough. By contrast, diffusion is a slow process in soil, typically measurable in millimetres per day. Where mass flow is insufficient to satisfy plant demand, therefore, ion concentrations at the root surface are reduced below that of the bulk soil solution, and marked zones of depletion occur. In such cases only a part of the soil volume is exploited, except for very mobile ions. The extent and development of these depletion zones are controlled partly by the plant, in creating the concentration gradient seen as a zone of depletion in autoradiographs, and partly by the soil, *via* the diffusion coefficient, D. This varies as follows (Nye, 1966):

$$D = D_l \times \theta \times f \times 1/b$$

where D_l is the diffusion coefficient in free solution; θ is the volumetric water content of the soil; f is an impedance factor; and b is the soil buffer power (dC/dC_l).

The drier and the more strongly buffered the soil, the lower the diffusion coefficient will be; clearly in a dry soil diffusion will be negligible and the wetter it is the nearer it approaches the free solution state. Impedance is a measure of the length of the diffusion path, which is affected by both moisture content and degree of compaction of the soil. The exact effect of these two interacting factors is complex (Fig. 3.14). The buffer power term accounts for adsorption of ions by soil and explains why the diffusion coefficients of non-adsorbed ions are relatively little affected by soil. Typical values of D in soil are shown in Table 3.6; the extremely low values for phosphate permit negligible movement and may severely limit supply. Even in poorly buffered soils phosphate diffusion is slow, and pronounced, though narrow, depletion zones develop (Fig. 3.15).

An important consequence of this, then, is that for ions for which the soil solution concentration is great enough to permit considerable movement by mass flow, little depletion occurs at the root surface, though the radius of the zone of influence is large. In contrast, ions supplied largely by diffusion typically have narrow (often less than 1 mm), but well-marked depletion zones. However, some

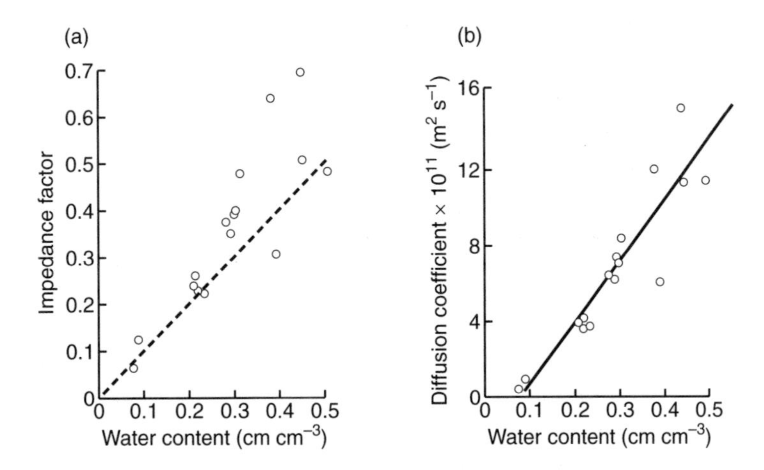

Figure 3.14

(a) Relationship between water content of a silt-loam soil and the impedance factor (f) for potassium; the dashed line is the 1:1 relationship. (b) The relation between potassium diffusion coefficient in the same silt-loam soil and soil water content; the line is fitted by eye. Redrawn from Kuchenbuch, K., Claassen, N. and Jungk, A. (1986) *Plant and Soil* **95**, 221–231

Table 3.6

Range of measured values for the effective diffusion coefficient ($m^2\ s^{-1}$) of various ions in soil (summarized from several authors by Barber, 1974; Fried and Broeshart, 1967; Tinker and Nye, 2000)

	Ion	Minimum	Maximum
Cations	K^+	2×10^{-11}	2×10^{-10}
	NH_4^+	4×10^{-12}	1×10^{-10}
	Ca^{2+}	3×10^{-12}	3×10^{-11}
	Zn^{2+}	3×10^{-14}	2×10^{-11}
Anions	NO_3^-	5×10^{-11}	1×10^{-9}
	Cl^-	3×10^{-11}	1×10^{-9}
	$H_2PO_4^-$	1×10^{-18}	4×10^{-13}

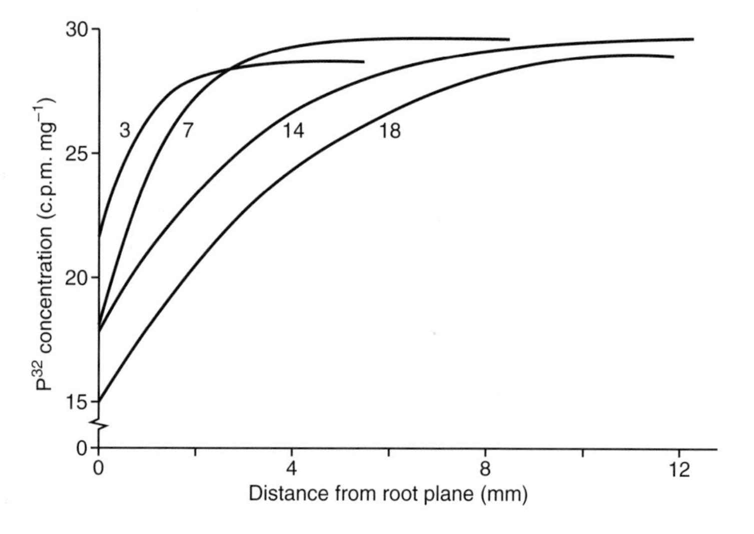

Figure 3.15

Development of depletion zones for phosphate around an actively absorbing root (figures on curves represent days from start of the experiment) (from Bagshaw *et al.*, 1972)

ions with high diffusion coefficients, such as nitrate, are present at low concentrations in most soil solutions, and diffusive supply is important. Since the area of diffusive supply is much greater for these ions than for ions such as phosphate, the depletion zones that develop will be wide and shallow, and this will mean that the whole soil volume will rapidly be exploited. In other words, the entire soil pool of nitrate is available to the plant, and where several roots are involved, competition will inevitably ensue. For phosphate, on the other hand, with an effective diffusion coefficient several orders of magnitude less than that for nitrate, the immediately available amount is that within a small radius of each root. Consequently the chances of depletion zones overlapping and so reducing the concentration gradient from bulk soil is much less at any root density. The extent of competition for ions in soil therefore depends in large measure on the diffusion coefficient, and for phosphate it will only occur at unusually high root densities or in soils with very low buffering capacities, and hence higher diffusion coefficients. Figure 3.16 shows that ion uptake per unit amount of root is much more sensitive to root density for K^+ (D usually between 10^{-10} and 10^{-12} $m^2 s^{-1}$) than for $H_2PO_4^-$ (D usually between 10^{-13} and 10^{-15} $m^2 s^{-1}$).

6. Limiting steps

Ion transport in soil can be very slow, but it will limit the uptake process only where it is slower than the flux across the root surface. That flux (F, mol $m^{-2} s^{-1}$), expressed on a surface area of root basis, is proportional to ion concentration at the root surface:

$$F = \alpha C_{lr}$$

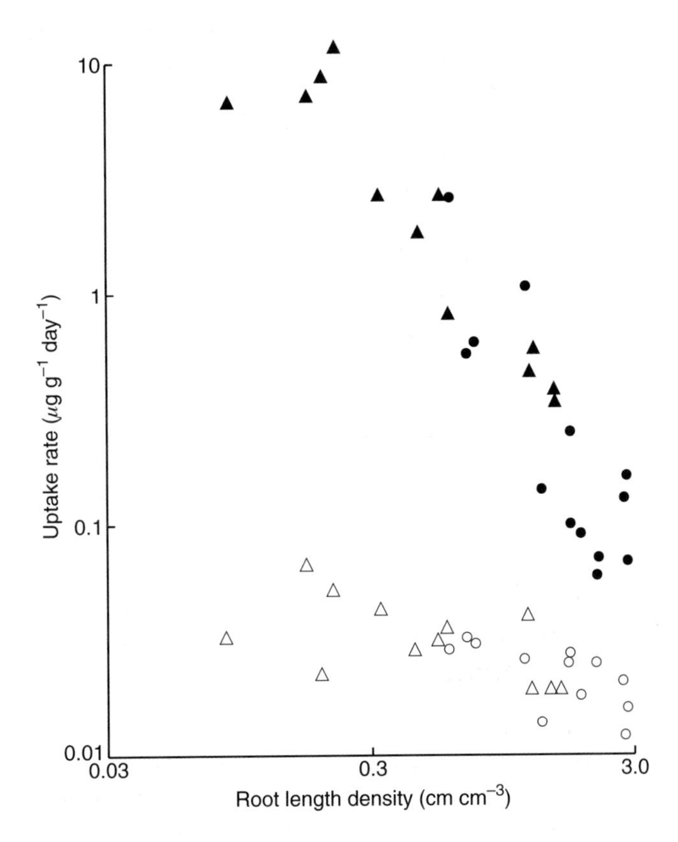

Figure 3.16

Effect of root length density on uptake of phosphorus (open symbols) and potassium (closed symbols) by *Lolium perenne* (circles) and *Agrostis capillaris* (triangles) per unit root weight. K uptake was much the more sensitive to root length density, because of its greater mobility (from Fitter, 1976)

Table 3.7

Root demand coefficients ($\overline{\alpha a}$) for four species grown in flowing culture at different potassium concentrations over 42 days (Wild *et al.*, 1974)

Species	$\overline{\alpha a} \times 10^{10}$ $m^2 s^{-1}$				
	[K] (μM)	1	3	10	33
Dactylis glomerata		11.9	8.0[a]	3.2	1.0
Anthoxanthum odoratum		7.9[a]	4.5	1.1	0.4
Trifolium pratense		7.2	5.2[a]	2.1	0.6
Medicago lupulina		30.1	19.9	6.9	2.3

[a] Concentration above which no further growth increases occurred.

where the proportionality coefficient α is referred to as the root absorbing power, with units of m s^{-1} (Tinker and Nye, 2000). The uptake can also be expressed on a root length basis, when it is known as an inflow (I, mol m^{-1} s^{-1}):

$$I = 2\pi(\overline{\alpha a})C_{lr}$$

The term $\overline{\alpha a}$, averaged over the root system (a is root radius), is the root demand coefficient of Nye and Tinker (1969). It is in effect a transport coefficient, in the same units (m^2 s^{-1}) as a diffusion coefficient (D). Since the two processes – diffusion through soil and transport across the root surface – are in series, the slower of the two will limit the rate of uptake. Values of $\overline{\alpha a}$ can be determined experimentally (Table 3.7) and vary with internal and external ion concentrations, plant growth rate and root morphology, as predicted by Nye and Tinker (1969). Where $\overline{\alpha a}$ is less than D, transport across the root surface will be limiting, and this will often be the case for ions such as NO_3^- and Ca^{2+}. Conversely the values of $\overline{\alpha a}$ for K^+ in Table 3.7 are very much greater than typical values of D for K^+ (around 10^{-11} $m^{-2} s^{-1}$: see Table 3.6), so that diffusion will normally be the rate-limiting step here.

3. Physiology of ion uptake

1. Kinetics

Plant root cells generally contain much higher ionic concentrations than their surrounding medium and are normally electrically negative with respect to it. This potential, of the order of -60 to -250 mV, is created by the presence of large organic anions which cannot move across the membrane, and by the activity of proton pumps, which transport H^+ across the membrane. These pumps are enzymes that convert ATP into ADP (ATPases), and so the maintenance of the potential difference across the membrane involves energy expenditure. As a consequence of this gradient, some cations can enter the cell passively, down the electrochemical gradient, although usually against the chemical concentration gradient for that ion; in some cases, certainly for Na^+, cations are actively exported back into the medium. By contrast, anions must be transported actively into root cells in company with protons, but against the electrochemical gradient and in most cases the concentration gradient too. Whether active or passive, however, transport is energy-dependent and so necessarily influenced by a number of external and internal factors, in particular the concentration of the ionic substrate and variables such as temperature that affect respiration. When energy supply is adequate, the dominant influence on the rate of ion uptake is the concentration gradient.

There is a characteristic relationship between uptake rate and concentration which, because of its similarity to that between the rate of an enzyme-mediated reaction and its substrate concentration, has been analysed in terms of Michaelis–Menten kinetics. The typical hyperbolic curve is shown in Fig. 3.17 together with two plots based on enzyme kinetics which produce linear plots. The enzyme analogy works because there are carrier molecules in the cell membrane, more or less specific for the ion, which can transport ions at a particular rate (v) under given conditions, so that at extremely high substrate concentrations all carriers are operating at maximum rate, and saturation uptake rates (V_{max}) are reached. The Michaelis–Menten equation can then be applied:

$$v = (V_{max} \times C_{ext})/(K_m + C_{ext})$$

where K_m represents that external concentration (C_{ext}) at which v is half V_{max}. This Michaelis constant, K_m, therefore describes a property we may call the 'affinity' of the uptake system; a low K_m implies high affinity, the system becoming half-saturated at low concentrations.

This high affinity uptake system operates at very low external concentrations, and can move ions even against the electrochemical gradient. At higher concentrations a distinct isotherm (response curve at constant temperature) can be observed (Epstein, 1973); this 'low affinity' uptake occurs at concentrations high enough for uptake to be down the electrochemical gradient. These two uptake systems, operating over different concentration ranges, correspond to distinct molecular systems within the plasma membrane, at least for K^+ (Maathuis and Sanders, 1996) and NO_3^- (Forde and Clarkson, 1998). The high affinity uptake system for K^+ uses a carrier molecule that co-transports K^+ and protons (known as a symporter), whereas the low affinity system operates through ion channels, which are transport proteins embedded in the membrane and capable of high passive fluxes of ions, either inwards or outwards. In ecological terms, the high affinity system is the most important, because the concentrations that plant roots normally experience are in that low range (see Table 3.4), and the role of the low affinity system in roots awaits clarification.

Anion uptake always occurs against the electrochemical gradient, but appears to involve a symporter, just as with K^+, with anions crossing the membrane in company with a proton. Antiporters, which move OH^- ions in the opposite direction, may also be involved.

Many of these ion transporters are now being identified at a molecular level. For example, a family of high affinity phosphate transporters (PHT) is now

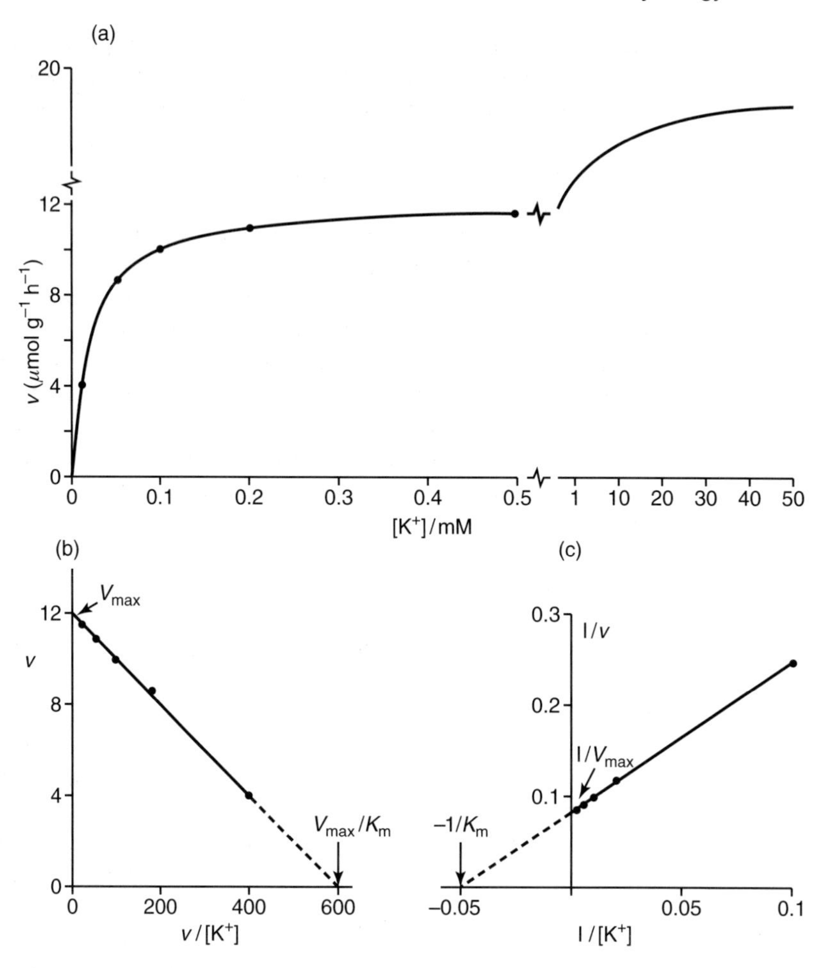

Figure 3.17

Absorption isotherm for potassium ions showing the two phases of uptake in relation to concentration (a); and two plots to obtain linearity from the asymptotic first phase (high affinity transport system): (b) Hofstee plot and (c) Lineweaver–Burke plot. $V_{max} = 12$ μmol g^{-1} FW h^{-1}; $K_m = 20$ μM. Hypothetical data

known from a wide range of plant species, including potato *Solanum tuberosum*, tomato *Lycopersicon esculentum* and *Arabidopsis thaliana*. These PHTs show very great DNA sequence homology and clearly represent a common system, in plants and in fungi such as *Saccharomyces*, *Neurospora* and *Glomus versiforme* (a mycorrhizal fungus, see p. 120). There are at least six PHT genes in *Arabidopsis*, although only two of these seem normally to be transcribed. Some of these transporters, for example *LePT1* and *LePT2* (both from tomato, *Lycopersicon esculentum;* the '*Le*' derives from the species from which the gene was identified) are expressed specifically in root epidermal cells (Daram *et al.*, 1998), which is strong evidence that they are involved in the acquisition of phosphate from the soil. When plants are grown with high phosphate supplies, the transcription of PHTs may decline (Liu *et al.*, 1998). This shows that plants are able to regulate the activity of the transporters to match the ion flux at the root surface, and that the transporters are short-lived, their appropriate concentration being maintained by a balance between destruction and synthesis. The molecular biology of P acquisition has been reviewed by Raghothama (1999).

Nitrate transporters are even better characterized. There are two distinct gene families involved in nitrate transport (*NRT1* and *NRT2*), many of which have been identified by finding mutants of *Arabidopsis thaliana* that are resistant to the herbicide chlorate: chlorate and nitrate are transported by the same mechanism, and hence chlorate-resistant plants may have disabled nitrate transport. Most of the genes in both the NRT families are induced by nitrate, but several are known that code for constitutive transporters (Crawford and Glass, 1998). This matches the known kinetics of nitrate uptake, for which three systems are known: constitutive low and high affinity systems, and inducible high affinity systems.

2. Interactions

Roots in soil absorb ions from a complex solution containing a wide range of ions, both inorganic and organic. Many of these ions have the ability to alter ion uptake. Because of the key role of proton pumps in maintaining electrochemical potentials in cells and energizing ion transport, it is unsurprising that altering pH has profound effects. Typically, cation uptake is depressed at very low pH values (below 4–5), when external H^+ concentrations are high.

Other ions interact because they share the same transport molecules, usually because they have similar physical characteristics (charge, ion size etc.). For example, arsenate ions, which are toxic to plants, are sufficiently similar to phosphate ions that they compete for the same binding sites. Arsenate tolerance is achieved by suppression of the high-affinity phosphate uptake mechanism (Meharg and Macnair 1992; cf. Chapter 6). K^+ and Rb^+ behave so similarly that ^{86}Rb is used as a convenient surrogate for K^+ ions in studies of potassium nutrition. These interactions are competitive, since the two ions compete for the transport sites, and their relative fluxes are a function of their relative concentrations. In other cases, the interaction is more one-sided: the binding affinity for one ion is much greater than for the other; for example, Cs ions bind weakly to the potassium transporter, and significant Cs uptake occurs only at very low K^+ concentrations (Erdei and Trivedi, 1991).

Calcium ions play a particularly important role in modifying ion uptake. In addition to typical competitive interactions with similar ions (e.g. with Sr^{2+}), Ca^{2+} ions play a key role in the structure of membranes and in counteracting the effects of high H^+ concentrations. In ecological terms, calcium is of interest because of the enormous range in free Ca^{2+} concentration that can be found in soil solutions, depending on whether or not the rocks the soil was formed from were calcareous (i.e. containing calcium carbonate – chalk or limestone).

Caution is therefore needed in extrapolating results from physiological experiments to understand the behaviour of ion uptake systems in soil. Even when whole plants are grown in nutrient culture, only essential elements are normally used (Hewitt, 1967), and very little is known of the effects of non-essential ions (with the notable exception of Al^{3+}, see Chapter 6). In many soils silicate ions are dominant and these can interact with manganese (Rorison, 1971), and in calcareous soils bicarbonate ions are abundant and powerfully inhibit iron uptake by non-adapted species, such as *Deschampsia flexuosa* (Woolhouse, 1966). To complicate matters further many soil solutions contain a wide variety of organic ions, such as phenolic acids derived from the breakdown of lignin, which are present in some quantity in the soil around roots, and these can inhibit phosphate and potassium uptake by barley roots (Glass, 1973, 1974).

3. Regulation

Ion uptake is affected by a wide range of external factors beyond those already described, including temperature and radiation flux, both of which determine the availability or rate of utilization of energy substrates needed to drive the transport process. Nevertheless, when measurements are made of ion uptake by intact plants, especially those growing under more or less realistic conditions, it is generally found that plants are able to regulate their uptake rates in such a way as to buffer many of the more obvious environmental constraints. For example, Clement *et al.* (1978) grew ryegrass *Lolium perenne* in a flowing culture system that maintained constant nutrient concentrations, over a range of three orders of magnitude in external nitrate concentration (0.2–200 $mg\,N\,l^{-1}$, i.e. 0.014–1.4 $mol\,m^{-3}$), but the total N content of the plants varied by only 7 and 28% in two separate experiments. Plant dry mass increased by a similar amount, so that nitrogen concentration in the tissue was barely altered. Such stability demonstrates a degree of regulation by the plant, and shows that it is not possible to extrapolate from short-term physiological studies to field conditions.

One control mechanism is induction of the uptake system. Barley *Hordeum vulgare* plants grown without nitrate in the medium have very low rates of nitrate uptake at low external nitrate concentrations, but high rates at high concentrations (Glass, *et al.*, 1990; Fig. 3.18). Pretreatment of the plants for 1 day with nitrate at 0.1 mol m^{-3}, however, resulted in a 30-fold increase in nitrate influx from very low concentrations (0.02 mol m^{-3}), but only a 2-fold increase from high concentrations (20 mol m^{-3}). This is because there is a constitutive low affinity nitrate uptake system in barley, and an inducible high affinity system (see above). Such induction is commonplace in biochemical systems where substrate availability is temporally variable and reduces the cost to the plant of maintaining non-functioning transporters or other molecules.

Nitrate uptake is well characterized (Crawford and Glass, 1998). Uptake appears to involve a number of distinct transport systems, but all use symport of two protons for each nitrate ion. There are both constitutive and inducible high affinity transporters, with a V_{max} of 20–100 μM, and also a constitutive low affinity system, which operates at nitrate concentrations from 0.25 mM upwards. Molecular genetic analysis of nitrate uptake mutants has revealed two families of nitrate transporter genes, *NRT1* and *NRT2*. As with the phosphate transporters, these gene families are widespread, with members found in fungal and animal cell membranes, as well as in plants. By measuring

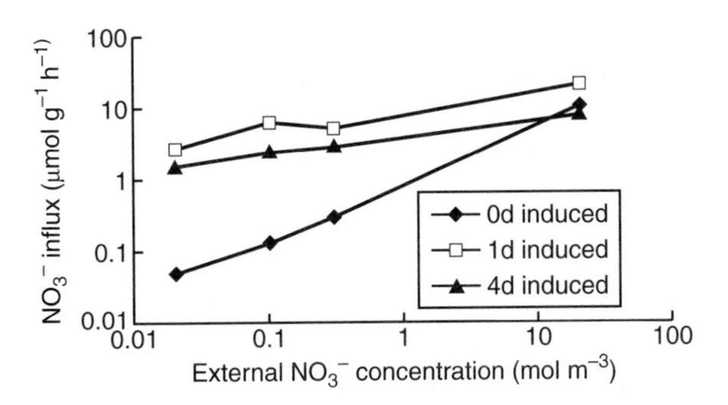

Figure 3.18

Nitrate influx to nitrate-starved barley roots as a function of the external nitrate concentration (from Glass *et al.*, 1990). The influx at low concentrations was greatly enhanced by the induction effect of growth in the presence of 0.1 mol NO_3^- m^{-3} for 1 or 4 days prior to measurement.

mRNA expression, it can be shown that some of these transporters (especially in the *NRT2* family) are induced by nitrate, providing the molecular basis for the responses observed.

Induction is a control mechanism but not a regulatory system, since that requires negative feedback: increased uptake from low concentrations and reduced uptake from high. Evidence for such regulation can also be seen in Fig. 3.18: when the barley plants were pretreated with nitrate for 4 days, the nitrate influx was lower than in plants pretreated for only 1 day. This shows that the plant is able to down-regulate the transport system as growth rate declines and demand falls, and this down-regulation can be shown to occur by reduced transcription of the inducible transporter genes (Krapp *et al.*, 1998). These findings open the possibility of a demand-led uptake system in which uptake rate is regulated by the rate at which the plant is utilizing the nutrient in question. It is now well established that plants do operate in this way: there seems to be a general relationship between uptake rate and growth rate, perhaps the most obvious measure of plant demand (Fig. 3.19: Garnier *et al.*, 1989).

What remains unclear is the nature of the signals that link plant demand and uptake rate. One possibility is that it is the rate at which ions are metabolized that is important. Nitrate, for example, must be reduced first to ammonium before incorporation into amino compounds. The system responsible consists of two linked and substrate-induced enzymes, nitrate and nitrite reductase:

$$NO_3^- \xrightarrow[\text{NAD(P)H} \rightarrow \text{NAD(P)}]{\text{nitrate reductase}} NO_2^- \xrightarrow[\text{red. ferredoxin (in leaves)}]{\text{nitrite reductase}} NH_3$$

This reduction can take place in either roots or leaves, depending on plant species, age and nitrate supply rate. The concentration of both enzymes and therefore the capacity for reducing nitrate is determined by the rate of NO_3^- supply. When plants are deprived of nitrate, the nitrate reductase (NR) activity falls rapidly, in less than 24 h (Long and Woltz, 1972), because the activity of

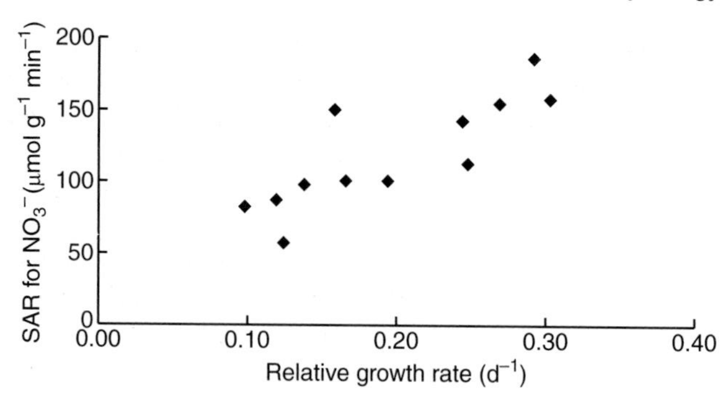

Figure 3.19

The specific absorption rate (SAR) for nitrate is a function of growth rate when measured in a range of grass species (from Garnier *et al.*, 1989)

NR is controlled by a reversible phosphorylation process (Huber *et al.*, 1996). Similarly, if nitrate is added to soil around plants growing in natural communities, they can show large increases in NR activity (Table 3.8). The level of NR activity in tissue may therefore be a good indicator of the nitrate availability in the habitat; the range of NR activity shown in Table 3.8 before induction suggests that species such as stinging nettle *Urtica dioica* were growing in more nitrate-rich soil than those in acid grassland (e.g. purple moor-grass *Molinia caerulea*) or calcareous grassland (e.g. small scabious *Scabiosa columbaria*). However, even though most plants show this inducible response, even after induction there are differences between species in NR activity, suggesting that there is genetic differentiation based on the fact that some habitats are

Table 3.8

Nitrate reductase activity in the field before and after induction by added nitrate, in species from various soil types (Havill *et al.*, 1974)

Soil	Species	Nitrate reductase activity (μmol NO_2^- g^{-1} FW h^{-1})	
		Before induction	After induction
Calcareous	*Poterium sanguisorba*	1.11	3.80
	Scabiosa columbaria	1.26	2.12
Neutral	*Urtica dioica*	7.93	16.10
	Poa annua	4.05	8.40
Acid	*Molinia caerulea*	0.52	1.50
	Galium saxatile	0.72	3.06
Acid peat	*Vaccinium myrtillus*	<0.1	<0.1
	Drosera rotundifolia	<0.1	<0.1

inherently more nitrate-rich than others. The most extreme difference is seen in plants from acid heathland, such as bilberry *Vaccinium myrtillus* and in insectivorous plants such as sundew *Drosera rotundifolia*. These species do not exhibit NR activity even after they are supplied with nitrate; they are therefore unable to use nitrate as an N-source. A possible explanation is that the acid soils they inhabit have low rates of nitrification, with the result that ammonium rather than nitrate ions are the main form of inorganic N. However, they can share these soils with species such as heath bedstraw *Galium saxatile*, which do show NR induction. More significantly both bilberry and sundew have alternative N supply routes, the former through its association with a particular type of mycorrhizal fungus that can obtain N directly from organic matter (see p. 120), and the latter, more obviously from its insectivorous habit. Indeed lack of NR activity is characteristic of several entire families of plants, including Ericaceae (heaths, mainly a northern hemisphere family), Epacridaceae (a southern hemisphere family ecologically similar to Ericaceae), and Proteaceae (Stewart and Schmidt, 1999).

The inducibility of the nitrate reductase system gives flexibility to the metabolic systems of many plant species, and is an economical response to the wide variation that occurs in nitrate supply in natural soils. However, it does not seem to be a key element of the control system that regulates nitrate uptake. One possible signal would be the accumulation of ions in the roots; if ions were not being used metabolically, they might accumulate in root cells and be the signal to down-regulate the uptake system. In practice, this seems unlikely. In maize roots, the external phosphate concentration could be varied between 0 and 0.5 mol m^{-3} without any effect on cytoplasmic concentrations of inorganic phosphate (P_i), which remained at 0.24 μmol cm^{-3} in an experiment by Lee and Ratcliffe (1993). At the same time, the vacuolar concentration, measured by NMR techniques, declined from around 7 μmol cm^{-3} at 0.5 mol m^{-3}, to 3.5 at 0.01, 0.5 at 0.001 and to undetectable levels when no P was supplied. However, the P uptake rate was not affected by these large differences in vacuolar P_i concentration, suggesting that there is no direct link between them.

The answer to the question of how uptake rates are regulated seems to lie in an understanding of the nutrition of the whole plant. For example, the concentrations of amino acids and carboxylic acids in the phloem both affect nitrate uptake rate in soybeans: adding malate nearly doubled the uptake rate (Touraine *et al.*, 1992) whereas supplying arginine to the cotyledons (which increased the phloem arginine concentration 10-fold) almost halved the nitrate uptake rate (Muller and Touraine, 1992). Organic acids might exert an effect by virtue of the bicarbonate ions released when they are decarboxylated in the roots, the HCO_3^- ions being exchanged for nitrate in an antiporter system. The mechanism by which amino acids regulate nitrate transport is unknown, but the example demonstrates the need to consider whole-plant nutrition in order to understand the regulation of the ion transport system.

Insights into the regulatory system are likely to come from studies on *Arabidopsis*, because of the ability to generate mutants. Two key mutants that may be used here are *pho1* and *pho2*; *pho1* is unable to load P into the root xylem and consequently accumulates P in the roots, with very low shoot concentrations (Poirier *et al.*, 1991), whereas *pho2* accumulates high concentrations in the shoots (Delhaize and Randall, 1995). The fact that these mutants maintain P uptake even though the root and shoot P concentrations, respectively, increase to abnormally high levels, shows that the feedback control mechanisms that regulate uptake do not depend simply on detecting local P accumulation.

4. Morphological responses

For many ions the rate-limiting step in uptake is transport through soil and not across the root (see p. 91). Changes in uptake kinetics may, therefore, have only a small impact on uptake, and the most effective response a plant can make is to alter the morphological properties of the roots. In other words, to have greater activity per unit amount of root will not increase uptake, but to have a greater amount of root will.

1. Root fraction

The simplest index of allocation to roots is the root:shoot ratio ($W_R : W_S$) or the root fraction (W_R / W_{RS}), which represents the proportion of total plant biomass allocated to roots. Root fraction is very plastic and generally increases with:

(i) low soil water supply (Chapter 4);
(ii) low nutrient supply (Ericsson, 1995);
(iii) low soil temperature (Davidson, 1969);

and shows less predictable responses to other environmental variables such as soil oxygen, light intensity and photoperiod (Aung, 1974). On the whole, plants seem to put more of their resources into root production in environments where growth is limited by soil-derived resources or factors.

Perhaps surprisingly, however, there is little evidence to support the often-stated view that plants from different habitats respond to nutrient deficiency differently in terms of the allocation of resources to root and shoot growth. When Reynolds and d'Antonio (1996) surveyed 77 published studies of the allocation responses to nitrogen availability, covering 129 species, they were unable to find any patterns among species categorized by habitat, life form or inherent growth rate. In other words the tendency to allocate more resources to root growth when nutrients are limiting is a general plant response, and not part of the adaptive mechanism to particular habitats or soil types. Plants therefore seem to be able to maintain some balance between the acquisition of energy and carbon by leaves and of water and nutrients by roots. Unsurprisingly the mechanism for such a general response has excited much interest and numerous possible signal molecules have been suggested, one of the most convincing being sucrose (Farrar, 1996). When nutrient deficiency develops in a plant, the rate of utilization of non-structural carbohydrate falls, allowing sucrose to accumulate. The full mechanism, when elucidated, will involve other signals and triggers, but sucrose is likely to be a key component.

2. Root diameter and root hairs

All except the oldest roots possess some uptake ability, but the segment behind the root tip, in most species, possesses a special importance by virtue of its root hairs (Plate 6c). The effect of root hairs is to increase the effective diameter of the root and hence the volume of soil exploited, with only a small additional cost in terms of tissue construction, compared with an actual increase in root diameter. Increasing effective root diameter will have little effect on the uptake of mobile ions such as nitrate, but should promote uptake where diffusive supply is important, as for phosphate. In the latter case depletion at the root surface will be extensive and the whole root hair zone may be depleted, as shown by autoradiographs of soil mixed with $H_2{}^{32}PO_4{}^-$ from which roots have been allowed to absorb P (Bhat and Nye, 1973). Further, in field conditions where soil moisture varies greatly and the solution contact with the root may be broken, root hairs can penetrate very small pores (diameter <5 μm) and by secreting mucilage maintain a liquid junction between root and soil without which both ion and water uptake would cease (Barley, 1970; Plate 8). Experimentally, however, the role of root hairs is not fully confirmed. P uptake by white clover was directly proportional to root hair length in a range of genotypes (Caradus, 1981) and where only root hairs were allowed access to ^{32}P in soil, they could still satisfy 60% of plant P demand (Gahoonia and Nielsen, 1998). However, Wheeler (1995) found that clover plants with long root hairs grew better than those with short hairs even in solution culture where transport limitations do not occur, which suggests that there may be other differences between root hair genotypes. Bole (1973), using chromosome substitution lines of wheat differing in root 'hairiness', could find no relationship between P uptake from soil and root-hair development, whereas Baon *et al.* (1994) found that short-haired rye (*Secale cereale*) plants were less able to obtain P from soil than long-haired plants. Most probably root hairs become increasingly important as the supply of an ion becomes more limited by diffusion; part of the confusion may occur because root hair growth may be sensitive to P supply: plants of *Arabidopsis* grown at low P (1 mmol m^{-3}) had root hairs three times longer than those grown at 1000 mmol P m^{-3} (Bates and Lynch, 1996).

Root hairs increase the effective diameter of roots, but root diameter itself can also vary and may be an important factor when depletion around the root surface is not extreme so that the concentration of ions in the root cell walls (which are in equilibrium with the rhizosphere soil) is high; uptake will then occur into all the cortical cells, not just the epidermis, and uptake will be related more to root volume than to surface area, as shown theoretically by Nye (1973) and found experimentally by Russell and Clarkson (1976). Mean fine root diameter varies remarkably between species. Some groups, including many grasses, have roots approaching the effective minimum (around 70 μm, set by the need for an epidermis, cortex and stele), whereas others are at least an order of magnitude coarser. The fine roots of the New Zealand tree *Podocarpus totara*, for example, are over 1 mm in diameter (Baylis, 1975). An order of magnitude difference in root diameter translates to two orders of magnitude difference in root volume (and hence presumably root mass) per unit root length, since volume is proportional to the square of the radius. One of the unexplained questions, therefore, is why some plants have such expensive roots.

Part of the answer lies in the role played by symbiotic mycorrhizal fungi in acquiring nutrients (see p. 120), but some inherent features of roots are also important. Fine roots are 'cheaper' for the plant to construct, but they grow more slowly and generally less far, are more vulnerable to predation and other damage, and probably normally live less long. All these may be important elements of the cost–benefit equation. On the other hand, fine roots are more likely to be able to respond to local variation in nutrient availability (see below).

Root diameter is not a fixed characteristic of a species. In most plants, the conditions under which roots develop determine diameter, just as they affect root hair growth. For example, Christie and Moorby (1975) subjected three Australian grasses to P concentrations ranging from 0.003 to 30 $mg\,l^{-1}$ and recorded increases in mean root diameter of 17, 22 and 41% at high P supply (cf. Table 3.11).

3. Root density and distribution

When a single root grows in an unlimited volume of soil, uptake is influenced by soil factors and plant demand. In real situations, however, the depletion zones around adjacent roots will frequently overlap. There will then be interference to supply patterns by one root upon another – in other words competition for nutrients (Nye, 1969). This interference will result from the overlapping of the depletion zones which effectively reduces the bulk soil solution concentration, resulting in the diminution of the diffusion gradient.

Competition will only appear, however, at high root densities for ions of low diffusion coefficients (such as phosphate), because the depletion zones are narrow and so unlikely to overlap. Figure 3.16 shows the different effect of root density on uptake of K^+ and $H_2PO_4^-$ by monospecific stands of *Lolium perenne* and *Agrostis capillaris* (Fitter, 1976). These effects are typical and can be shown equally with a single plant (Newman and Andrews, 1973). As a result an increase in local root density in areas where nutrient concentrations are high is an effective way to increase uptake of immobile ions such as $H_2PO_4^-$ or to a lesser extent NH_4^+ but not of mobile ions such as NO_3^- (Robinson and Rorison, 1983; Robinson, 1996).

When the roots of more than one species (or genotype) are interacting in this way, the result may not be just a general lowering in uptake rates for all roots, for the roots of one species may be able to lower the ionic concentration at the root surface more than the other. In this case nutrient ions will move down the resulting steeper concentration gradient more rapidly to the root surface of the former species (cf. Chapter 7, Fig. 7.5), which will therefore effectively exploit a greater soil volume. There is very little experimental evidence on this point, though it seems certain that this is what occurs in soil and it is well established that competitive ability for nutrients varies between species, with species balance dependent on nutrient levels. Welbank (1961), for example, showed that added N had a greater effect on the growth of *Impatiens parviflora* when its roots were competing with those of *Agropyron repens* than when they were in monoculture. Similarly, van den Bergh (1969), in a replacement series experiment in which the density of the two species was varied, found that at high soil fertility *Dactylis glomerata* always replaced *Alopecurus pratensis*, at pH 4.2, 6.2 or 6.7. At low fertility, however, *Alopecurus* was favoured at low pH and an equilibrium was established at the two higher pH values.

The mechanism of competition for nutrients and its interaction with competition for light and water, is discussed further in Chapter 7, but clearly the form, extent, and distribution of the root system will be critical. All these features are plastic characters and respond to various soil environmental factors such as texture (Kochenderfer, 1973) and pH (Fitter and Bradshaw, 1974). There are also dramatic responses to nutrient concentrations (see review by Robinson 1994), a fact exploited in agriculture by the practice of banding fertilizer so that roots can grow preferentially in the fertile zone (Duncan and Ohlrogge, 1958). Such effects can be mimicked in water culture (Fig. 3.20).

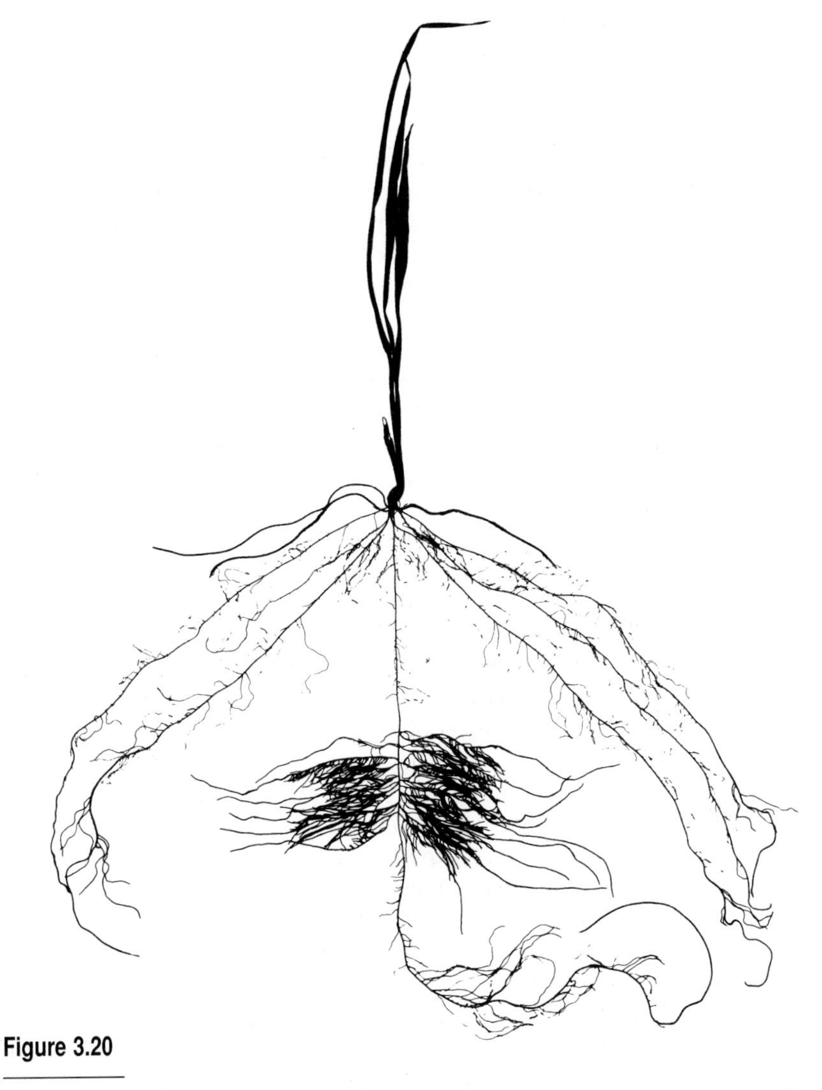

Figure 3.20

Silhouette of a barley plant grown with its root system in a nutrient solution containing 0.01 mM nitrate, except for the central section which received 1.0 mM nitrate. Note the elongation of the primary lateral roots and the proliferation of secondary laterals in this zone, compared with their absence elsewhere. (Courtesy Dr M.C. Drew)

Marked increases in root growth occur in response to local concentrations of, in particular, N and P, and can be ascribed to the fact that lateral root growth only occurs at a point on the root if adequate external nutrient concentrations exist there (Drew *et al.*, 1973). Growth of the main root axes of cereals is not concentration-dependent in this way, so that in infertile soil these will continue to grow, but lateral production will be inhibited until a fertile zone is reached. The root system is thus adapted to explore poor soil volumes but to exploit fertile ones.

These striking and dramatic changes are not accompanied by changes in shoot growth (Drew, 1975). There is a compensatory increase in uptake in the favourably placed segment with the result that nutrient uptake of the whole plant is scarcely affected in a plant only part of whose root system is exposed to high nutrient concentrations, as compared with one wholly in the high concentration (Drew and Nye, 1969; Drew and Saker, 1975). In other words, the plasticity of uptake kinetics observed between individuals in many species, for example, can also exist simultaneously between different parts of a root system.

A key question is whether this ability to proliferate roots in locally enriched patches in soil is a general response, in the same way as the allocation response that results in increased root biomass in plants grown in infertile soils, or whether it is an adaptive response, more marked in some species than others. Certainly it is not universal. Campbell *et al.* (1991) devised an ingenious experiment in which plants were grown in shallow basins of sand into which nutrient solutions were continually dripped from four nozzles. Two nozzles, at opposite sides of each basin, dispensed a low nutrient solution, and the other two a high nutrient solution. By adjusting the drip rate, they were able to create sharp boundaries of nutrient concentration between the four sectors in the sand (Plate 5d). When they planted seedlings in the basins, the plants were able to explore the soil without any artificial distribution of the roots, such as occurs in traditional split-root experiments. They showed that, although all the species they looked at placed more root in the nutrient-rich than the nutrient-poor quadrants, the proportion differed markedly. The species that placed the greatest proportion of their new root growth in the nutrient-rich quadrants were small and relatively slow-growing, such as harebell *Campanula rotundifolia*. In consequence the actual root densities that they were able to maintain in these nutrient-rich patches were lower than for larger, faster-growing species, such as stinging nettle *Urtica dioica* and false oatgrass *Arrhenatherum elatius* (Fig. 3.21). They suggested that there was a trade-off operating between scale and precision: large species foraged on a coarse-scale and so were bound to locate and exploit nutrient-rich patches, whereas small species foraged on a fine scale and were less likely to find the patches, which they then exploited by intensive root proliferation.

Such effects also occur in soil (van Vuuren *et al.*, 1996), but the complexity of nutrient supply from soil is much greater than from a nutrient solution, and it is not always the case that proliferation of roots results in increased nutrient capture. In the experiment of van Vuuren *et al.* (1996), the roots of wheat plants proliferated extensively in an organic patch in the soil, but only after they had taken up most of the nitrogen that was released by the decomposition of the patch. Root proliferation responses often occur over a time-span of 10–20

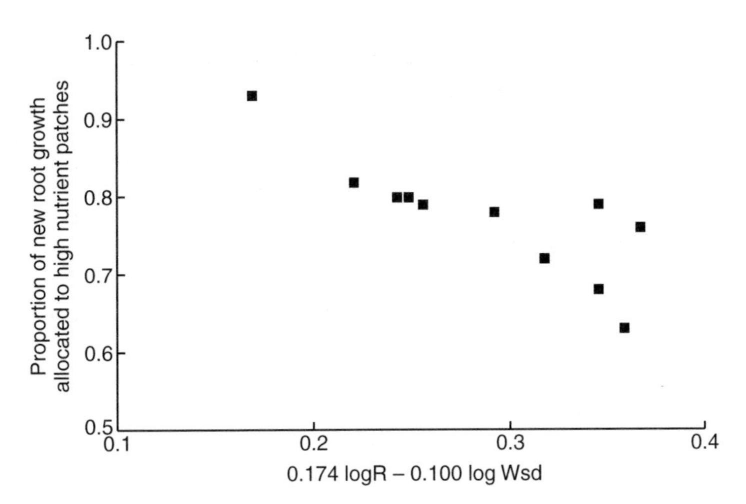

Figure 3.21

The precision with which plants develop new root growth into nutrient-rich patches is a function of plant size. The *y*-axis represents the fraction of new root growth that appeared in the two nutrient-rich quadrants in the experiment of Campbell *et al.* (1991), and therefore a value of 0.5 represents random placement, whereas 1.0 would represent all new root growth appearing in the rich quadrants. The plant size parameter on the *x*-axis was calculated from seed mass and relative growth rate using a regression model (figure from Fitter, 1994)

days, whereas soil heterogeneity may have a shorter time-scale; this paradox has not yet been resolved.

These apparently adaptive responses are under genetic control: a gene has been identified in *Arabidopsis thaliana* that controls the response to locally applied nitrate (Zhang and Forde, 1998). It seems puzzling, therefore, that they should not always result in increased nutrient uptake. Unfortunately, most experiments have been undertaken with single plants: in competition experiments, the plants that proliferate most do obtain the largest share of nutrients from an organic patch (Hodge *et al.*, 1999b), which is consistent with the idea of Campbell *et al* (1991) that species differ in whether they forage coarsely over a wide scale, or with precision at a fine scale.

Branching patterns are well-known to differ markedly between root systems, both within and between species. Herringbone-like patterns (Fig. 3.22a), which comprise a main axis with side branches but few or no further orders of branching, are more efficient in terms of the volume of soil explored for a given construction cost, but diffuse-branched systems (Fig. 3.22b) are more transport-efficient and generally minimize construction costs (Fitter, 1996). When plants are grown in nutrient-poor soils, they tend to produce root systems with a more herringbone architecture (Fitter and Stickland, 1991), and, for dicots at least, species from nutrient-poor habitats also have a more herringbone root architecture. The same responses occur when roots exploit nutrient-rich patches in soil: the roots that develop in these

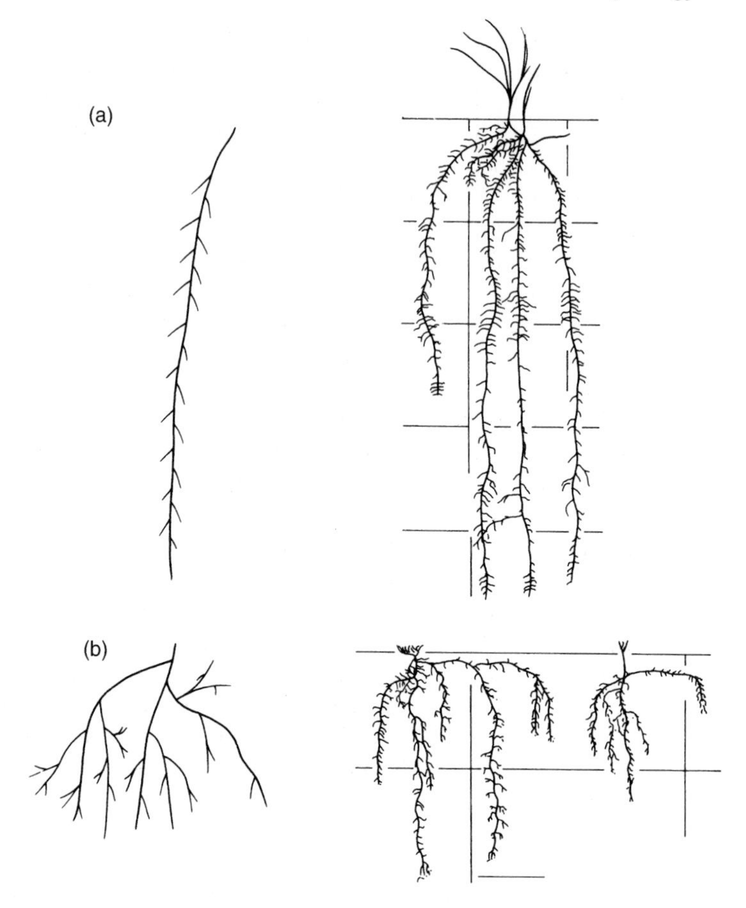

Figure 3.22

Idealized branching patterns (left, a and b) and actual examples which approach each pattern (right, a and b). (a) Herringbone pattern, *Calamovilfa longifolia*; (b) dichotomous pattern, *Phlox hoodii*. Root systems reproduced from Coupland and Johnson (1965), with permission from Blackwell Scientific Publications Ltd

patches are less herringbone in architecture (Farley and Fitter 1999b). This has the dual effect of reducing the cost (because herringbone systems are the most expensive to construct) and of keeping the new root growth localized within the patch.

4. Turnover

Ion uptake is proportional to surface area, though not necessarily the external surface, owing to diffusion into the cell walls. Not all parts of the root system are equally active in uptake, however, since old roots may in some cases become completely suberized, and Ca^{2+} in particular is taken up only by the youngest roots in which the endodermis does not constitute a barrier (Harrison-Murray and Clarkson, 1973). Evidence for other ions, such as $H_2PO_4^-$ and K^+, suggests that earlier assumptions that old roots are inactive are incorrect. Phosphate is readily taken up by barley roots 50 cm from the tip and can be translocated from there to the shoot (Clarkson *et al.*, 1968). At this distance from the meristem the endodermis is massively thickened and has a pronounced suberin layer. This layer, the Casparian band, prohibits movement of ions such as Ca^{2+} beyond the cortex, since they travel primarily in the free space of the cell walls (apoplastically), but provides no barrier to $H_2PO_4^-$ and K^+, which travel symplastically in the cortical cytoplasm, and the endodermal cells are traversed by many plasmodesmata (Clarkson *et al.*, 1971).

In ideal conditions, therefore, roots can continue absorbing ions for long periods, but in time that rate will diminish because of the depletion of the surrounding soil (i.e. transport through soil will become limiting) and because of the damage inflicted by pathogens and grazing soil animals such as nematodes. If nutrient levels elsewhere in soil are higher, there might be a benefit to the plant in allowing existing fine roots to senesce, and new roots to grow in unexploited soil zones. This is what happens above ground; as a plant such as wheat or sunflower grows, the lower leaves become progressively shaded and senesce, the nutrients in them being recycled to the growing tip. Similarly, deciduous trees are characteristic of productive environments in which the carbon investment in a leaf is rapidly repaid by photosynthesis. The leaves are not retained over unfavourable seasons (dry or cold) and senescence here is a controlled event, with resources, especially N and P, being withdrawn and used for new growth. On infertile soils, however, and in other unproductive environments (boreal and high-altitude forests, bogs, semideserts), many plants have small, resistant evergreen leaves, which can continue to produce, albeit at a slower rate, for several seasons (Small, 1972; Reich *et al.*, 1992; cf. Chapters 4 and 5).

Individual roots may live for a few days or for the entire lifetime of the plant. Rhizotron and minirhizotron techniques, in which roots are observed regularly either in underground chambers with perspex walls or in tubes inserted into soil and filmed using borescopes or endoscopes (Plate 5a, b), have shown just how extensive root turnover can be. Cohort analysis, in which a population of roots first observed on a particular date is then followed until all or most of its members have died (Fig. 3.23), allows the calculation of a half-life, which is the time taken for half the population of roots to die, or the median life span. Half-life data have been collected for a wide range of species and the result is surprising: most roots, even of long-lived plants such as trees, only live for a few weeks, typically 40–60 days (Fig. 3.24). The distribution of life-spans of individual roots is very skewed, which is why a half-life is the best way of expressing the data: a few roots live a very long time, even in grass root systems (see Fig. 3.23). Root systems are best thought of in similar terms to shoot systems, as consisting of a long-lived structural component (main roots; cf. branches) and a short-lived absorbing component (the fine roots; cf. leaves).

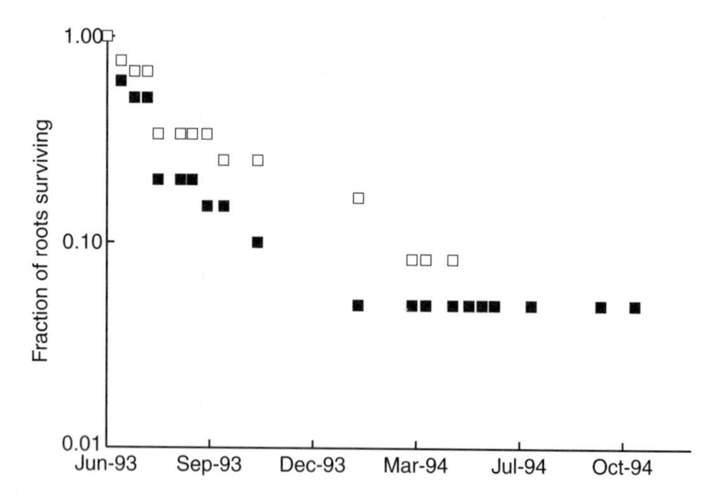

Figure 3.23

The survivorship of a cohort of roots (those 'born' at a particular time) in an experiment in which atmospheric CO_2 concentration was altered. Survivorship of the roots in elevated CO_2 (solid symbols) was shorter than in ambient CO_2 (open symbols), demonstrating faster turnover. It is common to find that, whereas most roots in a cohort die rapidly, following an exponential decay curve, a small fraction of the roots survive almost indefinitely. The latter roots represent structural elements of the root system, analogous to stems above ground. The time taken for half the roots to die is referred to as the half-life or median life span of the cohort (from Fitter *et al.*, 1997)

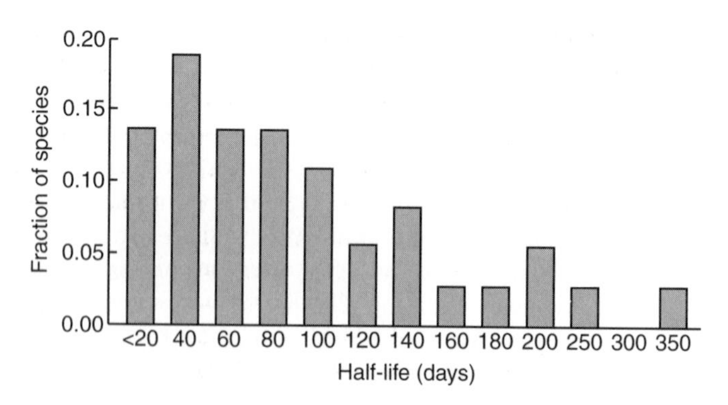

Figure 3.24

Root median life spans (half-life) vary greatly, and can be as short as a few days. This rapid turnover can only be measured using direct observation techniques. Data are from published studies (from Fitter, 1999)

5. Soil micro-organisms

1. The nature of the rhizosphere

The root is not the only living component of the root–soil system; soil ecosystems contain enormous numbers of bacteria, fungi, protozoa, nematodes and arthropods, most of which have yet to be described taxonomically (Tiedje, 1995). The microbial biomass in the soil of plots at Rothamsted Experimental Station which have been under continuous wheat and have received no fertilizer for over 100 years may be as high as 5 Mg fresh weight ha^{-1} (Brookes *et al.*, 1985), the equivalent of around 100 sheep! The supply of many nutrients depends on microbial degradation of organic matter, and soil animals play critical initial roles in this process, fragmenting the litter into pieces suitable for microbial attack. The traditional procedure for sampling soil microbes involved taking soil samples, suspending them in water, and plating them on agar to detect microbial growth. This gives a general view of the soil microbial populations, though it emphasizes the importance of the spore population and of rapidly growing micro-organisms. The set of organisms that will grow on an agar plate is also determined by the nature of the nutrient medium used.

A large part of the microbial population in soil is not recovered by these rather crude isolation techniques, as demonstrated by analyses of molecular and especially nucleic acid diversity, which have revealed the previously unsuspected taxonomic richness of the soil microbiota (Torsvik *et al.*, 1990). Molecular techniques reveal that traditional isolation methods typically recover around 5% of the species present. However, since about 95% are known principally from a DNA sequence, it is usually impossible to ascribe them any functional role. An alternative approach is to ignore the taxonomy of the micro-organisms and to concentrate on their biochemical and functional diversity. A widely used technique is Biolog (Winding, 1994), in which a soil extract is pipetted into a tray with a large number of small wells, each containing a distinct substrate. Colour changes in the wells indicate whether the microbial population can degrade each substrate. This technique is especially powerful for identifying changes in microbial communities, for example as the result of growing different plant species (e.g. Garland, 1996).

However, even techniques such as Biolog require the organisms to be able to grow in the plates, and we know that most microbes will not do so. In a very short time, all these techniques will be overtaken by the application of information about microbial genomes. These developing approaches will provide us for the first time with the ability simultaneously to characterize microbes taxonomically (by their nucleic acid sequences) and to determine their functional activity by expressed mRNAs. New technologies such as microarrays will result in an explosion of such data.

These microbes are not uniformly distributed; rather they are highly aggregated around energy sources, with the mineral matrix representing a microbial desert. Globally over half of primary production is allocated to roots, and in most soils, therefore, plant roots represent the major source of energy for soil microbes. Their influence on the soil microflora was first studied systematically by Starkey (1929), though the term 'rhizosphere', to cover the volume of soil in which the microbial populations are influenced by the proximity of a root, was coined by Hiltner in 1904.

The most striking feature of the rhizosphere is the stimulation of bacterial numbers and activity. The effect is often quoted in the form of an R (root) : S (soil) ratio (Katznelson, 1946), and typical R : S values range from 2 to 100, though they can be as high as 2000. The R : S ratio is based on the numbers of micro-organisms in the soil adhering to the roots relative to the numbers in bulk soil, and can be used for unicells such as bacteria as long as the problems of culturability are taken into account, but is almost uninterpretable for filamentous fungi, where spores can germinate on the plate and provide meaningless counts. A direct approach is to use scanning electron microscopy of root surfaces to examine rhizosphere (or strictly rhizoplane) populations (e.g. Rovira *et al.*, 1974). One of the most serious problems in studying the rhizosphere is that the conceptual definition (the volume of soil affected by the root) is not an operational definition for research scientists; it is impossible to separate this soil from the rest of the soil. In practice, researchers extract roots carefully from soil, and any soil adhering to the roots is collected and referred to as rhizosphere soil. Plate 7 shows why this works in practice.

The microbes in the rhizosphere are stimulated by increased concentrations of various chemicals that act as energy sources, and these are deposited there from a number of sources (Newman, 1985).

(i) By the sloughing of root cap cells as roots grow through soil. The few figures available here suggest that this is a surprisingly minor source of material, offering less than 10 mg g^{-1} root according to Newman's (1985) calculations.

(ii) By the secretion of compounds such as mucilage by the root cap and root epidermis, which facilitates the root's passage through the soil (Juniper and Roberts, 1966; Greaves and Darbyshire, 1972). Newman (1985) suggests that this may account for up to 50 mg g^{-1} root. Mucilage is a complex, dominated by polysaccharides, that is rapidly attacked by soil microbes, producing an even more complex mixture of mucilage itself, various breakdown products, microbial cells and mineral soil particles, that is refereed to as mucigel. Mucigel acts to bind the root to the soil and to absorb toxic ions such as aluminium (see also p. 261). Secretion refers to the transport of compounds across membranes under plant control and includes enzymes, signal molecules and ionophores that can bind to insoluble ions.

(iii) By exudation of compounds from intact cells, which is the passive loss of materials through leaky or damaged membranes. The compounds lost include amino acids, organic acids and sugars that are rapidly degraded by soil microbes. The quantity of carbon lost by roots in this way is difficult to estimate because it is hard to separate from secretions. Hodge *et al.* (1997) used a sterile microcosm and ^{14}C labelling to estimate total carbon loss from roots at 11–13% of net carbon assimilation. This is a high figure, greater than some other estimates, and most of the carbon probably originated from exudation. However, healthy roots can re-absorb many of the compounds that they exude; Jones and Darrah (1993b) found that maize roots could re-absorb 90% of exudates, albeit in a closed system that maximized that process. As a result, many early studies may have seriously underestimated C loss by roots.

(iv) The remaining carbon transfer to soil is therefore by the most obvious pathway: the death of roots or parts of roots. There is often very much more insoluble material in the rhizosphere than can be accounted for by the sources listed above, and the most likely explanation for the discrepancy is in the death of roots. Most roots are surprisingly short-lived (see Fig. 3.23) and therefore ultimately provide a major carbon input to soil.

In the light of such complexity and enrichment it is not surprising that the rhizosphere flora is qualitatively as well as quantitatively different from that of the bulk soil (Rovira and Davey, 1974). Bacteria requiring amino acids are abundant (Lochhead and Rouatt, 1955) and root-inhabiting microbes, both symbiotic bacteria (nitrogen-fixing *Rhizobium*) and pathogenic fungi (such as *Fusarium*), may show increased populations.

In addition, the rhizosphere is both physically and chemically distinct from the bulk soil, thanks to the direct activities of roots. Counter-ion movements of protons into the rhizosphere during cation (especially NH_4^+) uptake can reduce rhizosphere pH by up to 2 units (Marschner and Römheld, 1983), and conversely, where NO_3^- is the main source of N, rhizosphere pH can be increased by a similar amount (Plate 6). Roots can also both reduce the water potential of rhizosphere soil during the daytime, when transpiration is occurring, and increase it at night, when water may be lost from roots to the rhizosphere, a process that has been termed hydraulic lift (Richards and Caldwell, 1987; Caldwell *et al.*, 1998), since it can result in water from deeper layers in the soil being lost in surface layers.

Among the most important root exudates are those that alter the availability of ions in the rhizosphere. Changes in pH may have a profound effect, but the secretion of ionophores is the more specific. Many plants secrete compounds such as mugineic acid that form complexes with otherwise insoluble ions such as Fe^{3+}; they also have transport systems in membranes that can move the Fe–ionophore complex back into the cells (Römheld and Marschner, 1990). This mechanism allows plants to acquire sparingly soluble ions, and to do so in competition with bacteria that themselves produce ionophores. Another major impact on soil chemistry is achieved by the secretion of citrate, which exchanges directly with phosphate ions adsorbed onto the soil (see p. 87). A number of plant species produce cluster roots, which are dense clusters of tightly packed lateral roots in longitudinal rows along short sections of main root (Plate 8a); they are best developed in members of the Proteaceae, characteristic of nutrient-deficient soils in Australia and southern Africa, the Cyperaceae, and in lupins (Fabaceae). All these are non-mycorrhizal species (Lamont, 1993). Cluster roots often secrete large amounts of citrate (Gardner *et al.*, 1982) and the combination of the density of roots, the production of a compound that can exchange for adsorbed phosphate, and acidification of the rhizosphere (Dinkelaker *et al.*,1995) gives these species the ability to acquire phosphate in soils where other species suffer severe deficiency.

These changes in soil around roots create a distinct micro-environment for microbes, both in terms of resource availability and environmental conditions. The resulting complex microbial population in the rhizosphere can potentially influence nutrient uptake by roots in four ways:

(i) by altering root or shoot growth, or by directly damaging the root;
(ii) by altering the supply at the root surface and so competing with the root;
(iii) by direct impacts on nutrient uptake itself – inhibition or stimulation;
(iv) by mineralization of organic or dissolution of insoluble ions.

Evidence on most of these points is conflicting; for example, early work which suggested that roots stimulate the activity of phosphate (apatite)-dissolving bacteria and hence improve P supply, was later interpreted in terms of root growth stimulation by gibberellic acid produced by rhizosphere microbes (Gerretsen, 1948; Brown, 1975). It remains unclear whether microbes have significant effects under field conditions other than by directly affecting root growth (point (i) above), but since they are known to have such effects in laboratory studies, it seems likely that they do.

2. Nitrogen fixation

Only prokaryotes can convert atmospheric dinitrogen gas (N_2) into ammonium that can be incorporated into organic compounds; no eukaryote can do this. Among the many bacteria found in the rhizosphere are those capable of both free-living and symbiotic fixation of gaseous nitrogen. Several common genera of soil bacteria, such as *Azotobacter*, *Klebsiella* and *Clostridium*, can fix N_2 in the rhizosphere using energy obtained from root-derived materials. High levels of nitrogenase (the enzyme responsible for fixation) activity have been found in the rhizosphere of some tropical grasses, particularly *Paspalum notatum* (Döbereiner and Day, 1976). The bacterium most commonly responsible is *Azospirillum brasilense*, which may actually invade dead cortical cells, creating an endorhizosphere. For some time there was great excitement about the possibility that this associative symbiosis could be exploited agriculturally to produce strains of crop plants which would have very low fertilizer nitrogen requirements. Unfortunately, the energy costs of N_2 fixation are high (typically around 20–100 $g\,C\,g^{-1}$ N fixed) and the energy supply in the rhizosphere is usually inadequate (Giller and Day, 1985). In most circumstances, such associations could only supply around 5 kg N ha^{-1}, a few % of the N needs of a crop. To fix 5 $kg\,N\,ha^{-1}$, the associative fixers would require up to 0.5 $t\,C\,ha^{-1}$, which could represent as much as half the total below-ground productivity, and the rhizosphere bacteria responsible must compete with other microbes for this carbon. In most natural ecosystems where there is a net input of N from fixation, this derives either from symbiotic fixation or from photosynthetic, N_2-fixing cyanobacteria, which have their own energy source. There is some evidence that associative fixers may be responsible for a greater fraction of N uptake by sugar cane, possibly because one of the bacteria responsible (*Acetobacter diazotrophicus*) invades the root cortex where there will be less competition for carbon from other microbes (Boddey *et al.*, 1991). This association comes close to the better known and genuinely symbiotic nodule associations of legumes.

In symbiotic fixation, the energy supply is directly from the host. There are a number of types of symbiotic N-fixing association, but in the two most important types, the bacteria develop in a nodule on the host root that is supplied by the host vascular system (Plate 8b). The most important type in ecological terms is the rhizobial symbiosis. Rhizobia are bacteria that were once regarded as a single genus (*Rhizobium*), but are now classified into at least five, rather loosely related in the α sub-division of the Proteobacteria. It appears that the gene complex responsible for allowing symbiotic N-fixation has moved between rather distantly related bacteria by lateral gene transfer at several times in the evolution of the association (Young and Haukka, 1996).

Rhizobial associations are formed by members of a single plant family, the Fabaceae (and a single genus in the unrelated family Ulmaceae). Most species in two subfamilies of the Fabaceae, Papilionoideae and Mimosoideae, form nodules, but only a third of the largely tropical Caesalpinoideae. In other plant families, nodule formation is known in 158 species in 14 genera, including *Alnus* (alder), *Myrica* (bog myrtle), *Casuarina* (she oak) and *Hippophae* (sea buckthorn). All except one (*Trema* in the Ulmaceae, see above) are infected by a different symbiont, the actinomycete *Frankia*, and form larger, clustered nodules (Bond, 1976); these associations are termed actinorhizal.

The nodule acts as an important sink for fixed carbon. Over a 24 h period a pea plant (*Pisum sativum*) transported 11.02 mg of carbon to its roots, of which 6.29 mg (57%) were used as carbon skeletons for fixed nitrogen, and 4.34 mg (39%) for respiration (Minchin and Pate, 1974). Generally, however, the carbon efficiency of fixation is much greater for symbiotic than associative systems, and can be as high as 6 $\mathrm{g\,C\,g^{-1}}$ N fixed (Vance and Heichel, 1991). This means that to fix 50 kg N $\mathrm{ha^{-1}}$ (10 times as much as expected in an associative system) would require less carbon (0.3 $\mathrm{t\,ha^{-1}}$). Nevertheless, there is a carbon cost, and it seems likely that the plant must gain some ecological advantage from the extra nitrogen supplied to outweigh the carbon loss, for the association to persist. Many N-fixing plants are pioneer species on N-deficient soils. In the classic studies of vegetation succession in initially barren glacial moraines at Glacier Bay, Alaska (Crocker and Major, 1955), N-fixation by alders (an actinorhizal species) facilitated the invasion of spruce, by adding 35 kg N $\mathrm{ha^{-1}\,y^{-1}}$ over a period of 30 years (see above, p. 82). A similar role is played on abandoned china-clay waste heaps in Cornwall by gorse *Ulex europaeus* and tree lupins *Lupinus arboreus* (Marrs *et al.*, 1983). In agricultural systems, values as high as 200–400 $\mathrm{kg\,N\,ha^{-1}}$ per crop cycle have been recorded (Peoples and Craswell, 1992), which can satisfy the complete N requirements of a growing crop.

Where soil nitrogen levels are higher, however, the competitive advantage derived from fixation would be lost, and carbon used for nutrient uptake, defence or growth would lend more competitive ability than that used for fixing nitrogen. Legumes and other N-fixers are generally rare in mature communities and where they are found, as trees in tropical forest for example, they often belong typically to the little-nodulated Caesalpinoideae (Alexander, 1991). There is an alternative view of the evolutionary significance of nitrogen fixation. It may be that legumes have adopted an ecological strategy based on high leaf nitrogen concentrations. Nitrogen fixation can then be seen as a mechanism to achieve that state, in which the excess N may be used for non-metabolic purposes such as defence (McKey, 1994). In this view, legumes would not be expected to occur especially in soils of low N availability; rather, they would be found as minor members of a wide range of communities on a variety of soils, as is often the case. This would provide a distinct view of the ecological preferences of legumes, but would not affect our understanding of the importance of symbiotic fixation in raising N availability on N-poor soils.

3. Mycorrhizas

The rhizosphere is a quantitative phenomenon, defined by the increased microbial population. There is consequently a gradient of influence away from the root with the most enhanced area being the root surface, the rhizoplane. Microbes that occur in intimate contact with the root, in other words that live symbiotically with it, are likely to have the greatest impact. Mycorrhizas are symbiotic associations of a fungus and a plant root and can be viewed as a highly specialized development of a rhizoplane association, that has become at least partly invasive. There are several, quite distinct types of mycorrhiza, which evolved quite separately and function quite differently (Fitter and Moyersoen, 1996; Table 3.9). All have both an internal and an external phase, the fungus living simultaneously in these two different environments, and taking up materials from the soil that are transported back to the root.

Ectomycorrhizas (sheathing mycorrhizas) involve either a Basidiomycete or an Ascomycete fungus and a plant; the fungus forms a mantle of hyphae around the root and a network of intercellular hyphae in the cortex (the Hartig net) (Plate 8c). No intracellular penetration occurs and root morphology is typically altered, often with branches being short, thick and sometimes dichotomously branched. Endomycorrhizas are of several types, most of which are very specialized, such as those found on roots of plants in the orders Ericales (heathers) and Orchidales (orchids). By far the commonest type, formed with a single order of zygomycetous fungi (the Glomales), is variously known as the arbuscular (AM), vesicular–arbuscular (VAM) or even Glomalean mycorrhiza. The whole field is comprehensively reviewed by Smith and Read (1996).

Arbuscular mycorrhizas occur in most families of higher plants, as well as some bryophytes (liverworts but not mosses) and pteridophytes. Probably about two-thirds of all plant species normally form the symbiosis. This ubiquity suggests an ancient origin. Fossil evidence shows that Glomalean spores were present in the Ordovician (460 million years ago: Redecker *et al.*, 2000) and that Devonian plants (~400 million years old) contained fungi forming arbuscules and vesicles in their underground stems or rhizomes (Remy *et al.*, 1994). In addition, molecular evidence indicates that the Glomales as a group diverged from other fungi at about that time (Simon *et al.*, 1993). Both lines of evidence confirm that the arbuscular mycorrhiza evolved at or close to the time that plants first colonized the land. The relationship is apparently intracellular, the fungus producing haustoria that invaginate the plasmalemma and produce either arbuscules (which literally means 'little trees'), branching clusters of minute hyphae less than 1 μm in diameter (Plate 8d), or hyphal coils whose function is unclear. Most Glomalean fungi also produce vesicles, swollen structures containing lipids, which are presumed to be storage bodies and may have some function in propagation and persistence.

Most attention on both Glomalean and ectomycorrhizas has been focused on their ability to improve plant growth by enhancing P uptake. In P-deficient soils mycorrhizal plants typically grow markedly better than non-mycorrhizal ones (Fig. 3.25), but the reverse may be true in soils well supplied with phosphate (Fig. 3.26). Indeed in such soils plants normally show very low levels of colonisation. The advantage of the mycorrhizal plants cannot be explained on the basis of root morphology, since they take up phosphate faster per unit root length than non-mycorrhizal ones, at least when grown in ideal conditions, in growth rooms or in agriculture (Fig. 3.27). In fact mycorrhizal plants often

Table 3.9

Major types of mycorrhiza (from Fitter and Moyersoen, 1996)

	Ectomycorrhiza	Arbuscular mycorrhiza	Ericoid mycorrhiza	Orchid mycorrhiza
Taxa				
Fungi involved	Basidiomycotina, Ascomycotina, one genus in Zygomycotina	Zygomycotina (Glomales)	Ascomycotina esp. *Hymenoscyphus*	Basidiomycotina, Mycelia Sterilia, Deuteromycotina
Plants involved	Taxonomically diverse, few	Taxonomically diverse, many	Ericales	Orchidales
Morphology				
Diagnostic features	Hartig net	Arbuscule	Intracellular hyphal complexes in epidermis	Hyphal coils, pelotons in cortical cells
Other structures	Sheath	Vesicles, coils		
Functionality/changes in plant function				
P acquisition	Important	Important	Important	Exclusive[a]
N acquisition	Important	Marginal	Important	Exclusive
Water acquisition	?important	Marginal	Unknown	Unknown
Protection from pathogens	?important	?important	Unknown	Unknown
Others	?protection from ionic toxins	?micronutrients	Protection from ionic toxins	Carbon supply to plant

[a] 'Exclusive' is used to indicate that all the plant's resource acquisition is achieved by this route.

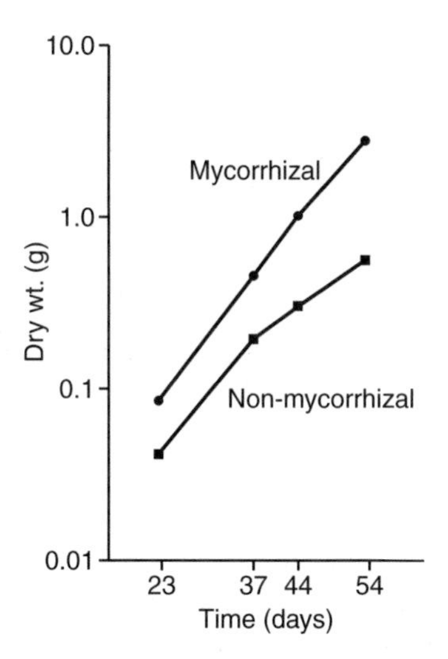

Figure 3.25

Growth rate of mycorrhizal and non-mycorrhizal onions, showing that non-mycorrhizal plants were unable to sustain the same rate. Note that these data are plotted on a log scale, and that the slope therefore represents relative growth rate; after 54 days growth the difference in dry weight was nearly 10-fold (data from Sanders and Tinker, 1973)

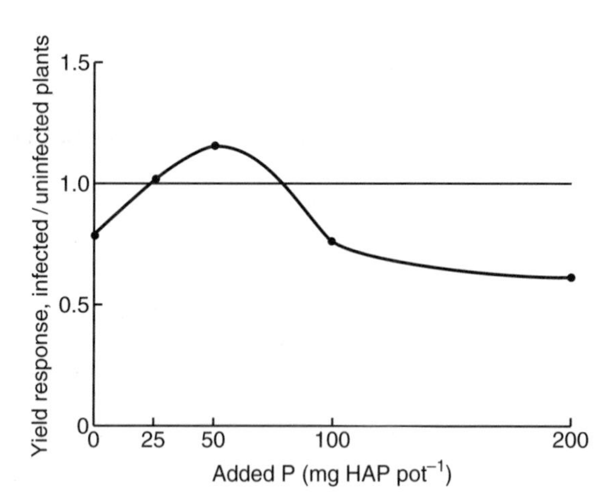

Figure 3.26

Yield response of soybeans *Glycine max* to infection by the AM fungus *Glomus fasciculatum* in sand culture amended with hydroxyapatite (HAP) as a phosphorus source. Note the marked depression of yield at high P supply rates. From Bethlenfalvay, G. J., Bayne, H. G. and Pacovsky, R. S. (1983). *Physiol. Plant.* **57**, 543–548

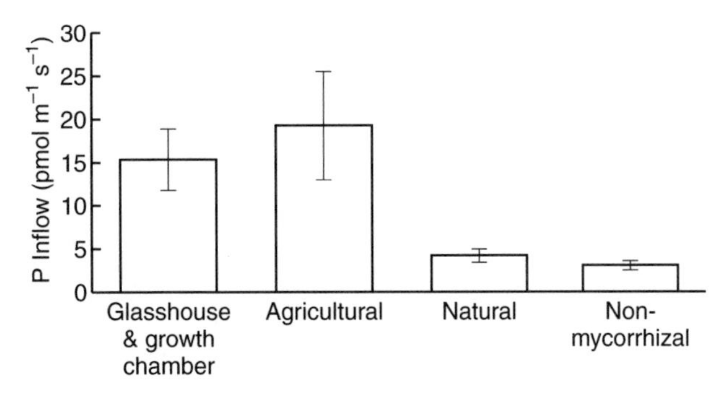

Figure 3.27

Phosphorus inflows (uptake rate per unit root length, with standard error) into mycorrhizal plants growing in controlled environments ($n=8$), in agricultural ($n=5$) and natural environments ($n=10$), and into non-mycorrhizal plants under all conditions ($n=9$). Inflows under ideal conditions were five times those into non-mycorrhizal plants, but this difference was not maintained under natural conditions. The data come from a range of sources.

have shorter root systems, visibly so in the case of ectomycorrhizas; it is possible that reported effects of rhizosphere microbes reducing root length may in part be due to endomycorrhizal infection (Crush, 1974; Fitter, 1977). For ectomycorrhizas (Harley and Lewis, 1969), orchid mycorrhizas (Alexander and Hadley, 1984) and arbuscular mycorrhizas (Rhodes and Gerdemann, 1975), but not for ericoid mycorrhizas, it is well established that phosphate travels from fungus to root, so the enhanced uptake must be due to the greater ability of the fungus to obtain phosphate. One obvious explanation of the increased P uptake of mycorrhizal plants is that the fungus might utilize sources of P not available to plants. However, when Hayman and Mosse (1972) grew arbuscular mycorrhizal plants on soil that had been enriched with ^{32}P, they found that infected and uninfected plants had the same specific activity as the soil, though the mycorrhizal plants had absorbed up to 30 times as much phosphate (Fig. 3.28). The added P was all in an available form, so that if the mycorrhizal plants had been using unavailable phosphate, the extra P they took up would not have been labelled, reducing the overall radioactivity of the absorbed P: they would therefore have shown lower specific activities.

An alternative explanation would therefore seem to be that the mycorrhizal root has different phosphate uptake kinetics. Although changes in kinetics have been reported, they are not great enough to account for the increased uptake, even for ectomycorrhizas, where such an explanation might be more attractive because the root is enclosed in the fungal sheath. For arbuscular mycorrhizas, Sanders and Tinker (1973) calculated the greatest inflow that onion roots in a sandy soil could achieve at the maximum rate of diffusive supply, irrespective of the uptake kinetics. They showed that this would be 3.5 pmol m^{-1} s^{-1}, which would be achieved when the root surface concentration was effectively zero. Under those conditions, the diffusion of phosphate ions to the roots surface would be maximal. Strikingly, this was approximately the rate achieved by

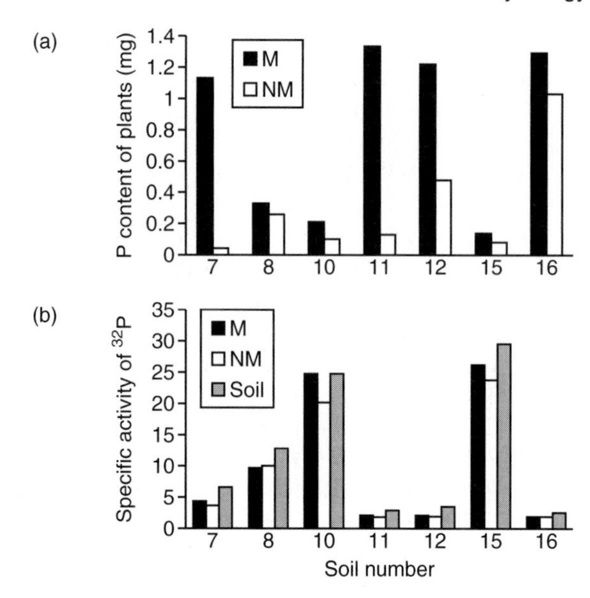

Figure 3.28

(a) Phosphorus content of mycorrhizal (M) and non-mycorrhizal (NM) onion plants grown in seven contrasting soils by Hayman and Mosse (1972); the plants captured very different quantities of P from the soils, and were wholly dependent on mycorrhizal P transport on some (e.g. soil 7). (b) Specific activity of ^{32}P in both mycorrhizal and non-mycorrhizal plants reflected that in the soil, irrespective of the degree to which the plants acquired P *via* mycorrhizal transport.

uncolonised roots (Table 3.10). Mycorrhizal roots had nearly five times higher rates than this calculated maximum, an impossible result unless the effective surface area for uptake had been incorrectly determined. Such an error would have been made if the fungal hyphae in the soil, attached to the mycorrhizal root, were exploring unexploited soil outside the depletion zone of the root, and translocating phosphate from there to the root. In that case, when inflow is

Table 3.10

Phosphate inflows of mycorrhizal and non-mycorrhizal onion plants (Sanders and Tinker, 1973). For growth data, see Fig. 3.25

Harvest interval			Inflow (pmol m^{-1} s^{-1})	
No.	Duration (days)		Mycorrhizal	Non-mycorrhizal
1	14		17	5.0
2	7		22	1.6
3	10		13	4.2
		Means:	17	3.6

calculated, the hyphae should be treated as part of the length of absorbing organ. Sanders and Tinker (1973) found about 80 cm of hypha per cm of root in the soil, which would adequately explain the superior uptake of mycorrhizal roots. The same effect is seen in Fig. 3.27, which shows data from a range of sources: P inflows to mycorrhizal roots are around five times greater than those to non-mycorrhizal roots. Importantly, however, inflows to mycorrhizal roots growing under natural conditions are only marginally greater than the non-mycorrhizal values, suggesting that many factors in natural soils (for example, grazing by the soil fauna) may reduce the effectiveness of the symbiosis (Gange, 2000).

It is generally accepted that this is how arbuscular mycorrhizas act. They explore soil that is inaccessible to the plant root, because of the limitations of diffusion. Mycorrhizal roots can exploit soil for phosphate up to 70 mm from the root (Rhodes and Gerdemann, 1975), whereas a figure of 10 mm or less would be more typical for uncolonized roots. This ability to acquire phosphate at a distance from the root probably explains why the symbiosis evolved at the time that plants colonized land. These early plants had no root systems and would have found it impossible to acquire such an immobile nutrient as phosphate; the mycorrhizal symbiosis was an essential requirement for a terrestrial existence (Pirozynski and Malloch, 1975). However, the mycorrhizal symbiosis can perform other functions. There are around 150 described fungal species in the order Glomales, and as far as is known, they exhibit very little specificity: in other words, most of these fungi will readily colonize the roots of most plants. That perception may be very strongly biased by the fact that most experimental work has been done with a very small number of species, which may indeed be non-specific. The other species are not popular experimental material because they are difficult to culture, possibly because they show some host preference. Of the few studied species, it is known that not all are equally good at facilitating P uptake. Jakobsen *et al.* (1992a,b) showed that three mycorrhizal fungi (*Acaulospora laevis, Scutellospora calospora* and a species of *Glomus*) differed markedly in their ability to transport phosphate from soil to a plant partner, *Trifolium subterraneum* (Fig. 3.29). Whereas *A. laevis* and the *Glomus* sp. were equally effective at facilitating P uptake from P sources close to the root (up to 10 mm), the *Glomus* was much less effective when the P source was more than 25 mm away. The reason was that *A. laevis* produced the greater hyphal length density in the soil at that distance from the root. The third species, *S. calospora*, took up P from soil, but retained it in the hyphae, and did not transfer it to the plant. These differences mean that it is necessary to ask why variation in what is apparently the key mycorrhizal activity (P uptake) should exist, and why plants do not form associations with fungi that can maximize the benefits they receive.

Glomalean fungi also provide other services to host plants (Newsham *et al.*, 1995), including protection from pathogens, increased micronutrient uptake (especially of Zn which, like phosphate, is immobile in soil), and improved water relations. The last is probably the result of the fungal hyphae binding roots to soil and so maintaining a water pathway as soil dries. It is possible that some Glomalean fungi have evolved improved abilities in some of these other functions, and that some of the diversity of the group can be explained on that functional basis. At present it is difficult to resolve this question because the

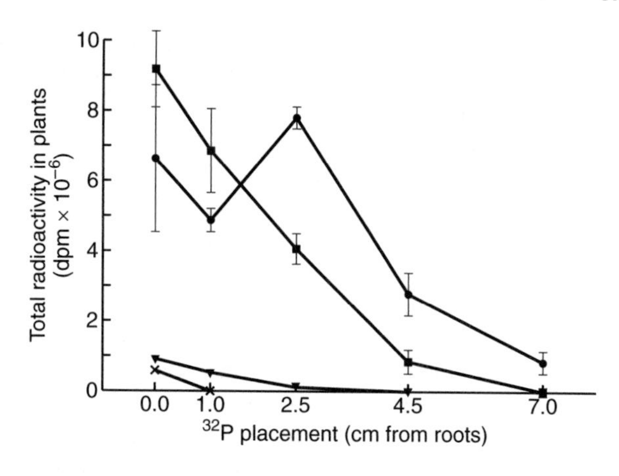

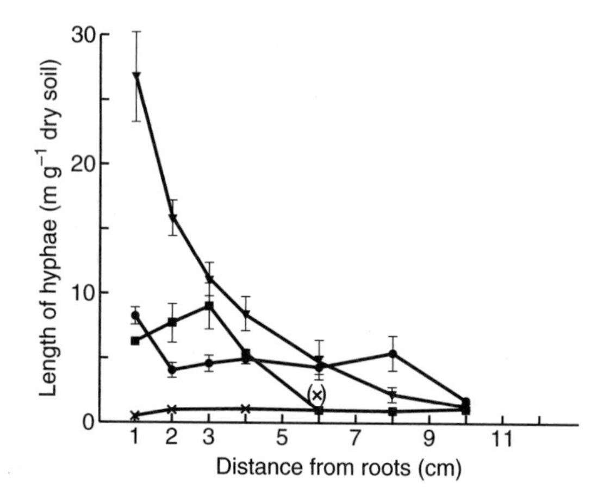

Figure 3.29

(a) The uptake of phosphate from soil by plants of *Trifolium subterraneum* depended upon the identity of the fungal partner in its mycorrhiza; the plants were grown with *Glomus* sp. (■), *Acaulospora laevis* (●) or *Scutellospora calospora* (▼), or with no AM fungus (×). (b) Differences among the same fungal species in hyphal growth at various distances from the roots of *Trifolium subterraneum*. The fungal species that produced most hyphal growth (i.e. *S. calospora*) did not transfer most P to the plant (cf. (a)) (from Jakobsen *et al.* 1992a, b)

taxonomy of the Glomales is confused (Morton and Benny, 1990), and molecular techniques are only just beginning to resolve it.

Ectomycorrhizal (EcM) fungi are more diverse (and their host plants much less so) than arabuscular mycorrhizas, and ectomycorrhizal associations may be specific. Many species of the genera *Boletus*, *Suillus*, *Amanita* and *Russula* are normally found only under particular tree species – the fly agaric, *Amanita muscaria*, for example, is characteristic of birch woods. By contrast over 100 different fungi can form associations with *Pinus sylvestris*, and these in turn with more than 700 other tree species (Molina *et al.*, 1992). EcM associations are certainly functionally complex. P uptake is important (Jones *et al.*, 1991), and the hyphae of ectomycorrhizal fungi can acidify soil and increase P availability in the same way that roots can (Cumming and Weinstein, 1990; Fig. 3.30, cf. Plate 7). This phenomenon appears to be due to the secretion by the fungus of oxalic acid into the hyphosphere (the soil around the fungal hypha, by analogy with the rhizosphere) (Wallender, 2000), and fungal species differ in their ability to do this. In addition, many EcM fungi can decompose organic matter (an ability common in the Basidiomycotina but not found in the Zygomycotina) and they can transfer N from these organic sources to the hosts. EcM mycorrhizas are commonest in boreal forests, where pines *Pinus* and spruces *Picea* are dominant trees; they are rare in tropical forests and only co-dominant with trees that have arbuscular mycorrhizas in temperate forests (oaks *Quercus* and beech *Fagus* are EcM, but maples *Acer*, for example, are AM). It seems that, as the rate of cycling of organic matter slows down, so N, which is supplied predominantly from the N cycle, becomes relatively more limiting than P, whose supply is dominated by inorganic equilibria. Consequently, the EcM mycorrhiza becomes progressively dominant at higher latitudes (Read, 1991). Beyond the tree-line, the vegetation is tundra on peat, where nutrient cycling has almost ceased. These communities tend to be dominated by species in the Ericales, which have a distinct type of mycorrhiza. Ericoid mycorrhizas, which involve a small number of Ascomycete fungi, are even more active in acquiring N from organic sources, and there is little evidence that P is important in this association.

The different types of mycorrhiza represent distinct evolutionary responses to nutritional problems. A small group of species (probably around 5% of all species) do not form mycorrhizal associations at all. Some whole families are either never or almost never mycorrhizal (Brassicaceae, Caryophyllaceae, Cyperaceae, Proteaceae); in other cases particular genera (*Lupinus* in the otherwise mycorrhizal Fabaceae) or even species may be non-mycorrhizal. All

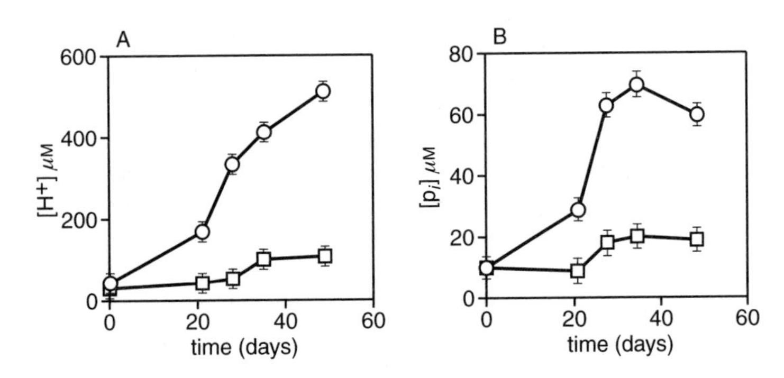

Figure 3.30

The release of protons into solution culture (a), and the consequent dissolution of inorganic phosphate (b) from insoluble aluminium phosphate was far greater in seedlings of the pine *Pinus rigida* colonised by the ectomycorrhizal fungus *Pisolithus tinctorius* (○), than in non-mycorrhizal seedlings (□) (from Cumming and Weinstein, 1990)

these species have evolved from mycorrhizal ancestors. They have lost the ability to form mycorrhizas for one of two reasons. Some have become habitat specialists (Peat and Fitter, 1994a) and occur in habitats where being mycorrhizal is either not possible (because the fungi are not there: mainly annuals in disturbed habitats) or not beneficial (because P is not limiting: some wetland species). Others have developed root systems that have alternative mechanisms for accessing P, notably cluster roots (see above, p. 117).

These non-mycorrhizal species are specialists: the norm for plants is to be mycorrhizal. Some of the mycorrhizal associations are specialist too; the ericoid association enables the species in the Ericales (not all of them, some have other types of association still, as described by Smith and Read, 1996) to grow successfully on acid peat soils where rates of N cycling are extremely slow. At the extreme is the orchid association which is apparently a parasitism by the plant on the fungus, since both mineral nutrients and organic carbon compounds pass from fungus to plant; in all other mycorrhizas, the carbon moves from plant to fungus. Some orchids and a few other plant species, however, have no chlorophyll, and survive by being 'mycoheterotrophic' (Leake, 1994). The mechanism by which the mycoheterotroph obtains all its nutritional needs from a fungus remains mysterious.

The key to the evolution of the various types of mycorrhizality (and non-mycorrhizality) is cost and benefit (Fitter, 1991). Except for the mycoheterotrophs, plants in mycorrhizal associations pay the cost of providing fixed carbon to the fungus. The benefit is increased acquisition of P or N, or protection from pathogens, or improved water relations, or probably other phenomena that have not yet been well documented. As long as the net selection pressure, over several generations, ensures that benefits outweigh costs, the symbiosis will persist. In the case of the arbuscular association, it has done so for 400 million years, suggesting that the selection pressures are strong.

6. General patterns of response to soil nutrients

The acquisition of nutrients from soil is fundamental to plant survival and fitness, and is a complex process. The various nutrients that plants require vary greatly in availability, both among themselves, and in space and time. In addition they have quite distinct physicochemical properties. For this reason there can be no single optimum strategy to acquire all nutrients. Most natural environments are nutrient-deficient, and so adaptations to nutrient-poor environments are numerous. These include adaptations that increase the amount of nutrient available to the plant, those that improve the plant's ability to acquire nutrients that are available, and those that optimize the use of the nutrients once captured. Examples of these include: exudation of citrate from roots that exchanges with phosphate (p. 117); mycorrhizal and root architectural responses to locally available nutrients; and increased lifespan of plant parts. The study of mineral acquisition was for a long time dominated by ion transport physiology. Variations in uptake kinetics are probably of greatest significance as short-term responses to local fluctuations in ion concentrations. Long-term adaptation is achieved by changes in demand for and use of nutrients, and by changes in root morphology (including symbionts) and distribution. All of these can be seen as elements of a conservative strategy (stress-tolerant in Grime's terminology: see Chapter 7).

If the demand for nutrients by a plant is reduced, there will be profound consequences for the plant. The most obvious is that there will be lower concentrations of nutrients in tissues and the need to make effective use of the nutrients that are acquired. Whenever plant parts die, a significant part (often around half) of the nutrients is lost to the litter. This means that the longer a leaf or root lives, the more efficiently will its nutrients be used to promote growth. One of the most important features of plants of nutrient-poor environments is the longer turnover times of both their leaves and their roots (Aerts and Chapin, 1999; cf. Figure 3.24). However, these characteristics typically involve a reduction in growth rate and the further consequence that longer lifecycles will be favoured; nutrient-poor soils tend to be dominated by perennial vegetation. Annuals are of course favoured by disturbance, which physically removes perennial vegetation, and where nutrient-poor soils are frequently disturbed (for example on mobile sand-dunes), few species can persist and a community of low diversity is found. Another consequence is the greater allocation of resources to unproductive fibrous tissue, which has a low nutrient content, as opposed to photosynthetic tissue. This in turn reduces palatability and provides protection from grazing, so preventing one possible loss of resources which might stimulate demand.

An analogous response is an increase in the root weight ratio, which again reduces growth rate by diverting resources from photosynthetic tissue, and makes possible changes in root morphology, which may improve nutrient acquisition. Nutrient-poor environments are often characterized by high levels of allocation of resources to roots: typically more than half of primary production goes below ground in natural ecosystems (Fogel, 1985). Nutrient deprivation also changes root morphology, with P-starved grass plants, for example, having main axis diameters of around 0.5 mm as compared with 0.7–1.0 mm in high P conditions (Christie and Moorby, 1975). Root diameter can be integrated for a whole root system by measuring the length per unit root weight or specific root length (SRL, $cm\,mg^{-1}$). SRL is typically increased by nutrient deprivation (Fitter, 1985). In field soils nutrient supply varies as much in time as in space, even or perhaps especially in fertile soils. Species capable of a good growth response to nutrients also have very much more plastic root morphology, including SRL (Christie and Moorby, 1975; Table 3.11).

Plants of infertile soils seem then to have a less flexible morphology and physiology. They are slow-growing, unpalatable, devote much photosynthate to root growth and metabolism, are usually mycorrhizal and tend not to respond morphologically to changes in nutrient concentration. These characteristics are well suited to the efficient acquisition and utilization of nutrients (and in many cases of water too, see Chapter 4) when they are scarce. In another context they are recognizable, too, as the characteristics of stress tolerance (Grime, 1979). As a consequence, such species are readily displaced from more fertile soils by species with opposite characteristics, which imply greater photosynthetic activity, shoot growth and competitive ability for light. However, it is difficult to generalize about the features of plants that are

Table 3.11

Responsiveness to phosphate supply, specific root length and root diameter of two arid-zone grasses (Christie and Moorby, 1975)

	Thyridolepis mitchelliana	*Cenchrus ciliaris*
Response to P[a]	3.0	29.6
Specific root length (cm mg^{-1})		
3 mg l^{-1}	16.2	13.6
0.003 mg l^{-1}	19.2	27.8
Ratio	1.10	2.04
Root diameter (μm)		
Nodal axes		
3 mg 1^{-1}	704	968
0.003 mg l^{-1}	552	460

[a] Response to P is the quotient of total dry weight of plants grown at 3 mg P l^{-1} to those grown at 0.003 mg P l^{-1}.

adapted to nutrient-poor soils. In any one habitat, there will be species with contrasting strategies. For example, each of a group of seven species from the ground flora of a temperate woodland had a unique pattern of response to the local enrichment of soil with nutrients (Farley and Fitter, 1999b), and Einsmann *et al.* (1999) found the same wide range of responses in ten species from a successional sequence in eastern North America. This should not surprise us; adaptations to nutrient deficiency are only one part of the overall suite of adaptations that each plant must deploy, and there may be multiple solutions to a given problem.

4

Water

1. Properties of water

Water is the major component of green plants, accounting for 70–90% of the fresh weight of most non-woody species. Most of this water is in the cell contents (85–90% water) where it provides an appropriate medium for many biochemical reactions, but water has other roles to play in the physiology of plants, and it is uniquely fitted, by its physical and chemical properties, to fulfil these roles.

The unusually strong intermolecular hydrogen bonds in water cause it to behave as if its molecules were very much larger. For example, the melting (0 °C) and boiling points (100 °C) of water (molecular weight 18) are anomalously high compared with those (−86 °C and −61 °C) of the closely-related compound, hydrogen sulphide (H_2S, molecular weight 34). Thus, unlike all other substances made up of small molecules, water normally remains in the liquid state under terrestrial conditions, although problems for living organisms do result from the freezing of water. For the same reason, the specific heat (4.2 J g^{-1}), the latent heat of melting (333.6 J g^{-1}) and the latent heat of vaporization (2441 J g^{-1} at 25 °C) of water are all exceptionally high, with important implications for the thermal economy of plants: its high specific heat buffers plant tissues against rapid fluctuations in temperature, whereas its high latent heat of vaporization facilitates leaf cooling by the evaporation of water. These thermal properties are also responsible for moderating the temperatures of moist soils, rivers, lakes and the oceans.

Water is an effective solvent for three groups of biologically important solutes:

(i) Polar organic solutes with which it can form hydrogen bonds, including amino acids, and low-molecular-weight carbohydrates, peptides and proteins, which carry hydroxyl, amine and carboxylic acid functional groups. Water also forms colloidal dispersions with higher-molecular-weight carbohydrates and proteins; the most important of these is the cytoplasm itself.

(ii) Charged ions, including the major plant nutrient ions (K^+, Ca^{2+}, $H_2PO_4^-$, NO_3^- etc.). Because of the partial polarization of their OH bonds, water molecules orientate themselves round ions to give larger and highly soluble hydrated ions. By the same mechanism, water molecules are attracted to fixed charges on the surfaces of plant cell walls, cell membranes and soil particles, giving tightly bound hydration layers a few molecules thick.

(iii) Small molecules, including the atmospheric gases (O_2, N_2), which, presumably, can fit into voids in the rather open structure of liquid water. Among the major atmospheric gases, CO_2 alone dissolves in water to give (bicarbonate) ions (although pollutant gases such as SO_2, NH_3 and nitrogen oxides also form true solutions in water).

Thus as well as acting as a solvent for biochemical reactions, water is also the medium for the transport and distribution of polar organic molecules (e.g. sucrose in the phloem), inorganic ions (nutrients from root to leaf in the xylem; CO_2 or bicarbonate to the site of photosynthetic fixation in the cell), and atmospheric gases (diffusion of oxygen to sites of respiration).

Two further physical properties, tensile strength and viscosity, are important in the long-distance transport of water and dissolved solutes. The high tensile strength (cohesion) of water, another consequence of the strong hydrogen bonds between its molecules, means that water can be drawn to the tops of tall trees by transpirational pull (the 'cohesion theory', considered in full on p. 185). On the other hand, liquid water has an unexpectedly low viscosity, facilitating rapid flow, for example in soil macropores, but this is less important in soil capillaries and plant apoplasts where the ratio of surface – bound to bulk water is high.

Water can move from the soil, through root and stem, to a transpiring leaf only if there is continuity of liquid throughout the pathway. Thus, in addition to continuous columns of water in the xylem, the plant also requires continuity of liquid water in the capillaries of the soil and the apoplasts of root and leaf. That this continuity exists is the consequence of the physical properties of water. First, as explained above, continuous films of water molecules are bound to hydrophilic surfaces of soil and cell wall capillaries; xylem walls are particularly hydrophilic (Tyree and Sperry, 1989). Secondly, since the surface tension of water is very high (73.5×10^{-3} kg s^{-2} at 15 °C, i.e. two to three times the value for most laboratory solvents), the filling of capillaries with water results in a reduction in the energy (and therefore increase in the stability) of the capillary/water system (reduction of the water surface in contact with air). These two forces retaining water in capillaries constitute the matric forces. For example, when the gravitational water drains from a water-saturated soil, the soil capillaries of diameter less than 60 μm remain filled with water retained by matric forces. As we shall see in later sections, problems arise for plants when the continuity of liquid water in the xylem is broken (cavitation/embolism, Tyree and Sperry, 1989).

Both the distribution and the morphology of green plants are influenced by the fact that water absorbs specifically in the infrared but is relatively transparent to short-wave radiation (p. 24). Thus aquatic plants can absorb photosynthetically active radiation at considerable depths in clear water; in land plants, the same optical properties have permitted the evolution of leaves in which the unpigmented epidermal cells allow passage of PAR to the underlying layers of mesophyll cells which are active in photosynthesis. However, the epidermis does not appear to act as a heat filter, protecting the mesophyll; indeed, under moderate irradiance, the epidermis can be at a lower temperature than underlying tissues owing to transpirational cooling.

In addition to its solvent properties, water is a biochemical reagent in, for example, hydrolysis reactions. Much of the chemical activity of water is a consequence of its dissociation into charged hydronium and hydroxide ions:

$$2H_2O \rightleftharpoons H_3O^+ + OH^- \tag{4.1}$$

where $K_w = 10^{-14}$
Thus even pure water is a 10^{-7} M solution of hydronium ions.

Because transpiration involves the diffusional loss of water molecules from the leaf in the gas phase, it is important to understand the different methods of expressing the water vapour content of air, and the interrelationships between the units used. The water vapour content of air in equilibrium with liquid water (referred to as the density of saturated water vapour or, more commonly, the saturation water vapour density, in g of water per m^3 of moist air) rises sharply with increasing temperature, as does the corresponding partial pressure exerted by water vapour (the saturation water vapour pressure, in kPa) (Table 4.1). The water vapour content of a volume of *unsaturated* air can also be expressed in terms of the water vapour density (g m^{-3}); however, since the air within the leaf is normally considered to be saturated, both the water vapour content of the bulk air, and the driving force for the diffusion of water molecules from the

Table 4.1

(a) Methods of expressing the water vapour content of air at different temperatures

Temperature (°C)	0	5	10	15	20	25	30	35
Saturation water[a] vapour density (g m^{-3})	4.9	6.8	9.4	12.9	17.3	23.1	30.4	39.6
Saturation water vapour pressure (kPa)	0.6	0.9	1.2	1.7	2.3	3.2	4.2	5.6

(b) Relationships between water vapour pressure deficit (VPD) and relative humidity (RH) of the air at different temperatures

VPD (kPa)/RH[b](%)								
0.1	83	89	92	94	96	97	98	98
0.2	67	78	83	88	91	94	95	96
0.3	50	67	75	82	87	91	93	95
0.4	33	56	67	77	83	88	91	93
0.5	16	44	58	71	78	84	88	91

[a] Note that, over the same temperature range, *the density of moist air* changes very little (from 1.3×10^3 to 1.1×10^3 g m^{-3}).
[b] The water vapour density of the air as a percentage of the saturation vapour density.

substomatal cavity to the bulk air, are commonly expressed in terms of the vapour pressure deficit (VPD: the difference between the actual vapour pressure of the bulk air and the saturated vapour pressure, in kPa). However, it is important to emphasize that a given VPD can correspond to a wide range of relative humidities, depending on the temperature. Interrelationships among these different expressions at different temperatures are presented in Table 4.1.

2. The water relations of plants and soils

1. Water potential

Water flowing downhill can turn a water wheel, yielding useful work. The water at the bottom of the gradient has lost part of its capacity to do work, and has a lower free energy content than it had at the top. The driving force for the flow of water is this difference in free energy, but thermodynamic analysis reveals only that flow *can* take place; it says nothing about the *rate* of flow (the kinetics), and, in the extreme case, an intervening dam could reduce the rate of flow to zero.

In a similar way, in accordance with the Laws of Thermodynamics, water moves in the soil–plant–atmosphere system in response to differences in its free energy content. Thus, in a well-watered, transpiring plant, the free energy content of water decreases progressively from the soil, *via* the xylem and the leaf apoplast, to the bulk air; and water flows from the soil through the plant to the air in response to this gradient in free energy. The existence of a gradient does not, however, guarantee flow, since the plant is able to establish barriers to flow, notably by closing stomata (see p. 149).

Plant physiologists have yet to develop a method of expressing the free energy content of water which satisfies all requirements, but it has become customary to use water potential (Ψ). In spite of some theoretical problems, water potential has proved to be a powerful tool in the exploration of plant and soil water relations (Johnson *et al.*, 1991). Water potential is defined as the free energy per unit volume of water, assuming the potential of pure water to be zero under standard conditions (usually ambient temperature and atmospheric pressure). Since energy per unit volume has the same dimensions as pressure, plant and soil water potentials are conventionally expressed in pressure units, correctly MPa, although the bar (1 bar = 10^5 Pa = 0.1 MPa) was widely used up to around 1980. Some authorities prefer to use free energy per unit of mass (J kg^{-1}, where 100 J kg^{-1} is approximately equal to 0.1 MPa, depending on temperature).

Because water potential increases with temperature, it is important to maintain constant temperature during a series of experiments; this is rarely achievable in the field. On the other hand, water potential is lowered below that of pure water by dissolved solutes and also by the binding of water to surfaces by matric forces. Since these effects are considered to be mutually independent, the water potential of a solution (Ψ) can be expressed as

$$\Psi = \psi_s + \psi_m \qquad (4.2)$$

where ψ_s, the solute potential, is the lowering of water potential caused by dissolved solutes, ψ_m, the matric potential, is the lowering of water potential owing to matric forces, and Ψ will be negative, since the potential of pure water is set at zero.

Since the water in the soil and plant is invariably accompanied by solutes, and acted upon by matric forces, water potentials in the soil–plant–atmosphere system are usually negative; water flows towards regions with more negative values, and the expressions 'lower' and 'higher' water potential indicate more and less negative values, respectively. For example, in rapidly transpiring Sitka spruce trees, water flowed upwards, but down a water potential gradient, from the soil (−0.04 MPa) to the terminal shoots (−1.5 MPa) at 10 m above the soil surface (Hellkvist *et al.*, 1974).

Because the water potential in plant tissues can be raised by hydrostatic pressure, equation 4.2 must be modified to give

$$\Psi = \psi_s + \psi_m + \psi_p \tag{4.3}$$

where ψ_p, the pressure potential, is the increase in water potential owing to hydrostatic pressure. Since ψ_p is positive, the resulting water potential is less negative than would be the case if only solute and matric effects were included. High hydrostatic pressure within plants can result in zero water potentials, or even positive values, for example in the xylem of guttating plants, whose hydathodes can release drops of liquid water.

2. The water relations of plant cells

1. Mature tissues

Most of the cells involved in plant–water relations are fully mature, with a large fraction of the cell water contained in a central vacuole. The thin layer of cytoplasm, together with its associated plasmalemma and tonoplast, can be treated as a complex semipermeable membrane, separating the vacuolar contents from the external medium.

In a leaf, the external medium is the water in the cell walls and intercellular spaces (the leaf apoplast), which is subject to atmospheric pressure (i.e. $\psi_p = 0$). Since the solute concentration is normally low, ψ_s is small, and the water potential is determined by the matric forces exerted by the cell walls. Thus

$$\Psi_{apo} = \psi_m \tag{4.4}$$

where ψ_m will normally be high (more positive than −0.1 MPa) in a leaf which is well supplied with water but not transpiring (e.g. at night). In the vacuole, matric forces are less important, and water potential is determined largely by the solute concentration. Thus

$$\Psi_{vac} = \psi_s \tag{4.5}$$

where ψ_s can vary from −0.5 to −3.0 MPa and even lower, according to species, environment and the degree of osmoregulation achieved (see p. 166).

Since Ψ_{vac} is lower than Ψ_{apo}, there is a driving force for water influx across the cytoplasm, which would raise the water potential in the vacuole by diluting the vacuolar sap, while increasing its volume. In the absence of cell walls, this influx of water would continue until either the cell burst or the difference in water potential had been abolished. However, the volume of a mature leaf cell is limited by its walls, and only a small influx of water can be accommodated by their elasticity (compare with the *plastic* deformation of the walls of cells which are still expanding; see below). Consequently, hydrostatic pressure (turgor pressure or, simply, turgor) builds up within the vacuole, pressing the cytoplasm against the inner surface of the walls, and thereby raising the water potential of the vacuole. As turgor increases, adjacent cells press against one another, with the result that an originally flaccid leaf becomes increasingly turgid. Ultimately, equilibrium is reached when the driving force for water influx (the difference in solute potential) equals the driving force for efflux (cell turgor pressure), and there is no net movement of water into the vacuole. At this point, cell turgor is at its maximum, and

$$\Psi_{apo} = \Psi_{vac} \tag{4.6}$$

$$\text{i.e.} \qquad (\psi_m)_{apo} = (\psi_s + \psi_p)_{vac} \tag{4.7}$$

The progressive increase in cell turgor caused by water influx is illustrated by Fig. 4.1 in which, for clarity, the water potential of the bathing medium is set at zero. The idealized, flaccid cell, with a cell volume of 1.0 units and a ψ_s of -1.6 MPa, takes up water from the bathing medium, thereby increasing cell volume and raising ψ_s by sap dilution. Since the cell walls resist expansion, ψ_p rises at an increasing rate until maximum turgor is achieved at $\psi_p = 1.2$ MPa, whereas ψ_s has risen to -1.2 MPa. At this point, since $\Psi_{vac} = \psi_s + \psi_p = 0$, the driving force for entry of water has declined to zero and the cell has reached maximum volume (1.3 units).

This treatment has concentrated on the water relations of vacuoles but, since virtually all plant biochemical and physiological processes take place in the cytoplasm or cytoplasmic organelles, the water relations of the cytoplasm are of much greater interest. It is customary to assume that the cytoplasmic water is in thermodynamic equilibrium with the very much larger volume of vacuolar water,

$$\text{i.e.} \qquad \Psi_{cell} = \Psi_{cyt} = \Psi_{vac} \tag{4.8}$$

However, even though the cytoplasmic and vacuolar potentials are equal, the relative contributions of the components of water potential need not be the same. Thus, because of high concentrations of colloidal particles, there will be a significant matric potential contribution to cytoplasmic water potential.

2. *Growing tissues*

The water relations of immature tissues are broadly similar to those of mature tissues except that the cell walls can undergo irreversible (plastic) deformation to permit cell and tissue expansion. The relative rate of volume increase of a cell can be described by the Lockhart equation:

$$1/V.\mathrm{d}V/\mathrm{d}t = \varphi(P - Y) \tag{4.9}$$

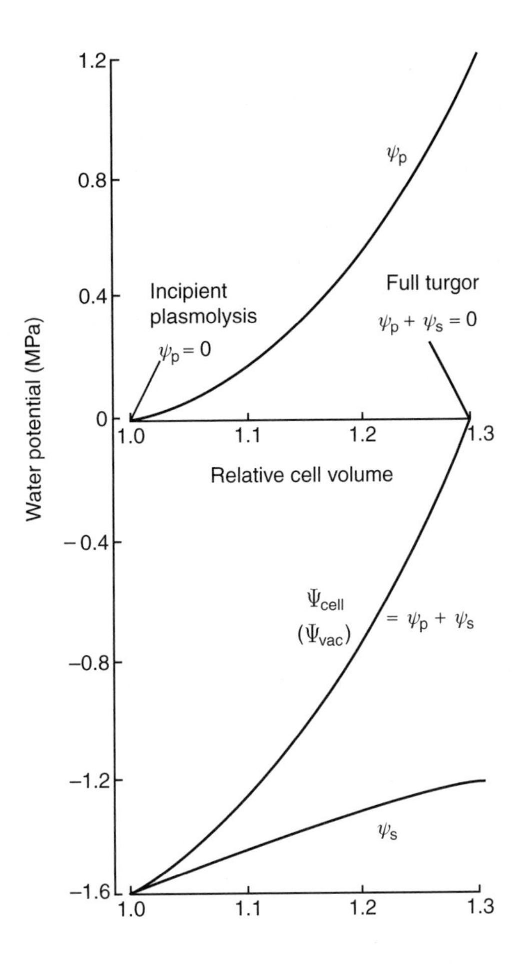

Figure 4.1

The relationships between vacuolar water potential (Ψ_{vac}) and its components, pressure potential (ψ_p) and solute potential (ψ_s), at different volumes of an idealised mature plant cell (redrawn from Meidner and Sheriff, 1976)

where V is cell volume, φ is the (volumetric) extensibility of the cell wall, P is cell turgor pressure and Y is the yield threshold pressure (i.e. the minimum turgor pressure to give cell growth) (e.g. 0.22 – 0.25 MPa in Fig. 4.2). Thus, in principle, the rate of expansion of an immature cell can decrease as a result of reduced turgor or extensibility, an increase in yield threshold, or a combination of such effects. As discussed below, the final volume of the expanding cell, and the organ of which it is an element, depends on the duration as well as the rate of expansion.

Measurements of single cells using pressure probe techniques (e.g. Spollen *et al.*, 1993) have shown that turgor pressure tends to be constant among the cells of the expansion zones of *unstressed* plant tissues, indicating that cell wall loosening is the primary cause of cell expansion. Although the biochemistry of such loosening is very complex, the disruption of calcium pectate complexes by

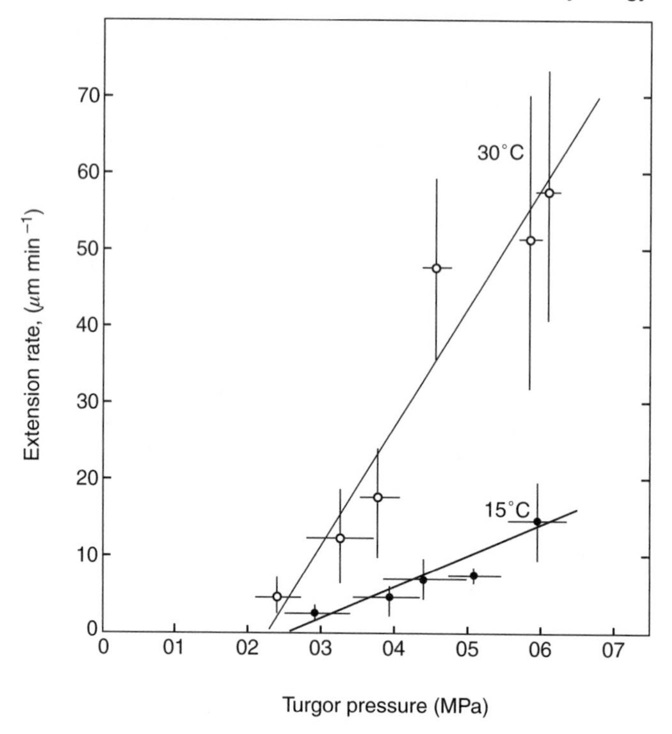

Figure 4.2

The effects of variation in the turgor pressure of cells of the growing zone (brought about by variation in the osmotic pressure of the growing medium) on the rate of extension of maize roots at two temperatures. The bars indicate s.d. Note that the yield threshold turgor pressure is unaffected by pretreatment or temperature, but the extensibility is greater at 30 °C (from Pritchard *et al.*, 1990).

lowered pH, the enzymic hydrolysis of linkages between cellulose molecules, and the action of expansins appear to be central events (Cosgrove, 1993, 2000; McQueen-Mason, 1995). The maintenance of cell turgor shows that the rates of uptake of additional water and solutes keep pace with cell expansion (Pritchard, 1994).

3. Plant water stress

1. Expression and classification of water stress

As noted in Chapter 1, because of the complexity of plant/water relations, there is no single index of water supply by the environment (soil water content; bulk air humidity etc.) which can be used to quantify the degree of water deficit stress (or water stress) to which a plant is subjected. In the absence of an *environmental* index, it is the convention to quantify water stress in terms of the extent to which tissue water content has fallen below that at full turgor (i.e. below the optimum water content for growth and function). The principal index is tissue water potential, although relative water content (RWC: water content as a percentage of the fully hydrated content), turgor (Jones and Corlett, 1992) and water deficit can be of value in some circumstances (see below). The use of *pre-dawn leaf water potential* as an index of stress is founded on the assumption that equilibration of water potential between leaf and soil has occurred overnight (but see p. 155).

Since the photosynthetic uptake of CO_2 *via* open stomata is inevitably associated with water loss to the atmosphere, and some loss of turgor (Fig. 4.1), nearly all plants are exposed to some degree of water stress throughout their lives during the daily period of illumination. Coping with water stress is a *routine* aspect of plant life, not simply a feature of species adapted to dry habitats. As we shall see later, the leaves of desert plants such as *Larrea tridentata* can survive water potentials as low as −11.5 MPa, with photosynthesis continuing at −5 to −8 MPa. On the other hand, species adapted to the understorey of moist forests are rarely exposed to (or equipped to deal with) values lower than −1 MPa. It is, therefore, misleading to refer to 'typical' levels of water stress.

Nevertheless, in reviewing the effects of water stress on plant growth and function, Hsiao (1973) found it convenient to use three, loosely defined, degrees of water stress, in relation to a 'typical mesophyte' (probably best represented by the crop and weed species of temperate agriculture):

Mild stress: Ψ_{cell} slightly lowered, typically down to −0.5 MPa at most;
Moderate stress: Ψ_{cell} in the range −0.5 to −1.2 or 1.5 MPa
Severe stress: Ψ_{cell} below − 1.5 Mpa,

This terminology is used in the remainder of this chapter. Lawlor (1995) has proposed an alternative, but broadly compatible, classification for mesophytes, based on RWC: values down to 90% are associated with effects on stomata and cell expansion; 80–90% with effects on photosynthesis and respiration; and below 80% (corresponding to water potentials of −1.5 MPa or lower) with the cessation of photosynthesis and the disruption of cell metabolism.

2. The influence of water stress on plant cells and plants

Interpretation of the effects of different degrees of water stress on plant physiology (i.e. the 'strain' caused; Chapter 1) can be complicated by the fact that responses can be evoked at the organ, tissue, cell or molecular level. For example, the stomata of mesophytic plants start to close at leaf water potentials in the range −0.5 to −1.0 MPa (Fig. 4.3) (or possibly even higher under the influence of intraplant signals), thereby reducing the flux of CO_2 from the bulk air to the photosynthetic mesophyll. Thus the rate of photosynthesis may be reduced by a whole leaf response before there are significant effects of water stress on individual cells, chloroplasts, membranes or reactions (Lawlor, 1995; Tezara *et al.*, 1999).

The primary effect of dehydration on plants is loss of turgor. For the idealized leaf cell illustrated in Fig. 4.1, the imposition of mild water stress is associated with a rapid reduction in turgor pressure, which continues at a declining rate per unit of water potential under moderate stress. Severe water stress involves a complete loss of turgor ($\psi_p = 0$), and leaf wilting. As the volume of the dehydrating cell contents decreases, there is a tendency for the

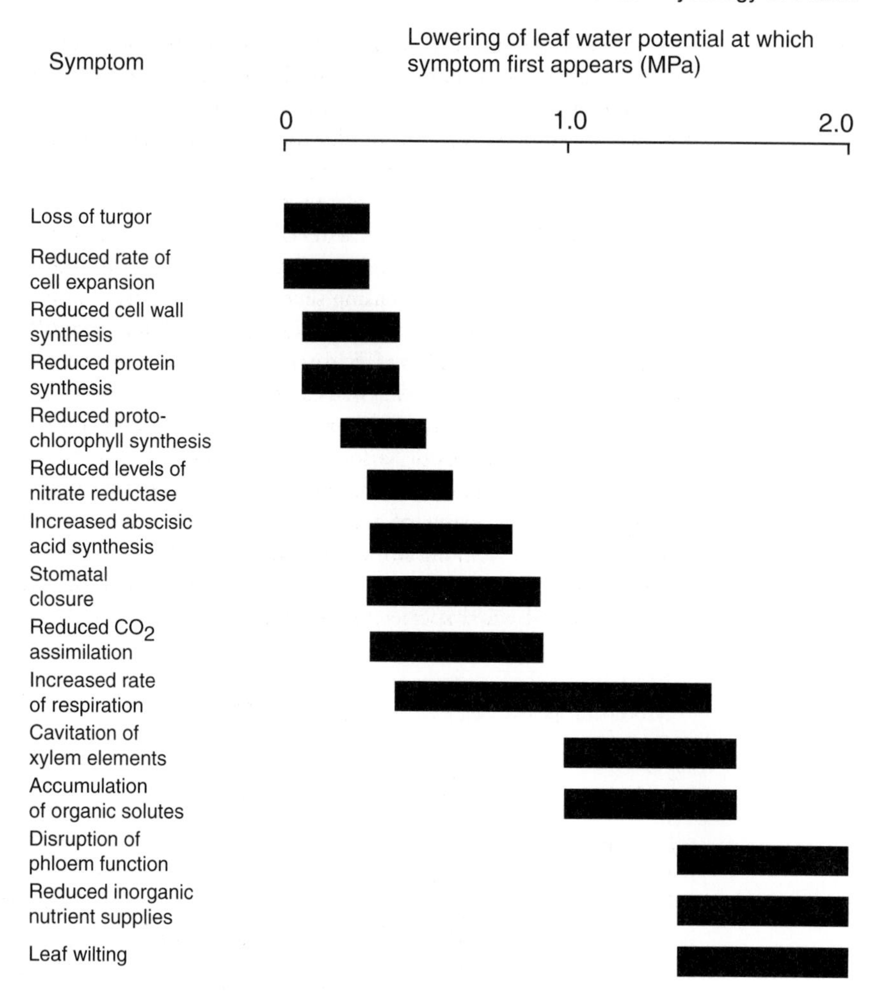

Figure 4.3

The influence of water stress on the physiology of mesophytic plants. The horizontal bars are guides to the level of stress at which the relevant symptoms first occur. The lowering of leaf water potential is in relation to a well-watered plant under mild evaporative demand (updated from Hsiao *et al.*, 1976)

plasmalemma to shrink away from the cell wall; nevertheless, plasmolysis seems to be a rare event in nature since it is difficult for water or air to move inwards to fill the space vacated between membrane and cell wall. The exposure of cells to severe water stress, therefore, implies mechanical stress as well as serious dehydration, which brings reactive molecules closer together. Resurrection plants (p. 182; Plate 11) provide examples of genotypes which can survive the extreme mechanical and chemical stresses of dehydration and rehydration.

Loss of turgor has a range of implications for plant leaves. The Lockhart equation (4.9) predicts that rates of expansion of cells in an immature leaf, and therefore of the entire leaf, will fall, but this does not take account of additional negative effects of drought on cell wall extensibility (e.g. Neumann, 1995; see also the effect of low temperature on extensibility, p. 198). On the other hand, the rate of cell division, and the duration of leaf expansion, are both relatively unaffected by mild to moderate stress, although both will be curtailed under severe stress. The net result of these different effects of loss of turgor is illustrated by the field measurements of droughted sunflower plants (Fig. 4.4). Comparing leaves of the same insertion, the rate, but not the duration, of leaf expansion declined with increasing severity of water stress, at least in the first 40 days of the treatments. Thus, the final area of leaf 3 fell from 69 cm^2 (Fig. 4.4a; fully irrigated) through 62 cm^2 (b; dry to bud formation) and 38 cm^2 (c; rainfed) to 31 cm^2 (d; dry to anthesis); the corresponding values for leaf 12 were 207, 200, 62 and 54 cm^2. However, as the degree of turgor loss increased, the duration of leaf expansion also began to fall (e.g. compare leaf 21 under treatments (a) and (c)); this is explained by the fact that growth was not possible during a substantial part of the potential duration as cell turgor had fallen below the threshold for expansion (equation 4.9). In the field, recovery of turgor at night can permit higher rates of expansion than during the day but, in general, mesophytic plants experiencing long-term loss of turgor will have

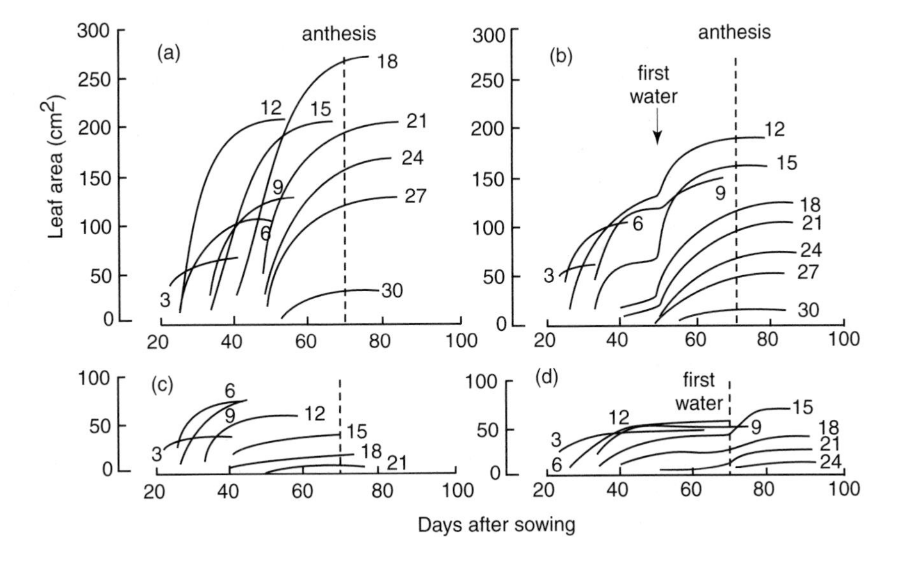

Figure 4.4

Time courses of leaf area expansion of sunflower (*Helianthus annuus*) cv. Sungold plants grown under different irrigation regimes in Victoria, Australia: (a) irrigated weekly, 450 mm; (b) no water up to bud formation, 330 mm; (c) rainfed, 121 mm; (d) no water up to anthesis, 193 mm. Individual leaves are numbered in order of appearance, and the curve for leaf 12 is displaced, for clarity, in (a) and (b). (Reprinted from Connor, D. J. and Jones, T. R. (1986), Response of sunflower to strategies of irrigation. 2. Morphological and physiological responses to water stress. *Field Crops Research* **12**, 91–103. ©1985, with permission from Elsevier Science)

smaller leaves and less extensive canopies; less *PAR* can be intercepted and the potential for dry matter production will be reduced. As we shall see in subsequent sections, many of the species adapted to drier environments show adaptations, notably osmoregulation, which tend to maintain turgor under water stress.

In addition to slowing growth, a lowering of leaf water potential by less than 0.5 MPa (mild stress) is associated with some disruption of biosynthetic activities including the generation of cell wall components, chlorophyll, enzymes and protein in general (Fig. 4.3). Under moderate stress there is further reduction in turgor, leading to the narrowing of stomatal aperture and a progressive reduction in photosynthetic activity. Increased respiration may also play a part in stomatal closure owing to an increase in CO_2 concentration within the leaf air spaces (see p. 149). With the onset of severe stress, photosynthetic exchange of CO_2 ceases and a general disruption of metabolism is signalled by high rates of respiration and the build up of proline and/or sugars in leaf tissues; in plants resistant to drought such accumulation of organic solutes, leading to osmoregulation, can occur at lower stress (see p. 166). There is increasing evidence that, under moderate to severe water stress (xylem water potential lower than -1.0 MPa), herbaceous plants begin to show symptoms of xylem cavitation and reduced xylem conductance (e.g. in maize, Tyree *et al.*, 1986; Borghetti *et al.*, 1993). As we shall see in section 5, cavitation of xylem elements in trees tends to begin at much lower water potentials, typically -2.0 to -3.0 MPa (Jones and Sutherland, 1991).

This stepwise development of (generally reversible) symptoms of water stress is accompanied by, and to some extent brought about by, changes in the synthesis and transport of growth substances (notably the enhanced biosynthesis of abscisic acid; see below). There can also be reductions in the supply of mineral nutrients to the leaf *via* the xylem, and, under severe stress, in rates of flow in the phloem (Schulze, 1991). However, the functional relationships between xylem (under tension) and phloem (whose contents must be under pressure for transport to occur), in close proximity within the stele, are far from clear. Thus, plants exposed to moderate to severe water stress will also be experiencing nutrient stress, a changed balance of growth substances and altered patterns of partitioning of photosynthate, as well as temperature stress.

Owing to a lack of standardization of the treatments imposed, and their severity, it is not yet possible to correlate the many effects of water stress on gene expression with the pattern of physiological and biochemical events outlined here. Drought brings about changes in the expression of some genes whose products are clearly involved in plant responses to water shortage, but these vary widely from short- to long-term responses (e.g. synthesis of aquaporins (see below), osmotic agents, cuticular waxes and cryoprotectant molecules, Ingram and Bartels, 1996). Some gene products are similar to high- and low-temperature shock proteins (Chapter 5), but most have no identified role in drought stress. Perhaps the most promising area of investigation centres on the role of certain gene products in signalling the onset and severity of water stress (Shinozaki and Yamaguchi-Shinozaki, 1997).

In summary, exposure of plants to even mild water stress can affect growth, and disturb metabolic processes. Depending on their severity, such effects can reduce the ability of the plant to survive and reproduce. Consequently, it is important for terrestrial species either to avoid water stress (e.g. by phenology), or to evolve anatomical, morphological or biochemical adaptations which lead to the amelioration or tolerance of water stress. Much of the remainder of this chapter is devoted to the study of these adaptations.

4. Supply of water by the soil

The interception of dew and rain by leaves can play a central part in the survival of some arid-zone species but, as far as most terrestrial plants are concerned, this is of negligible importance compared with the absorption of soil water by the root system.

1. Climate

The quantity of water held by a soil depends primarily on the climate, and specifically on the excess of precipitation over evaporation (P-E). Thus in 'extremely humid' regions such as the northwestern coasts of Europe and North America, where annual precipitation is at least twice evaporation, it is unusual for the availability of soil water to be a factor in the survival and distribution of plant species. Exceptions to this rule include habitats such as cliffs, and shallow or very coarse-textured soils with poor water-holding capacities. In contrast, in 'extremely arid' regions, where potential evaporation is more than twice precipitation, soil moisture levels are normally so low that vegetation is sparse or absent.

Between these extremes, there is a spectrum of climates, classically documented by Walter and Leith (1960), where the availability of soil water depends not only on annual P-E but also on the distribution of rainfall within each year. Many of the features common to seasonally dry climates are illustrated in Fig. 4.5, which compares the distribution of rainfall for two contrasting, but not extreme, years in a tropical savanna region. In 1970/71, the 'wet' season began (25 November) and ended (8 March) decisively, with a few scattered days of rainfall in early November, late March and early April. The continuity of precipitation ensured that significant soil moisture deficits did not occur until the end of the rains. In contrast, both the beginning and the end of the 1971/72 wet season were ill-defined, and a higher proportion of annual rainfall fell in the first half of November, and in April. Furthermore, there were at least two mid-season periods (26 December to 4 January; 26 January to 8 February) during which, according to agrometeorological calculations, annual crop plants would have been exposed to water stress (Hay, 1981). The months of May to October were entirely dry in each year.

To be successful under such conditions, a *wild* plant species must possess a set of characteristics, including the ability:

(i) to survive long periods without rain every year (six months, in this example);
(ii) to make full use of the period of soil water availability for the completion of its annual life cycle, without sustaining serious injury owing to water stress at the beginning and end of the wet season. Since the date of the beginning of the growing season varies from year to year, the plant requires a

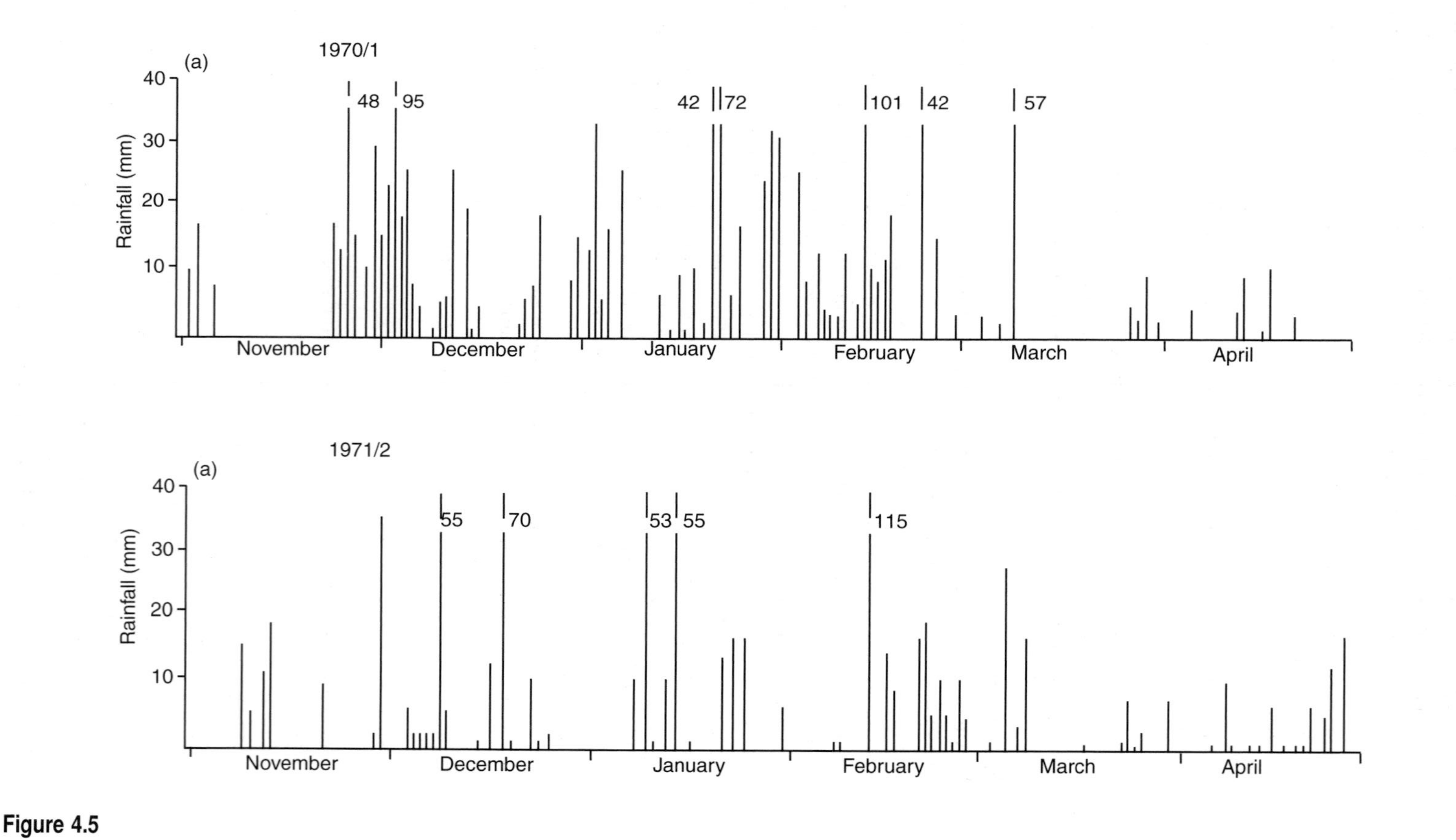

Figure 4.5

The distribution of rainfall in a tropical savanna region (Bunda, Central Region, Malawi) in (a) 1970/71 and (b) 1971/72. There was negligible rainfall during the remaining months of each year

mechanism to signal when it is 'safe' to start or resume growth. (Mechanisms of this kind are discussed below when considering the timing of germination of dormant seeds, and of leafing out of deciduous perennials);

(iii) to avoid, ameliorate or tolerate shorter periods (hours, days) of mild to severe water stress at different times and growth stages within the growing season; and

(iv) to survive occasional years of drought when either the total amount of precipitation or the length of the wet season is significantly reduced below the mean.

In more humid and more temperate regions, where rainfall is less seasonal, points (i) and (ii) are less important, but (iii) and (iv) remain essential characteristics for plant species in all but the most humid climates.

2. Soil/water relationships

Within a given climatic zone, the availability of water for plant uptake depends upon the water-storing properties of the soil. Soils consist of mineral particles of varying shape and diameter bound together into aggregates by organic matter, clay particles and other inorganic components such as iron oxides. Within and between these aggregates there is a three-dimensional network of interconnected spaces of diameters varying from a few centimetres (drying cracks, earthworm or termite channels), through a few millimetres (between aggregates) down to few micrometres or tenths of a micrometre (within aggregates). Although these spaces are irregular in shape, it is a useful first approximation to treat the network as if it were made up of a series of more regular cylindrical pores. When a soil is saturated with water after prolonged rainfall, this pore space becomes temporarily water-filled (i.e. the soil is saturated) but, as the following analysis shows, a freely draining soil cannot hold all of this water for plant use.

Since the concentration of solutes in soil water is generally low, the major forces retaining water in soil pores are the matric forces, which increase as pore diameter (d) decreases. The water potential in a water-filled pore is, therefore, inversely related to pore diameter by the expression:

$$\Psi_{\text{pore water}} = \psi_{\text{m}} = -0.3/d \qquad (4.10)$$

where water potential is in MPa and d in μm (Payne, 1988). For example, pure water held in soil pores of diameter 10 μm will be at a potential of -0.03 MPa such that a suction exceeding 0.03 MPa would be required to withdraw this water. Similarly, since gravity exerts a suction equivalent to about 5 kPa in temperate zones, then all pores wider than 60 μm will tend to drain spontaneously after a soil has been saturated with water.

Even in a freely draining soil, the loss of gravitational water can take two to three days. Once drainage is complete, the soil is at field capacity (FC: containing the maximum amount of water, normally expressed in g per 100g oven-dry soil, that it can hold against gravity). The term field capacity cannot strictly be applied to soils whose drainage is impeded.

3. Availability of soil water to plants

Transpirational loss of water from a leaf canopy establishes a gradient of water potential in the plant/soil system. If this causes the water potential in the xylem of a root axis to fall below that in the adjacent soil pores, then water can flow into the root and pass *via* the xylem to the site of transpiration. Soil water is, thus, available to plants only if the root xylem water potential (and, in turn, the leaf water potential) can be lowered below the soil water potential.

Significant quantities of water are held in the pores of finer-textured soils at potentials of −3 MPa or lower (Fig. 4.6), but few mesophytes can tolerate the lowering of leaf water potential required to absorb this water (see Fig. 4.3 for the implications of leaf water potentials of −2 MPa). Many of the species which have been studied in detail (principally temperate crop plants) can withdraw water from pores wider than about 2 μm, corresponding to a water potential of −1.5 MPa; once all the water in these pores has been exhausted, no more can be transported to the leaf, and the plant will wilt and ultimately die, unless the soil is recharged with water. For this reason, the property 'permanent wilting point' (PWP) has been defined as the moisture content (g per 100 g oven-dry soil) after the soil has come to equilibrium under a suction of 1.5 MPa. The available water content of the soil is then the difference between FC and PWP.

This internationally-accepted standard method of determining PWP and available water content is useful in comparing the water relations of different soils (e.g. Fig. 4.6), but it is no more than a useful laboratory convention. In practice, a given soil can release differing quantities of water to different species (and even to plants of the same species raised under different conditions) according to the minimum leaf and root xylem water potentials

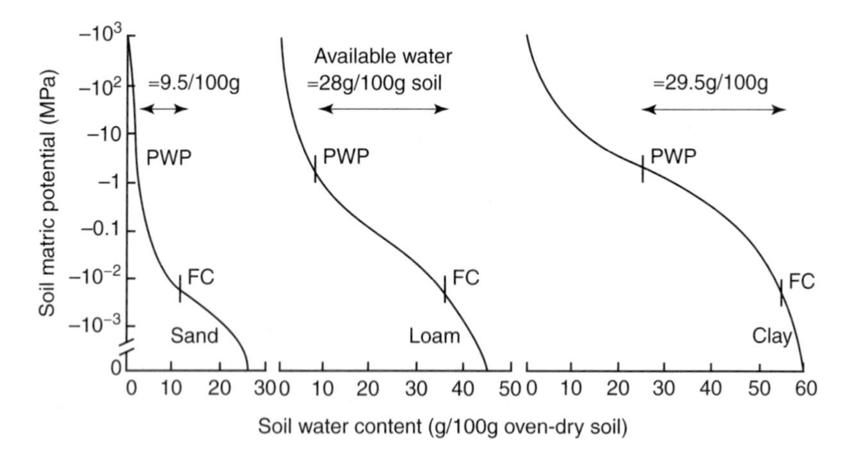

Figure 4.6

Representative soil water release curves for coarse (sand), medium (loam) and fine (clay) soils. FC and PWP indicate field capacity and permanent wilting point, as defined in the text. Note that although the available water contents of the loam and clay soils are very similar, in the loam soil a higher proportion of this water is held at matric potentials above −0.1 MPa (i.e. in wider pores) (redrawn from Brady, 1974)

which can be tolerated (e.g. Table 4.5), and the distribution of the roots within the soil (see p. 156).

The amount of soil water that is available for uptake by a plant, thus, depends on the size distribution of the pores within the soil, which, in turn, depends on the size distribution of the soil particles (the soil texture) and the soil structure. In general, medium- to fine-textured soils, and those with a good aggregate structure, tend to hold more water for plant use than coarse-textured and poorly structured soils (Fig. 4.6). Temperature also influences soil water availability through its effect on the viscosity of water; in the foregoing discussion of FC, it was concluded that only those pores wider than 60 μm drain under gravity, but this is based on field measurements in temperate zones. In tropical and subtropical zones, the lowering of water viscosity under warmer conditions, coupled with very free drainage, can lead to the draining of pores as narrow as 10 μm. Thus soils of similar texture and structure can hold different amounts of available water in temperate and tropical zones.

Since any increase in the solute content of soil water lowers soil water potential, root xylem potential must be lowered further to extract the same amount of water; this effect is particularly pronounced in saline soils (see Chapter 6). Finally, in some environments, soil texture and structure can have a large influence on the availability of water to plants through their effects on soil surface properties and water infiltration. For example, in direct contrast to the normal conclusion about the influence of texture (e.g. Fig. 4.6), coarse-textured soils in arid regions tend to retain a higher proportion of intense, episodic precipitation than do finer-textured soils, whose pores cannot accept water at a sufficient rate to avoid run-off. Furthermore, because of run-off, water accumulates in soils below depressions, irrespective of soil properties.

So far, this treatment of soil water availability has assumed a uniform branching of roots within and between aggregates throughout the soil volume. This is clearly not the case for younger plants, whose roots exploit the water stored in successive soil horizons as they extend and ramify, or even for more mature root systems, whose axes tend to be concentrated in larger pores, with laterals exploring only the adjacent aggregates (see Chapter 3). Although root distributions of this kind lead to less than uniform patterns of soil water depletion (e.g. Smucker and Aiken, 1992), plants do not draw water only from the immediate vicinity of their actively absorbing roots: as extraction of the water adjacent to a root proceeds, a depletion zone (analogous to those for mobile ions, see p. 93) develops, causing water to flow towards the root surface over distances of at least several millimetres.

At the same time, the withdrawal of water from the larger pores reduces the volume of the soil through which flow can take place, and increases the length/tortuosity/impedance of the pathway between the remaining available water and the root surface (see p. 95). These effects combine to give a progressive and drastic reduction in the hydraulic conductivity (reciprocal of resistance to water flow) of a soil as it dries. For example in Gardner's (1960) classic study of a Pachappa sandy loam soil, hydraulic conductivity fell from 6 cm day^{-1} at field capacity to less than 10^{-6} at permanent wilting point

(−1.5 MPa). Consequently, the maintenance of a steady flow of water into a root from a drying soil requires:

(i) a progressive lowering of root xylem water potential to maintain a potential gradient between the xylem and the remaining available soil water;
(ii) a progressive increase in the steepness of the gradient to overcome the increasing resistance to water flow offered by the drying soil. This effect is explored further on p. 155.

In conclusion, as we shall see in later sections, it is common for the rate of flow of available water from the soil to be insufficient to meet the short-term demands of the canopy during periods of rapid transpiration; as a result, it is possible for transpiring leaves to be exposed to severe water stress before bulk soil PWP has been reached.

5. Loss of water from transpiring leaves

Within a leaf, the walls of the mesophyll cells adjacent to the substomatal cavity must remain moist to permit the dissolution and uptake of carbon dioxide for photosynthesis. Consequently, as long as the stomata are fully closed, and the temperature is stable, the air contained in the leaf will be saturated with water vapour (Table 4.1). Under such conditions, water molecules can escape from the leaf to the surrounding air only by diffusing across the hydrophobic cuticle covering the outer surface of the epidermis. The rate of cuticular transpiration depends on the thickness, continuity and composition of the cuticle, being low in all species, and particularly low in young leaves with undamaged surfaces, and in most plants adapted to dry habitats.

When leaf stomata begin to open in response to diurnal rhythms or environmental cues (see below), the air outside the leaf is normally not saturated with water vapour. The vapour pressure deficit (VPD), between the substomatal cavity and the bulk air, is the driving force for the outward diffusional movement of water molecules, *via* the stomatal pores, according to the equation:

$$\mathrm{F} = \frac{d_{\mathrm{l}}^{\mathrm{s}}(T_{\mathrm{l}}) - \mathrm{RH}.d_{\mathrm{a}}^{\mathrm{s}}(T_{\mathrm{a}})}{R_{\mathrm{l}}} = \frac{K\rho_{\mathrm{a}}}{P} \cdot \frac{\mathrm{e_l} - \mathrm{e_a}}{R_{\mathrm{l}}} \qquad (4.11)$$

where

F is the rate of transpiration;
$d_{\mathrm{l}}^{\mathrm{s}}(T_{\mathrm{l}})$, $d_{\mathrm{a}}^{\mathrm{s}}(T_{\mathrm{a}})$ are the saturation water vapour densities at the temperature of the leaf (T_{l}) and the bulk air (T_{a});
RH is the relative humidity of the bulk air;
R_{l} is the leaf diffusive resistance (the resistance to the diffusion of water molecules offered by the pathway between the leaf mesophyll and the bulk air);
K is a constant;
ρ_a is the density of the bulk air;

P is atmospheric pressure;

e_l, e_a are the water vapour pressures in the substomatal cavity and the bulk air, respectively;

$e_l - e_a$ is the VPD of the bulk air if the air in the substomatal cavity is saturated, and the leaf is at the same temperature as the bulk air (see also Table 4.1 and equations 5.1 to 5.4).

Thus, for a leaf with a plentiful water supply, the rate of water loss by transpiration is proportional to the VPD of the bulk air, and inversely proportional to the leaf diffusive resistance. The diffusion pathway can be resolved into a number of components in series: the distance moved within the leaf air space; entry into, passage through, and exit from the stomatal pore; and the boundary layer outside the leaf surface. However, as long as the leaf apoplast is well supplied with water, the rate of stomatal transpiration at a given VPD is determined by the sum of the stomatal and leaf boundary layer resistances (normally expressed as s cm^{-1}; but note the widespread use of the reciprocal of resistance, conductance, with dimensions of velocity, e.g. cm s^{-1}).

1. Stomatal resistance

The resistance to the diffusion of water molecules offered by the stomata of a leaf is proportional to the number of stomata per unit of leaf surface (stomatal frequency/density), and inversely proportional to the diameter of the stomatal aperture, which, in turn, is dependent in a complex manner on a number of intrinsic and environmental factors (See below). Across the plant kingdom, the variation in stomatal frequency (e.g. 5–260 per mm^2; Fig. 4.7) is much greater than that of pore dimensions (e.g. fully open diameter varying from 15 to 30 μm; Meidner and Mansfield, 1968), and there is increasing evidence that stomatal frequency can vary with growth environment. For example, study of herbarium specimens and plant fossils has revealed a broad inverse relationship between stomatal frequency of a species and the prevailing atmospheric concentration of CO_2 (Woodward, 1987; Beerling and Woodward, 1997). There is a clear relationship between stomatal frequency and the degree of shading in the species niche (Chapter 2), but not, in general, with its water relations (Peat and Fitter, 1994b) (but see p. 174).

In considering the many interrelated responses of stomata, it is important to establish some basic phenomena. Stomata normally open in the light and close in the dark, either as a direct response to *PAR* or blue light (Zeiger and Zhu, 1998), or under the control of endogenous circadian rhythms. There are, however, a number of important exceptions, including plants employing the CAM pathway of photosynthesis (see p. 59 and 177), and other species, whose stomata remain open at night (e.g. the potato). Stomata also respond to lowering of the CO_2 concentration in the substomatal cavity, opening at a threshold value, whose magnitude is related to the photosynthetic pathway used (C_3, C_4 or CAM, see p. 176 and Chapter 2); they tend to close progressively, but not completely, with increase in the concentration of CO_2 in the bulk air (Assmann, 1999). It is likely that stomatal closure at high temperature is also a response to enhanced levels of intracellular CO_2 as cell respiration rises (see p. 201).

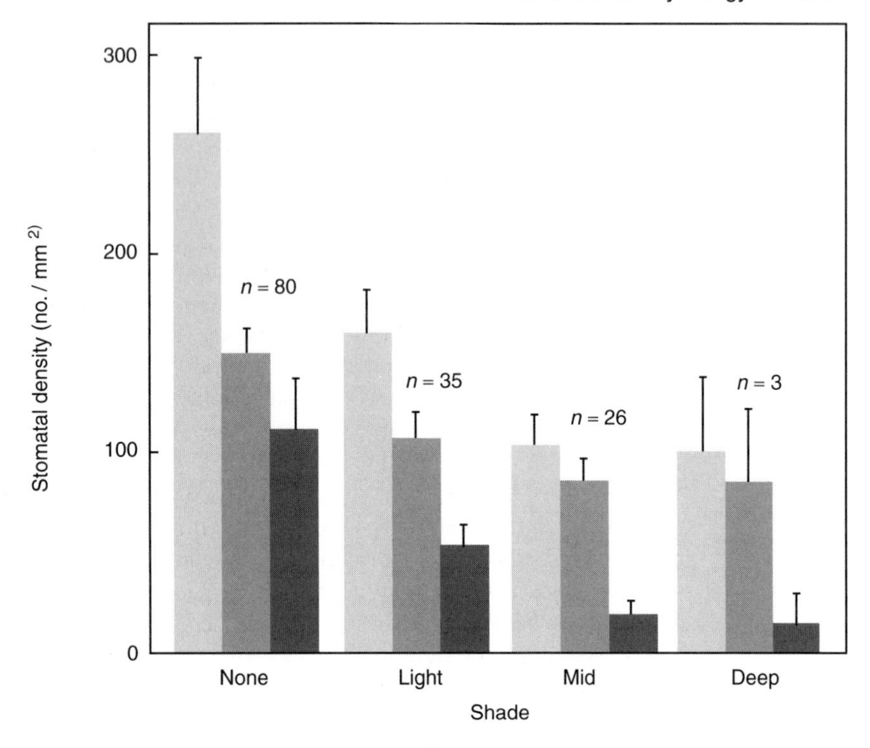

Figure 4.7

The density of stomata on the surfaces of leaves of plant families in the British flora which are adapted to different levels of shade. Each set of histograms refers to the total density (expressed over one surface), the density on the abaxial surface, and the density on the adaxial surface. The bars indicate s.e.m. These families can be classed as weakly hypostomatous (preponderance of stomata on the abaxial surfaces) (from Peat and Fitter, 1994b).

The evolution of these two patterns of response, to light and CO_2, has acted to facilitate the supply of CO_2 to the photosynthetic machinery when *PAR* is available. Stomatal resistance plays an important role in controlling the rate of loss of water *and* the rate of gain of CO_2, and, to a certain extent, their relative magnitudes. Here we concentrate on the basic physiology of stomatal function, laying the foundation for understanding of the role of stomatal properties in determining water use efficiency (no. of moles of CO_2 fixed per mole of water lost), which is of central importance in the interpretation of plant responses to climate change (see Fig. 4.15, and Chapter 2).

The mechanism of stomatal opening is understood in broad terms. Exposure of leaves to a solar radiation or CO_2 signal initiates a cascade of biophysical and biochemical events at the plasmalemma and tonoplast of the guard cell, causing a lowering of its solute potential (ψ_s) by an influx of solutes (principally K^+, Cl^- and/or organic acid ions) (Assmann, 1999). The resulting influx of water causes the turgor of the guard cells to rise above that of surrounding subsidiary, and other epidermal, cells, and the stomatal pores open as a consequence of the mechanical effects of this difference in turgor (Talbott and Zeiger, 1998). The degree of opening of the stomatal aperture, and the resistance to gaseous diffusion through the pore, depend, therefore, on the magnitude of the difference in turgor pressure.

If the turgor of all the cells of a leaf is affected by water loss, stomata can respond *directly* to the water status of the plant: they reach full closure at relatively low thresholds of leaf water potential, which are broadly related to the water supply in the natural range of the species (e.g. -1.0 MPa for *Vicia faba*; -1.8 MPa for *Zea mays*; -2.0 MPa for *Sorghum bicolor*; -4.3 MPa for *Gossypium hirsutum* (cotton); and -5.8 MPa for the desert evergreen *Larrea tridentata*; Ludlow, 1980; see p. 183). This type of response can be seen as a crude protection against low leaf potentials, xylem cavitation and metabolic disruption.

Nevertheless, stomatal closure can be initiated in leaves which have not yet experienced significant loss of turgor, for example, in plants growing in soils dried to very modest soil moisture deficits (e.g. -5 kPa), or with intact root systems divided between wet and drying soil (Davies *et al.*, 1994; Tardieu *et al.*, 1996; Davies and Gowing, 1999). These findings are the basis of the new technique of 'partial rootzone drying' which has been shown to enhance the performance of crops such as grapes (Loveys *et al.*, 2000). Detailed study of the effect of growth substances on membrane behaviour indicates that such closure is caused by abscisic acid (ABA)-induced efflux of cations and anions from the guard cell vacuole, leading to a loss of turgor (e.g. MacRobbie, 1997); indeed some consider ABA to be a universal mediator in the control of stomatal aperture (Tardieu *et al.*, 1996). ABA synthesis is stimulated in droughted leaves and roots, and transported in the xylem, but it is not yet clear whether stomatal closure is caused by changes in the concentration in the leaf, or in the rate of delivery to the leaf, of ABA; changes in xylem chemistry, including variation in pH, may also play a role (Davies and Gowing, 1999). Stomata which close in response to ABA can become unresponsive to other factors for prolonged periods, even when the plant is not experiencing water stress (Mansfield, 1983). As pointed out by Meidner and Mansfield (1968), such a delay in stomatal opening is of particular value to mesophytes, which normally experience only short periods of drought; by the time that normal responses are re-established, it is likely that the soil water will have been recharged with rain.

The suggestion that stomata can respond to plant water status in a third way – by sensing air humidity directly, before it can affect leaf water potential (Schulze, 1993) – has been challenged (Monteith, 1995). It now seems likely that the observed effects can be explained in terms of complex interactions between stomatal resistance and rate of transpiration, which tend to optimize water use efficiency (e.g. Jarvis *et al.*, 1999) (see p. 173).

In summary, stomatal resistance is determined by the interplay of a complex array of factors – irradiance, light quality, CO_2 concentration, temperature, leaf water status, wind, growth substances, soil water status, xylem chemistry, endogenous rhythms, and complex feedback and forward interactions with transpiration rate (Jarvis *et al.*, 1999). There are several pathways of signal sensing and transmission at the cellular level (Blatt and Grabov, 1997), and there is evidence that there can be different 'patches' of stomatal response on the surface of a single leaf (Mansfield *et al.*, 1990). As we shall see in Section 4, the success of a plant species in a given environment depends on the manipulation of the responses to these various cues to give a favourable balance between the photosynthetic uptake of CO_2 and water conservation.

2. Boundary layer resistance

Even in the absence of wind, the bulk air surrounding a leaf is turbulent, owing to convective heat exchange. Consequently, the air is thoroughly mixed, and water molecules move rapidly from a transpiring plant canopy into unsaturated air by mass flow rather than, much more slowly, by molecular diffusion. However, at the leaf surface there is a relatively still layer of air, the boundary layer (e.g. Fig. 4.8), through which water vapour must diffuse before entering the turbulent bulk air (see Chapters 2 and 5). The thickness of the boundary layer (δ) depends on leaf shape and size, and on wind velocity; for example, for the simplest leaf shape (grass leaf with parallel sides):

$$\delta = Kd^{0.5}/u^{0.5} \tag{4.12}$$

where K is a constant, d is the (downwind) dimension of the leaf (length or breadth), and u is wind velocity (see Fig. 4.14).

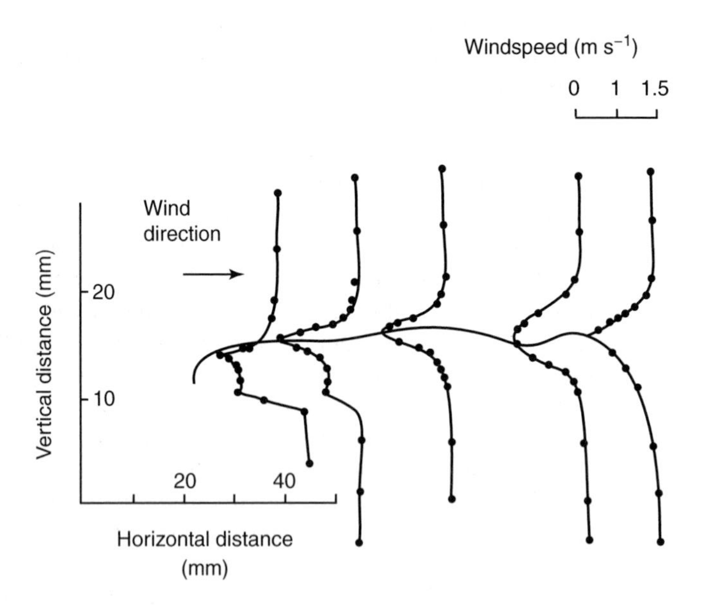

Figure 4.8

Profiles of mean windspeed at the surfaces of a *Populus* leaf. The distance axes define the position of each measurement point on the leaf; and the vertical and windspeed axes define the windspeed profiles at each point. The boundary layer increases, in the downwind direction from near 0 to around 4 mm (from Grace and Wilson, 1976, with permission from Oxford University Press)

The following are representative values of resistances to the diffusion of water vapour across the cuticle, through the stomatal pores, and across the boundary layer of the leaves of a mesophyte:

Cuticle (r_c) 20–80 s cm^{-1} (much higher values for some species);
Stomata (r_s) 0.8–16 s cm^{-1} (depending on degree of opening; when fully closed, $r_s = r_c$);
Boundary layer (r_a) 3.0 s cm^{-1} (at 0.1 m s^{-1} wind velocity, Beaufort Scale Force 0); 0.35 s cm^{-1} (at 10 m s^{-1}, Force 6).

Since the stomatal and cuticular pathways are in parallel, such high values of r_c indicate that cuticular transpiration will be a negligible fraction of total transpiration as long as the stomata remain open. Leaf diffusive resistance (R_l) is, therefore, the sum of r_s and r_a, and, since r_s is normally much the larger component, stomatal aperture determines the rate of transpiration. However, as wind speed diminishes, the thickness of the boundary layer increases (equation 4.12; Fig. 4.14) and r_a constitutes an increasingly larger fraction of R_l. At very low wind velocities or in still air (which are relatively uncommon occurrences in the field), boundary layer resistance can determine the rate of stomatal transpiration irrespective of the degree of stomatal opening. For the same reason, the rate of exchange of gases between air and leaf can be very low if measurements are made in still air under controlled conditions; this can lead, for example, to underestimates of the rate of photosynthesis or of the rate of uptake of gaseous pollutants (see Chapter 6).

6. Water movement in whole plants

In a freely transpiring plant rooted in a soil at field capacity, water evaporates from the moist cell walls of epidermal and mesophyll cells in the interior of the leaves and is lost to the atmosphere *via* stomata according to equation 4.11. As water loss proceeds, the water potential in the leaf apoplast falls below that of the leaf cells, and also below the water potential in the xylem and the soil. Water is, therefore, withdrawn from neighbouring leaf cells, causing a lowering of cell water potential (Fig. 4.1). In contrast, although there is continuity of liquid water between leaf and soil *via* the xylem, rapid equalization of water potential throughout the plant cannot occur because there is a resistance to hydraulic (mass) flow in the soil/plant system. Instead, the transpiration of water from the leaves sets up a gradient in water potential, down which water tends to flow from the soil to the leaf apoplast.

The pathway of water movement from the root surface to the site of evaporation in the leaf is predominantly extracellular. Under low transpirational demand, water flows radially inwards through the cell walls and intercellular spaces of the root epidermis and cortex up to the endodermis, where further apoplastic movement is blocked by Casparian strips in fully differentiated young roots. Thereafter, water passes through the cells of the endodermis before entering the lumina of the xylem elements by way of the stelar parenchyma apoplast. The route is less certain for older and less permeable roots with a suberized exodermis, or which have undergone secondary thickening; it is considered that water enters the apoplast of such roots *via* lenticels or cracks.

There is increasing evidence that, under higher transpirational demand, the resistance of the pathway across root cell membranes and through the root symplast can be lowered owing to the activity of aquaporins (highly selective water channels in cell membranes, under metabolic control; Steudle and Henzler, 1995). Such changes not only lower the hydraulic resistance of the symplastic pathway, but also increase the overall capacity of the root for water movement to the xylem (Steudle and Peterson, 1998).

The pathway then follows the root and stem xylem into the leaf where bundle sheaths and branching networks of veins deliver water to the apoplast or symplast within a few cells of the site of evaporation. In some species, the flow within the leaf may encounter suberized cell walls, and be diverted into the symplasm (O'Dowd and Canny, 1993). Overall, most of the water in a plant, including the content of most cells, is effectively 'off-line' and not part of the pathway, but exchange does occur with living cells bordering the pathway. The participation of such 'off-line' stores of water in the hydraulic flow within the plant can be demonstrated by measuring diurnal changes in leaf thickness and stem diameter; as we shall see in Sections 4 and 5, storage of this kind can play a major role in the survival of plants in arid environments, and can contribute a substantial proportion of the daily transpiration of trees (Tyree and Ewers, 1991) and succulents.

The highest resistance to hydraulic flow within herbaceous plants appears to reside within the root system, although this can vary with demand, and the conductance of the stem can fall sharply owing to cavitation in xylem elements (Fig. 4.3) (hydraulic flow in trees is discussed in more detail in Section 5). Nevertheless, resistances to hydraulic flow within the plant are generally small compared with the resistance to the diffusion of water vapour from the leaf into the bulk air. Consequently, the rate of transpiration at a given VPD is determined by leaf diffusive resistance and, more specifically, by stomatal resistance, except in still air. The hydraulic resistance within the plant will not normally limit the rate of transpiration, but the resistance to water movement towards a root system from a drying soil can dominate plant water relations under certain circumstances, as outlined below.

A modified version of the classic drying curves of Slatyer (1967) can give some insight into the relative importance of the different resistances in the soil/plant/atmosphere system. Figure 4.9 shows changes in leaf, root surface and soil water potential during six days of soil water depletion by a *model plant whose roots have explored the entire, uniform, soil volume.* At the start of the first day, the stomata open progressively over a period of several hours, in response to external cues (solar radiation) or endogenous rhythms, causing a progressive rise in the rate of transpiration. Because of the hydraulic resistance in the plant and soil, water begins to move from the soil to the leaf only after a gradient in water potential has been established. Since the soil is at field capacity ($\Psi_{soil} = 0$), an adequate flow can be maintained without lowering leaf water potential below -0.6 MPa, thereby exposing the leaves to only mild water stress for a few hours. Because of the low hydraulic resistance of the wet soil, water flows to the root in response to a very small difference in water potential (< 0.1 MPa).

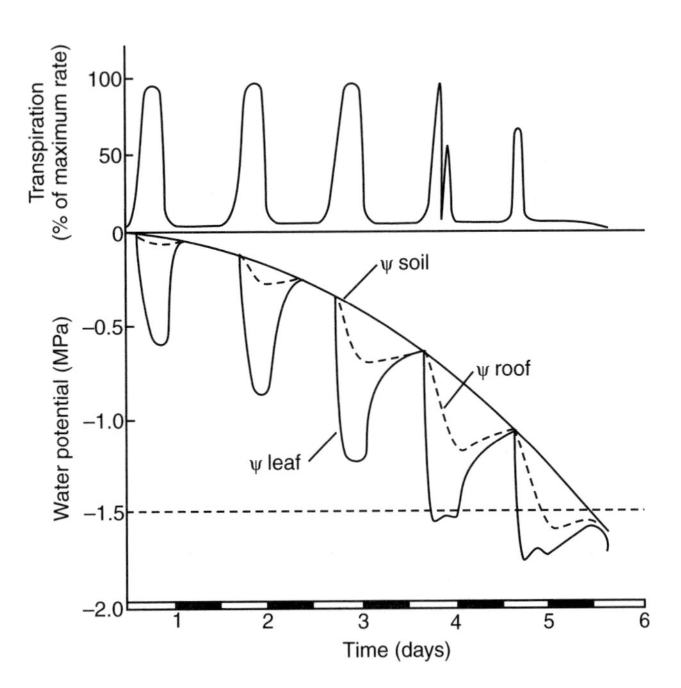

Figure 4.9

Schematic representation of the changes in leaf, root surface, and bulk soil water potentials, and in the rate of transpiration, associated with the exhaustion of the available soil water over a five day period. See text for full description (adapted from Slatyer, 1967)

In the evening of the first day, the rate of transpiration falls with progressive closure of the stomata, and the upward movement of water to the leaf begins to exceed the rate of loss. Consequently, the leaf apoplast and cells are rehydrated, and the difference in water potential between soil and leaf are abolished overnight. However, because the stored soil water has been depleted, the equilibrium soil and plant water potential at dawn on day 2 is approximately −0.1 MPa. Plant/water relations during the second day are essentially similar to those on the first except that it is now necessary to lower leaf water potential to about −0.9 MPa to maintain the gradient necessary to support the same rate of transpiration (which is determined by stomatal resistance). Meanwhile, the soil hydraulic resistance is beginning to rise as the soil dries, so that a water potential difference of 0.1−0.2 MPa is now required to maintain flow to the root surface.

At the start of the third day, the equilibrium water potential in soil, root and leaf has fallen to −0.4 MPa and it becomes necessary during the course of the day to lower leaf water potential to −1.2 MPa. The progressive increase in soil resistance has two effects: first, the potential difference between root and soil required to ensure the same supply of water as in previous days has increased to 0.3 MPa but, what is more important, the slower movement of water through the soil is now delaying the overnight equilibration of water potential. Consequently, the leaves are exposed to mild to moderate water stress during most of day 3.

These developments become more serious on day 4 when leaf water potential falls below −1.5 MPa, partly as a result of decreased xylem conductivity owing to embolism. The stomata close completely for a short period at mid-day in response to turgor effects, and the diurnal pattern of transpiration is disrupted for the first time. Finally, by the end of day 5, when the soil water potential has fallen to −1.5 MPa (PWP), there is no available water left, the plant wilts on day 6 and eventually dies if the soil is not rewatered.

This idealized account of plant/water relations during a drying cycle provides only a basic framework for considering plant/water relations. It serves to underline the general conclusion that *in the presence of available soil water*, the rate of transpiration is largely determined by stomatal resistance, whereas the hydraulic resistance of the plant determines the lowering of leaf water potential that is required for water to flow from a soil at a given matric potential, and the hydraulic resistance of the soil controls the rate of rehydration at night. However, at least three major modifications are required.

First, depending on species, habitat and growing conditions, partial closure of stomata would normally occur earlier in the cycle, at least from day 3, in response to partial loss of turgor, alteration in the ABA supply to the leaf and/or other signals (Davies and Gowing, 1999). Secondly, it appears that full overnight equilibration of water potential within the plant may not be as universal as formerly thought (Johnson *et al.*, 1991). Thirdly, in the field, soil drying does not occur uniformly as in Fig. 4.9, but from the surface downwards. Thus, as exploitation of the available soil water proceeds, with roots withdrawing water from progressively deeper horizons, root xylem water potential need not fall towards −1.5 MPa until water supply becomes limiting in the lowest horizon tapped by the root system. In fact, the pattern of root water potential displayed in Fig. 4.10 suggests that root water relations are a

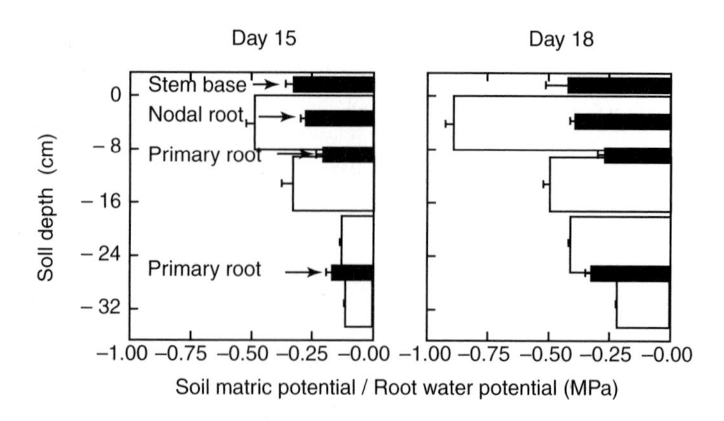

Fig. 4.10

Soil matric potentials (□) and predawn root water potentials (■) measured at different depths under a maize crop in Switzerland, after 15 and 18 days of soil drying. Horizontal bars indicate s.e.m (from Schmidhalter, 1994)

poor guide to the status of the surrounding soil (Schmidhalter *et al.*, 1998) an observation which needs to be reconciled with the proposed role of roots in signalling the onset of drought to shoots.

3. Adaptations favouring germination and seedling establishment in dry environments

It is important for plants growing in an environment where water is available for only a limited part of each year to be able to make full use of this period of favourable conditions to intercept solar radiation, accumulate dry matter and invest in reproductive development. Seeds or other propagules of adapted species tend to be ready for dispersal at the end of, or just after, the wet season, at a time when conditions may still be favourable for germination but will rapidly become very unfavourable for the subsequent growth of seedlings. It is, therefore, not surprising to find that the seeds of many successful species from drier habitats are innately dormant when first shed from the mother plant, and will not germinate until the start of the next prolonged period of favourable conditions. This is analogous to autumn dormancy which prevents the germination of temperate species until after winter (see p. 224).

For example, in tropical savannas, plant growth and development are not normally restricted by water supply from the beginning of each wet season up to 20–30 days after the end of the rains (e.g. Fig. 4.5), but thereafter the available water in the top 1–2 m of soil is progressively depleted by evapotranspiration. Seeds dispersed over the dry soil surface could experience good conditions for germination during isolated late rains (e.g. April 1972 in Fig. 4.5), but the resulting seedlings would soon run out of water. Similarly in deserts, where rainfall is generally less seasonal and predictable, it is essential that the seeds of ephemeral species do not germinate until the soil contains sufficient water to enable the resulting plants to complete their life-cycles. Seed dormancy is, therefore, not a reliable adaptation unless it is associated with a system which can signal the onset of conditions which are favourable for seedling establishment and growth (Mayer, 1986; Gutterman, 1998).

If the seasonal pattern of precipitation is reasonably regular then such a system need not be elaborate. For example, the seeds of several tropical grassland species simply require a few weeks of after-ripening before they will germinate; in their natural range, the risk of isolated late rains will normally have passed before the after-ripening period has been completed. In other species, the hilum of the drying seed can act as a one-way valve, opening in dry air to permit the loss of water vapour but closing as the humidity of the air rises; the rehydration of the seed, leading to germination, can then occur only after prolonged immersion in liquid water, which would normally signal the start of the next wet season (Koller and Hadas, 1982).

Alternatively, the need for environmental monitoring can be avoided by a 'random or opportunist strategy', which can also be effective in areas of irregular rainfall (Gutterman, 1998). For example, many species exploit the advantages of seed dimorphism, releasing varying proportions of soft-coated, non-dormant seeds and hard seeds whose coats are impermeable to water

(Kelly *et al.*, 1992). Within a given generation of seeds, the non-dormant will germinate whenever conditions permit, whether subsequent conditions are favourable for seedling establishment or not. Since the rupture of impermeable seed coats by decay, abrasion or fire (see Chapter 5) can take from a few months to several years, dormant seeds of one generation become ready to germinate over several subsequent growing seasons. To maintain the population of a perennial species, it is sufficient for the germination of only one seed of this, or another, generation to coincide with appropriate conditions for establishment.

For many species, the existence of adequate water is not a sufficient condition for germination, and other environmental factors must also be favourable. This is clearly demonstrated in desert regions where rainfall at different times of the year causes the germination of different groups of ephemeral plants, even though the full range of potential species is represented in the seedbank (e.g. in the Sonoran Desert; Shreve and Wiggins, 1964). Experiments under controlled conditions have shown that, for species native to arid zones of Australia, this effect can be attributed to differing temperature requirements for germination (Mott, 1972). Other desert species need a specific solar radiation signal before germination can begin (Koller and Hadas, 1982; Mayer, 1986; see also Thomas, 1992).

The complex interrelationships between plant and environment which can be required for successful reproduction in arid environments are well illustrated by *Mesembryanthemum nodiflorum*, an annual of the Aizoaceae from the most extreme desert areas of Israel (Gutterman, 1980/81). Although the mature seeds are released from the mother plant by wetting of the dry capsule by raindrops, germination cannot normally start until leaching has reduced the soil surface salinity, and the seeds have undergone years of primary seed dormancy. Thereafter, germination is controlled by a regular seasonal variation in innate dormancy, which ensures that only a small fraction of the seeds germinates outside the winter wet season. This endogenous rhythm, which can persist for many years, does not appear to be synchronized with the seasonal variation in temperature (e.g. Table 4.2). Furthermore, there are additional variations in germination potential, related to the position of the seed in the capsule (Gutterman, 1994), that act to spread germination over a longer period. Finally, the mechanism of dispersal, which deposits the seeds close to the, now senescent, mother plant, ensures that any resulting seedlings start life in a microhabitat which has already proved suitable for the full life cycle of the species.

With a few notable exceptions such as groundnuts and subterranean clover, the reproductive propagules of terrestrial species are scattered over the soil surface, and germination normally takes place within the surface layers of soil. Since, even in humid zones, these layers can be subject to frequent and rapid cycles of wetting and drying, it has been assumed that plants adapted to arid habitats possess features which favour successful establishment under such hostile conditions. For example, arguing from correlations between seed size and habitat characteristics for the native Californian flora, Baker (1972) proposed that large seed size was adaptive in xeric habitats: substantial stores of assimilate in seeds would be important in the early development of an extensive

Table 4.2

The germination of seeds of *Mesembryanthemum nodiflorum*, collected near the Dead Sea, Israel, between 1972 and 1974, and subjected to different temperature regimes under continuous light in the laboratory in 1978 (% of seeds germinated after 9 days) (from Gutterman, 1980/81).

Month of germination test	Constant temperature			Alternating temperature	
	15 °C	25 °C	35 °C	15–35 °C	35–15 °C
April	77	69	89	58	99
June	12	17	7	19	31
September	12	15	6	14	16
December	67	81	–	64	85

root system, exploring deeper, less variable soil horizons. It has proved difficult to confirm this hypothesis experimentally (e.g. in species adapted to semiarid areas of Australia; Leishman and Westoby, 1994).

For those species which colonize *bare* soil (burned or cultivated areas, coastal sand dunes and cliffs), seed stores are probably less important than the ability to lodge in 'safe sites' where the water relations are more favourable than the extreme surface conditions. For example, a wide range of temperate weed species produce very small seeds which are able to lodge in cracks in the soil surface where better contact can be made with soil moisture, and evaporative loss from the seed is reduced by an undisturbed, humid boundary layer (Harper *et al.*, 1965; see p. 170). In a number of species from several families (Geraniaceae, Gramineae, Ranunculaceae etc.), each dispersal unit is equipped with a hygroscopic awn or similar structure whose twisting movements, in periods of fluctuating humidity, drive the seed into the soil (e.g. Fig. 4.11). In *Avena fatua*, a common weed of temperate cereal crops, the awn first acts as a flight, enabling the seed to lodge upright in the soil, and then screws the seed into deeper layers of soil where the water supply is less variable and evaporation is reduced (Thurston, 1960). In the dry tropics and subtropics, this mechanism may also serve to place seeds at depths where they cannot be damaged by fire.

In a study of several composites, Sheldon (1974) found that the pappus ('parachute') attached to the fruit not only aids the wind dispersal of the fruit, but also ensures that it lodges, in soil cracks and crevices, with the attachment scar downwards. Consequently, good contact can be made between soil moisture and the micropile, which is adjacent to the scar, and through which water enters the fruit. In other species, the secretion of mucilage improves contact with soil water.

The rate of germination of the seeds of temperate species tends to decline significantly if the surrounding soil is dried from field capacity to a matric potential of only −0.1 MPa (e.g. Harper and Benton, 1966), whereas inhibition of germination occurs at a soil water potential which is broadly related to the water supply of the species in its natural range (e.g. Table 4.3). Thus the seeds of sorghum can germinate at lower matric potentials than can the seeds of maize (which is adapted to moister zones than sorghum), and halophytes can germinate in soils and sediments with very low solute potentials. However, the range of minimum values for successful germination is modest when compared with the matric potential of dry seeds (about −50 MPa); even at a soil water potential of −3 MPa (Table 4.3), there is still a very substantial difference in water potential between seed and soil, tending to drive water into the seed, at least at the start of seed hydration.

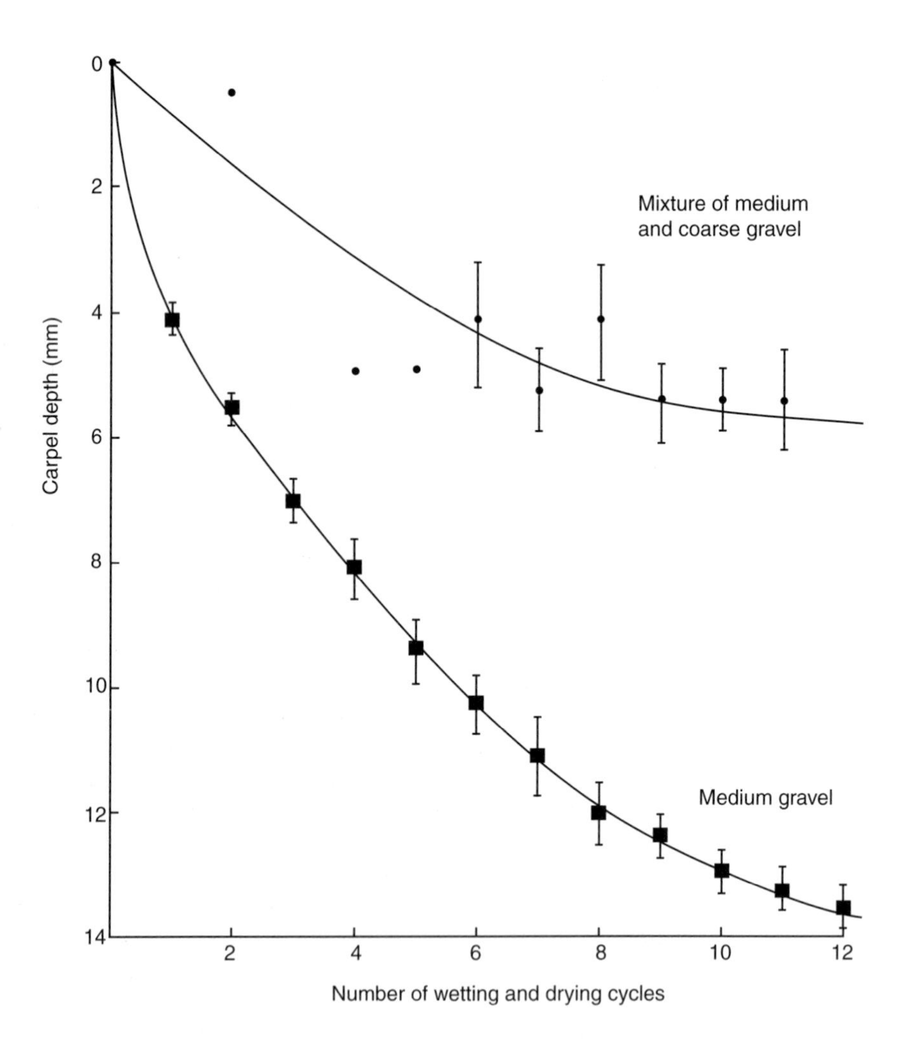

Figure 4.11

Vertical movement of carpels of *Erodium cicutarium* (Geraniaceae) into gravel substrates over twelve cycles of wetting and drying. The beak of the carpel coils and uncoils with varying humidity, driving it into the soil (redrawn from Stamp, 1984)

Table 4.3

Minimum bulk soil water potentials at which the seeds of selection of plant species can germinate (compiled from a range of sources)

Species	Water potential (MPa), mainly at 25 °C
Trifolium repens	−0.35
Pisum sativum	−0.66
Triticum aestivum	−0.80
Cicer arietinum	−1.20
Zea mays	−1.25
Sorghum bicolor	−1.5 to −2.0
Salsola kali (temperate halophyte)	−3.0

This apparent anomaly can be explained by the original experiments of Owen (1952) who found that wheat seeds could absorb water and germinate in a gaseous environment corresponding to a water potential at least as low as −3.2 MPa. This was much lower than the wheat value in Table 4.3 (−0.8 MPa) for two reasons. First, Owen's wheat seeds had access to an unlimited quantity of water whereas, in soil, the moisture content is finite and falls rapidly as matric potential is lowered (see Fig. 4.6). Secondly, the resistance to the flow of water through the boundary layer around each seed in Owen's experiment would have been much lower than that offered to the movement of liquid water through a drying soil (see p. 147). These conclusions are confirmed by Wuest *et al.* (1999), who found that, at a given soil water potential, the germination of wheat seeds was little delayed by the interposing of a barrier to the movement of liquid water (but not to water vapour). Thus it is likely that the rate of movement of water into seeds in soil is governed by the resistance to hydraulic flow *and* by the amount of water available in the soil. Since these are *soil* and not seed characters, it is surprising to find that pronounced differences in the ability of seeds to extract water do exist between species.

One possible explanation has been proposed by Hegarty and Ross (1980/81) who found that the minimum water potential for seed germination in a range of crop species was substantially higher than that for subsequent growth of the established plant, in spite of the fact that much less water was required for germination. The mechanism of this differential sensitivity to water supply between processes in the same plant is not understood, although growth substances are implicated. Nevertheless its value to plants establishing in drier environments is clear, since germination will not begin unless a substantial quantity of water is available for seedling establishment and growth.

4. Adaptations favouring survival and reproduction under conditions of water shortage

To grow and reproduce successfully in all but the most humid environments, plants must be able to survive periods of exposure to water stress varying in length from hours to years. In the driest zones (deserts, semideserts, etc., with discontinuous vegetation cover), the emphasis tends to be on survival, rather than growth and reproduction, since rainfall is episodic. For example, each individual plant of the herbaceous perennial, *Agave deserti*, sets seed once only at the end of its lifetime of 10–30 years (Schulze, 1982). Similarly conservative life cycles are found in very cold environments (see p. 222).

In moister climates, such as tropical savannas, with more reliable seasonal patterns of precipitation and complete vegetative cover, survival and reproduction can also involve intense interspecies competition for resources. For example, variation between species in the sensitivity of cell and leaf expansion to mild water stress (Fig. 4.3) could have important implications for the interception of solar radiation. The life cycles of mesophytes, the typical plants of humid areas, are not normally in phase with the seasonal variation of precipitation, but are more likely to be limited by temperature (Chapter 5); nevertheless, they do have to endure unpredictable episodes of water stress (hours to weeks) during droughts, and in dry habitats within generally moist environments (shallow soils over impermeable rock, sand dunes, cliffs etc.). Thus, the terms xerophyte and mesophyte are not precisely defined; there are few sharp discontinuities between xerophytic and mesophytic communities as we move from rainforest through savanna grassland to desert. Furthermore, apparently xeromorphic characters can turn up in some unexpected habitats (e.g. small leaves, slow fall of leaf RWC with declining Ψ_{leaf}, and resistance to high temperature, in the temperate *wetland* shrub *Erica tetralix*; Bannister, 1971).

Problems of water shortage and the maintenance of turgor are, therefore, universal among terrestrial plants. The physiological and morphological characteristics and life-cycles which have evolved in response to water deficit can be divided into three main classes (considered in turn in the following sections, although a given species can possess adaptations of each class):

(i) adaptations leading to the acquisition of the maximum amount of water (avoidance of water stress, and the amelioration of its effects);
(ii) adaptations leading to the conservation and efficient use of acquired water (amelioration and tolerance, but also avoidance in the case of those species which restrict their activities to periods of water availability);
(iii) adaptations (mainly biochemical and ultrastructural) which protect cells and tissues from injury and death during severe desiccation (tolerance).

It should, however, be emphasized that plants in arid habitats are normally exposed to a range of interacting stresses (principally water shortage, high irradiance and high temperature, as well as interference with mineral nutrition and assimilate partitioning). Thus, leaf characters and behaviour must be assessed in terms of the balance struck among conserving water, maintaining CO_2 influx and optimizing the interception and loss of radiant energy so as to avoid photochemical and/or thermal damage; increase in temperature alone tends to cause an increase in the rate of transpiration through its effect on saturation water vapour density (Table 4.1). This chapter concentrates, where feasible, on water relations, making extensive cross referencing to Chapters 2 and 5.

1. Acquisition of water

1. Root system morphology and distribution

In humid zones, plants do not generally require deep and widely-spreading root systems for water uptake because supplies of soil water are not limited and transpirational demand can be satisfied from a relatively small soil volume. Root: shoot ratios tend to be low; for example, it is estimated that roots account for only 21–25% of the total biomass of boreal/temperate coniferous forests (Lange *et al.*, 1976). In drier, tropical savanna woodland, the proportion rises to 30–40%, whereas the root systems of some prairie and desert species, extending to great depths, can represent 60–90%. However, conclusions based on simple ratios of this kind can be misleading because a variable fraction of the root system is active in water uptake, no account is taken of the contribution of mycorrhizal tissues, and the allocation of dry matter to underground organs, in cold and arid environments, can represent the recharging of carbohydrate or lipid reserves, rather than a stimulation of root growth (see p. 222). Furthermore, an extensive root system may be necessary for other purposes such as anchorage or uptake of phosphate (see p. 106). Measurement of the ratio:

$$\frac{\text{length (or surface area) of absorbing roots}}{\text{area of transpiring leaf}}$$

would be much more useful (than proportions of biomass), but it is not practicable at present to discriminate between absorbing and non-absorbing root tissues in the field (McCully, 1999).

In contrast, there is a wealth of qualitative and semiquantitative information on the distribution of plant roots in soil profiles, in relation to soil water availability, although much of the available information comes from very labour-intensive studies before 1970. For example, since much of the rain falling in deserts drains rapidly into the soil from depressions and river courses, the level of the regional water table can remain stable from one year to the next, even in very arid regions (e.g. at about 2 m in the Sonoran Desert; Shreve and Wiggins, 1964). Ephemeral annuals with shallow fibrous root systems can flourish briefly in moist depressions (Plate 9), whereas adapted perennials (phreatophytes), at low plant population density, have few roots in the hot dry upper horizons but deep, relatively unbranched root systems tapping groundwater. In a review of woody species native to mediterranean and desert areas of Israel and the Americas, Kummerow (1980) found maximum rooting depths of up to 9 m, and there have been more extreme reports (18 m for *Welwitchia mirabilis* in South Africa; 30 m for *Acacia* spp. in the Middle East). The ability of such root axes to conduct water to the shoot at reasonable rates depends on the diameter of their xylem elements – a problem analogous to the transport of water to the crown of tall trees (see p. 185). In assessing the competitiveness of such perennials, it is also necessary to evaluate the role of root axes in absorbing water from deep horizons and releasing it during the night to more superficial layers (Caldwell *et al.*, 1998).

In the North American prairies, where precipitation is higher and more seasonal, the root systems of native grassland species tend to be deep, but with profuse branching in the top 0.8–1 m of soil. For example, the maximum rooting depth of the dominant perennial grass *Agropyron smithii* varies from 1.5 to 3.6 m in the USA, and from 0.6 to 1.5 in the moister Saskatchewan prairies (Coupland and Johnson, 1965). This pattern of root distribution, which is also a feature of tropical savanna grasslands, leads to efficient absorption of water and nutrients in the superficial soil horizons during wet periods, and the extraction of stored moisture from deeper horizons during drought or in the dry season. Genera such as *Agropyron* can maintain their dominance by absorbing water from horizons below the maximum rooting depth of competing grasses. In this type of habitat, some shrubs and herbs have a root system morphology similar to that of desert plants, with long, unbranched tap roots absorbing water from below the grass roots (e.g. 1–3 m in Saskatchewan).

One important feature of prairie species is the morphological plasticity of their root systems, in response to differences in soil conditions (Kummerow, 1980). Fig. 4.12 illustrates variation in the spread and maximum depth of root systems of *Agropyron smithii* which are associated with differences in soil moisture status between habitats of similar soil type. However, results of this kind should be treated with caution since they may also reflect differences in above-ground growth. A much more remarkable example of root system plasticity is provided by the herbaceous perennial *Artemisia frigida* which relies on a deep tap root in dry soil but, under moister conditions, the tap root senesces, to be succeeded by a mass of fibrous roots in the top metre of soil (Weaver, 1958; Coupland and Johnson, 1965).

Deep rooting is clearly an important feature of established perennial plants in desert, mediterranean, prairie and savanna regions, as well as other dry habitats such as sand dunes, where groundwater is a major source of supply. However, it is difficult to understand how individual plants of such species can become established. For some, this can be by vegetative reproduction, producing rhizomes which can rely on the parent root system until their own roots have extended sufficiently to reach groundwater.

Alternatively, as discussed above, large quantities of seed can be released, a very few of which may germinate and establish under unusually wet conditions. Even so, it is essential that the first roots extend rapidly to make contact with stored water while the surface layers remain moist. This provides an explanation for selection for large seed reserves (Baker, 1972) and very high rates of root growth in some species from dry areas; for example, in their natural range in the Namibian Desert, seedlings of *Welwitchia mirabilis* can produce a tap root of 35 cm in length within ten weeks of germination (von Willert, 1985). Rapid root growth at the start of the growing season can also give a competitive advantage to annual plants. In particular, since the seminal roots of *Bromus tectorum* and *Taeniatherum asperum* can extend more rapidly at low

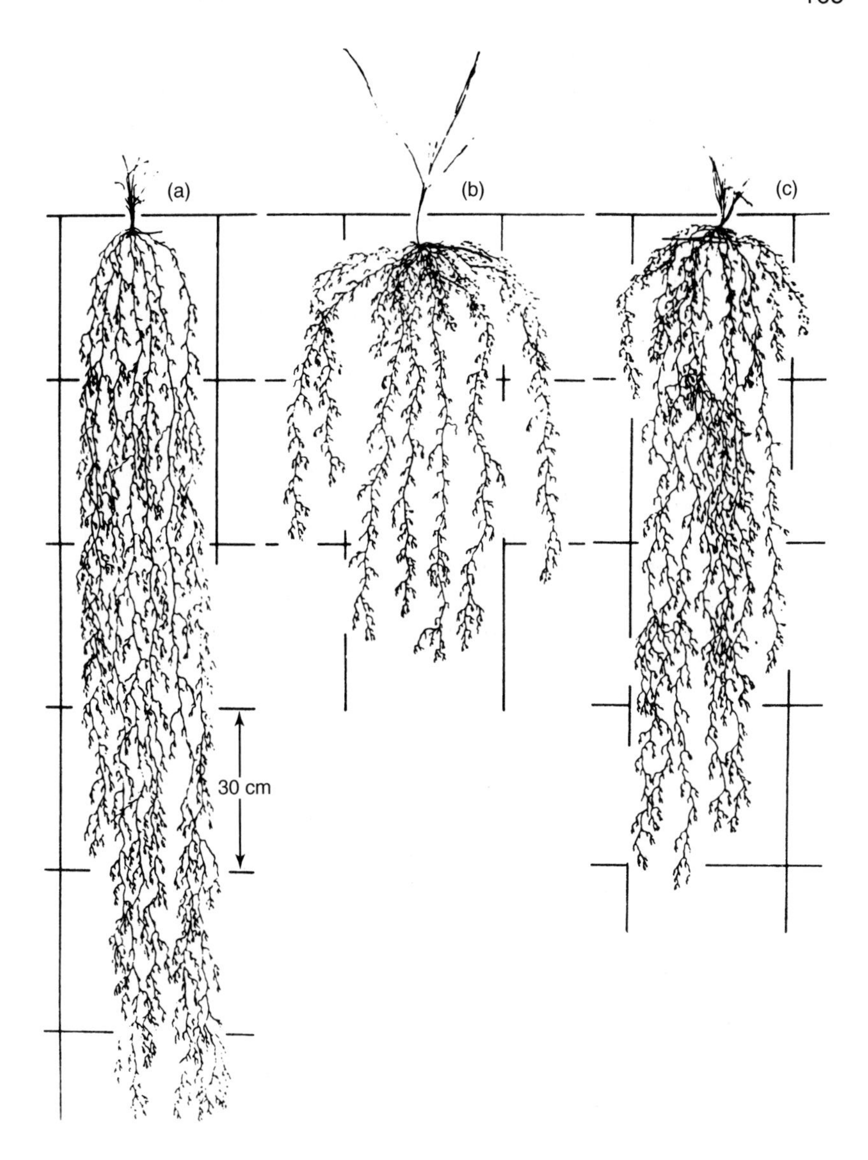

Fig. 4.12

Results from the classic era of root system evaluation in the field: root systems of *Agropyron smithii* growing in the Saskatchewan Great Plains. Within the dark brown soil zone, rooting tends to be deeper on more xeric, south-facing slopes (a) than on level sites (b); rooting is also deeper in the drier climate of the brown soil zone (c) (from Coupland and Johnson, 1965)

temperature than the roots of competitors in the western rangelands of the USA, these grasses absorb a disproportionate fraction of the limited supply of stored water in early spring by sending roots into successive horizons ahead of other species (Harris and Wilson, 1970).

2. Leaf characteristics

In areas of very low or irregular rainfall, the morphology and anatomy of some plant species have evolved to favour the interception and absorption of dew or rain directly by the shoot rather than *via* the soil (e.g. *Anthyllis henoniana* in Tunisia; Ourcival *et al.*, 1994). Although there has been considerable debate over the role of dew in the water economy and distribution of xerophytes, it does appear to contribute to *survival* in certain circumstances, possibly as a result of the temporary suppression of transpiration, and by direct leaf cooling. The collection of water by leaf canopies is, however, not restricted to plants from dry regions; in particular, the blade and sheath structure of the Gramineae and Cyperaceae, both temperate and tropical, leads to the accumulation of liquid water at the junction of the sheath with the culm, and between sheaths. More spectacularly, the extracellular leaf 'water tanks' of epiphytic bromeliads can supply plant requirements for several days (e.g. Zotz and Thomas, 1999).

In addition to the enormous variation in their root morphology and distribution, plant species can also vary in the ability of individual roots to extract water from soils. The lowest soil matric potential at which a plant can extract water is determined primarily by the most negative *leaf* water potential that it can tolerate (in order to set up a sufficiently steep gradient of water potential). This lower limit is, of course, equal to the solute potential of the leaf cells (Fig. 4.1) and because the relevant values for mesophytes, including many crop species, fall within the range −1.0 to −2.0 MPa, their roots can dry soils down to matric potentials within the same range. This is the reason for the widespread use of −1.5 MPa as the standard measure of permanent wilting point (see p. 146).

However, since the solute potentials of xerophytes and halophytes tend to be significantly lower (commonly in the range 3.0–4.0 MPa, and even lower for certain halophytes), the true PWP (which is really a characteristic of the plant and not the soil) will be much lower than −1.5 MPa; the quantity of additional water which can be extracted by their roots at matric potentials lower than −1.5 MPa will depend on soil texture and structure (e.g. Fig. 4.6). Nevertheless, there is little advantage to be gained by the plant by lowering leaf water potentials much below −4 MPa because the minute quantities of capillary water left in the soil are distributed discontinuously in the finest pores, and hydraulic resistance is very high. Where extremely low leaf water potentials are tolerated (e.g. down to −16 MPa in the xerophyte *Artemisia herba-alba*, Richter, 1976; see also *Larrea tridentata*, p. 183), they are the consequence of severe desiccation and salinity, and not an adaptation for the acquisition of more water from the soil, because the leaves cannot function actively at such low values. Conversely, some woodland herbs can absorb much less water from a given soil than mesophytic crop plants, having permanent wilting points corresponding to soil matric potentials and cell solute potentials in the range −0.5 to −1.0 MPa.

This analysis has concentrated upon the lowering of leaf water potentials required to withdraw water from drying soils, without taking account of the need to maintain leaf turgor for cell function and growth. Furthermore, for a wide range of species, the solute potential of the cell is not a fixed property of the species because the reversible accumulation of solutes (predominantly organic ions, sugars, amino acids and quaternary ammonium compounds; see Chapter 6) can result in significantly lower values. For example, arid-zone grasses have shown osmotic adjustment of up to 0.7 MPa in the field (e.g. Wilson and Ludlow, 1983), and Morgan (1984) reviewed diurnal variations in leaf cell solute potential of up to 2.0 MPa in wheat.

Osmotic adjustment (osmoregulation) is induced by exposure of the plant to water deficit over a period of days, although variation in irradiance and CO_2 concentration can affect the response. The capacity for osmoregulation, which tends to decline once a leaf is fully expanded, is not universal; grasses and cereals have provided many of the best examples in the field and laboratory, but more limited evidence suggests that dryland legumes do not show osmotic adjustment to the same degree (McNaughton, 1991). Bennert and Mooney (1979) found that for the desert halophyte *Atriplex hymenelytra* and the desert xerophyte *Larrea tridentata*, diurnal changes in solute potential ensured full turgor maintenance at leaf water potentials as low as −3 MPa, whereas the desert ephemeral *Camissonia claviformis* showed little osmoregulation and wilted at a leaf water potential of −1 MPa (see p. 184).

Clearly, the lowering of leaf solute potential by osmotic adjustment permits cells to maintain turgor and function at lower leaf water potentials but these changes appear to be associated with a reduction in cell wall extensibility (Spollen *et al.*, 1993; see equation 4.9). Thus osmotic adjustment should be seen as a short-term adaptation, whose rapid deployment can lead to prolongation of the functional life of a leaf under stress, rather than contributing to leaf expansion. Furthermore, the extent to which cells in this condition can function normally will depend on the metabolic compatibility of the secreted solutes at such high concentrations in the vacuole (e.g. Fig. 6.11).

2. Conservation and use of water

1. Water use efficiency

The assimilation of CO_2 is inevitably associated with loss of water vapour through open stomata. However, this is not a simple exchange of one CO_2 molecule for one water molecule; since the diffusion pathway for water is shorter (no mesophyll/residual component, see below) and the concentration gradient driving water out of the leaf tends to be steeper than that driving CO_2 inwards, the amount of water transpired is greatly in excess of the amount of CO_2 fixed. Raschke (1976) has estimated that, for an idealized leaf, a minimum of 20 water molecules are lost per molecule of CO_2 taken up (20 °C, 70% RH), rising to 430 at 50 °C and 10% RH; Carlson (1980) calculated a theoretical minimum of 109 molecules of water per molecule of CO_2. Ratios for real leaves can be much higher than these values.

Clearly, where water supply limits plant growth, there may be benefits to be gained from improved water use efficiency (ratio of photosynthesis to transpiration, see Fig. 4.15; Jones, 1993), although profligate use of water coupled with rapid growth and development is a possible alternative approach to competitiveness (e.g. in mediterranean species; LoGullo *et al.*, 1986). The occurrence of substantial genetic variation in this efficiency amongst plant species was demonstrated by the pioneering measurements of Shantz and

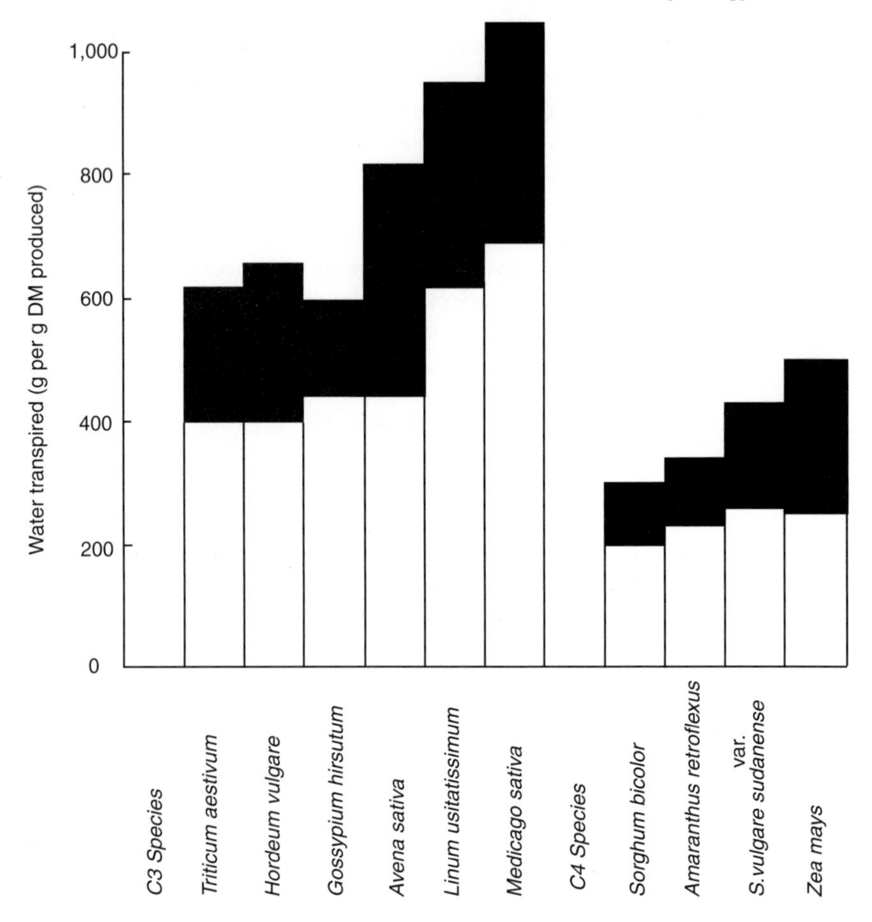

Figure 4.13

Relationships between water use and dry matter production for nine species grown in the open in pots in the dry rangelands of Colorado, in two contrasting seasons (unshaded columns: dry season, evapotranspiration = 848 mm; unshaded plus shaded columns: wetter season, 1198 mm) (from data of Shantz and Piemesal, 1927)

Piemesal (1927). The data in Fig. 4.13 demonstrated a broad relationship between water loss per unit of dry matter produced and the water supply in the natural range of the species; in particular, they gave the first indication of the profound differences in physiology between species using the C_3 and C_4 pathways of CO_2 assimilation (see p. 59, 176). The water-use efficiency of a plant, crop or plant community is the culmination of a set of (commonly-interrelated) biochemical, physiological, anatomical and morphological features rather than a single adaptation. A range of such adaptations and combinations of adaptations is examined in the following sections followed by consideration of the use of stable carbon isotope discrimination as an index of water use efficency. The implications for water use efficiency of the rise in atmospheric CO_2 concentration are considered in more detail in Chapter 2.

2. Canopy and leaf characteristics

As has been noted already, perennial species from drier habitats tend to develop deep and/or extensive root systems, giving high values of the ratio:

$$\frac{\text{length (or surface area) of absorbing roots}}{\text{area of transpiring leaf}}$$

High values of this ratio can also be achieved by reduced canopy leaf area either as a set of plastic responses to water stress (loss of turgor during expansion, resulting in smaller leaves, and early senescence; e.g. Squire *et al.*, 1984) or as a permanent morphological feature (Begg, 1980).

Xerophytes characteristically have smaller, thicker leaves than those of mesophytes, giving a higher ratio of photosynthetic mesophyll to transpiring leaf area (e.g. Abrams *et al.*, 1994). This ratio can be expressed by the specific leaf area (dm^2 of leaf area per g of leaf dry weight, see p. 46); thus, as we pass from the extremely arid Sahel zone of North Africa to more humid areas of Central Europe, the mean specific leaf area of woody perennials rises from 0.36 through 0.70 (N. Sahara) to 1.10, which is very similar to the value of 1.03 recorded in Ivory Coast rainforests (Stocker, 1976). Similarly, Larcher (1975) tabulated values greater than 1.5 for species adapted to mesic conditions (*Fagus sylvatica*, 1.4–1.6; *Oxalis acetosella*, 1.8; *Impatiens noli-tangere*, 2.2), but much lower specific leaf areas for xerophytes (*Sedum maximum*, 0.12; *Opuntia camanchica*, 0.026). The adaptive significance of such differences should be treated with caution since they are, in part, a response to temperature and irradiance (see p. 46), and in some xerophytes, one consequence of extreme succulence, as discussed below.

Although reductions in leaf dimensions and in specific leaf area tend to reduce leaf area for transpiration, such changes are also associated with the thinning of the leaf boundary layer (Fig. 4.14) and, in many species, with increased stomatal frequency. These changes tend to *reduce* both the boundary layer and stomatal resistances, resulting in potential *increases* in water loss by transpiration rather than decreases. The effect of leaf blade dissection, a feature common to plants adapted to xeric environments, on boundary layer resistance depends on whether or not the boundary layers of individual leaflets overlap and merge (see p. 48). If they do, then the leaf can benefit from a higher resistance to water loss per unit of photosynthetically active leaf area. In general, in mesic and xeric habitats, changes in leaf size, dissection and specific leaf area appear to be more important in the *thermal* economy of the leaf than in the water economy (although, of course, the specific effects of drought and high temperature stress on the plant are difficult to disentangle in the field) (see p. 231). In contrast, in cold environments, large leaves with thick boundary

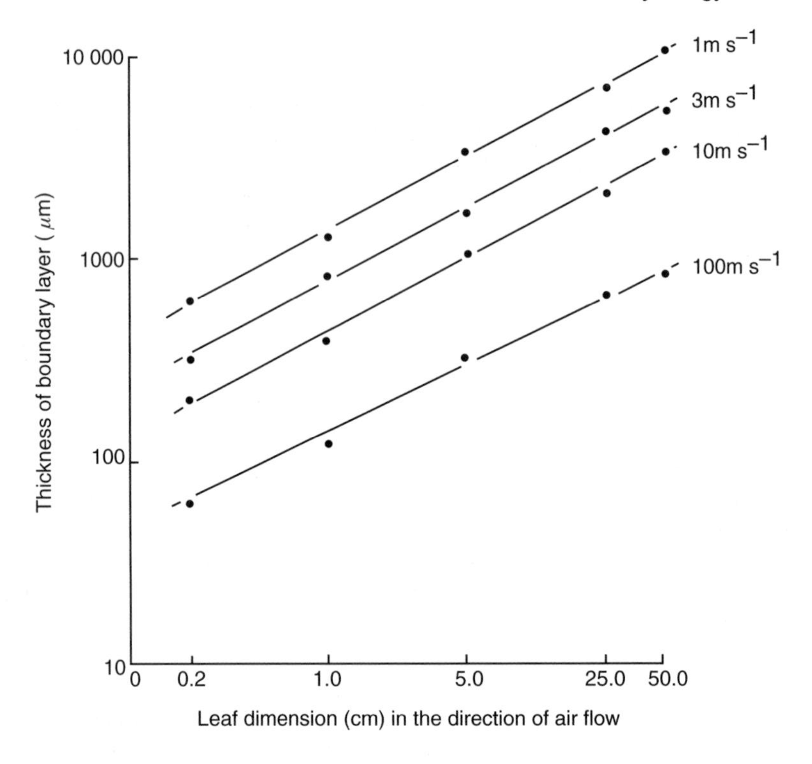

Figure 4.14

Thickness of the boundary layer over an idealized leaf as a function of wind velocity and leaf size (from data of Nobel, 1974)

layers can be beneficial to both the thermal and water economies (e.g. Geller and Smith, 1982; see p. 215).

The wide range of genotypes of *Geranium sanguineum* native to contrasting sites in Europe provides a useful illustration of the interaction between leaf morphology and temperature/water relations. As we move from moister woodland sites, through progressively drier habitats to the most xeric (shallow, well drained soils over limestone pavement), leaves of adapted populations are smaller, thicker and more dissected (Table 4.4). Similar (plastic) changes can be induced by transplanting populations from wetter to drier sites (Lewis, 1972). In parallel with these changes, stomatal frequency rises and the leaf boundary layer thins, giving substantial, but not statistically significant, reductions in leaf diffusive resistance to the exchange of water and CO_2, when the stomata are open (Table 4.4).

Although such variation in leaf morphology does not, in these habitats, appear to be of major significance in terms of water economy, the lowering of boundary layer resistance can increase the flux of heat from the leaf surface, especially when the stomata are closed and there is no transpirational cooling (see p. 194). The observed leaf characteristics will, therefore, provide some protection against the higher irradiances and temperatures experienced at drier

Table 4.4

Differences in leaf anatomy and diffusive resistance to water vapour loss, among populations of *Geranium sanguineum* from European habitats of differing water supply. Components of the leaf diffusive resistance (R) are indicated by r_a (boundary layer at 0.15 ms^{-1}), r_c (cuticle) and r_s (stomata) (simplified from Lewis, 1972).

Habitat type (in order of increasing water supply)	Leaf anatomy		Diffusive resistances (s cm^{-1})				
	Thickness (μm)	Stomatal frequency (per mm^2)	r_a	r_c	r_s	R stomata open	R stomata closed
Limestone pavement	297	279	0.20	13	0.9	1.2	13.6
Dry grassland	276	308	0.19	58	1.4	1.7	58.0
Woodland fringes	252	255	0.31	27	2.4	2.7	27.4
Woodland	237	187	0.45	14	3.0	3.3	14.4
Probability[a]	**	*	***	***	***	n.s.	n.s.

[a] Probability of significant difference between habitat types estimated by analysis of variance: *5%; **1%; ***0.1%; n.s. differences not statistically significant.

sites. For example, Lewis (1972) calculated that at midday irradiances typical of temperate zones in summer, unshaded 'woodland' leaves would be between 10 and 20 °C above air temperature compared with 2–10 °C for 'limestone pavement' leaves. Effects of this kind, explained in greater detail on page 194, would tend to reduce the photosynthetic efficiency of the 'woodland' leaf if the temperature optimum were exceeded (see p 201). Furthermore, larger, less dissected 'woodland' leaves growing in open sites under water stress (stomata closed) and high irradiance, would be exposed to leaf temperatures within the thermal death zone for this species (47–50 °C). In direct contrast, the bulky shapes and daytime closure of stomata of desert succulents appear to favour water conservation to the detriment of the thermal economy of the plant, although some protection is afforded by the high specific heat of water (see below and p. 235).

The damaging effect of water stress on actively growing plant tissues can, in the case of some perennials, be avoided by bearing leaves only when the supply of water is adequate for normal function. At the onset of prolonged drought, water loss can be reduced dramatically by leaf abscission, thereby avoiding very low leaf water potentials and xylem embolism. Responses of this kind, triggered by water stress, seasonal changes in environmental factors other than water supply, or by endogenous rhythms, can be observed in a variety of dry environments from deserts to mediterranean zones (e.g. *Salvia mellifera* in arid coastal areas of California; Kolb and Davis, 1994); however, when leaf loss is brought about directly by drought in more mesic areas (e.g. in walnut; Tyree *et al.*, 1993), it can cause premature and damaging defoliation of trees, thereby threatening subsequent growth and survival. In tropical savannas, many woody species show a regular seasonal pattern of abscission, shedding leaves soon after the beginning of the regular dry season (e.g. Fig. 4.5), but reforming the canopy several weeks before the main rains begin. Since differences in photoperiod are small, it is likely that leaf production responds to increasing temperature, with the result that the entire wet season is available for completion of the annual life cycle.

On the other hand, the seasonal dimorphism (large leaves during the wetter winter replaced by smaller leaves in the dry summer), which is a feature of dominant shrubs of plant communities of the Middle East (Orshan, 1963), facilitates photosynthetic activity throughout the year. In other species, CO_2 assimilation can continue after abscission, albeit at a lower rate, by photosynthetic stem tissues (e.g. Smith and Osmond, 1987); this can be particularly effective in those plants whose leaves use the C_3 pathway of photosynthesis during cool conditions but whose stems and other tissues can exploit the C_4 or CAM pathways during hot dry conditions (Osmond *et al.*, 1982; see p. 176).

3. Stomatal responses

Without doubt, closure of stomata is the most important process in the protection of plants from exposure to severe water stress. The rapid responses of stomatal aperture to changes in the rate of transpiration, or in the concentration of abscisic acid (see p. 149), can be considered as 'first lines of defence', protecting the leaf from dehydration even before low leaf water potentials have developed. Such responses, involving complex signalling within the plant, are not restricted to species from dry habitats, but are common to plants from deserts through mesic temperate zones to the Arctic.

A 'second line of defence' can also be deployed: closure in response to lowered leaf water potential. Many investigations have shown that stomata can remain fully open until a critical or threshold leaf water potential is reached; from this value, the aperture begins to narrow with further water loss, and closure can be complete, cutting off both photosynthetic exchange of CO_2, and water loss by stomatal transpiration, within 0.5 MPa of the threshold. There is considerable variation in both threshold and full closure values among species and growing conditions; in particular, the leaf water potential at which complete closure occurs is generally much lower for field-grown plants than for those raised under controlled conditions. At least part of the reason for this difference is that the onset of drought takes longer in the field, permitting some degree of adaptation (e.g. by osmoregulation see p. 167).

It has been proposed that some species rely on the rapid response/xylem signal strategy whereas others employ responses related to leaf turgor. It is difficult to confirm this theory without an extensive re-evaluation of critical/ threshold water potentials for a wide range of species and habitats. However, such re-evaluation may be unnecessary if, as Tardieu *et al.* (1996) claim, ABA mediates both types of response.

Threshold and full closure water potentials, and the magnitude of the difference between these values, are broadly dependent on the water relations in the natural range of the species. As extreme examples, the stomata of *Larrea tridentata* closed at leaf water potentials between -4.0 (controlled conditions) and -5.8MPa (field), whereas the corresponding values for *Vicia faba* were -0.6 and -1.0 MPa (Ludlow, 1980). In a comprehensive review of existing data,

Hsiao *et al.* (1976) proposed generalised threshold values of −0.5 to −1.0 MPa for mesophytes and −1.0 to −2.0 MPa (and lower) for xerophytes; these values remain broadly valid. Thus mesophytes, which are exposed to water deficits for relatively short periods, use stomata to conserve water and avoid the effects of moderate to severe water stress; furthermore, since partial closure has a greater influence upon water loss than CO_2 uptake (see below, Fig. 4.15, and Chapter 2), water use efficiency improves progressively between threshold and closure water potentials, albeit at a cost in terms of total CO_2 fixed.

In contrast, xerophytes must continue to assimilate CO_2 under drier conditions and for longer periods of time. Osmoregulation can make it possible for turgor to be maintained and for stomata to remain open at much lower leaf water potentials, but these characteristics must be associated with a greater ability to tolerate desiccation and the disturbance of cell metabolism by high concentrations of organic solutes. However, the relationship between leaf water potential and leaf water content is not unique, but varies widely among species and growing conditions. For example, Bannister (1971) found that the threshold water potential in the dwarf shrubs *Erica cinerea* (from dry heathland) and *Calluna vulgaris* (from moister zones within the same plant community) were very similar (−1.8 to −2.0 MPa), but that this corresponded to water contents of 88% for *Erica* and 75% for *Calluna*. Thus in *Erica*, as in other species from dry habitats, rapid lowering of leaf water potential is associated with relatively modest losses of water from the leaf tissues. In a range of South American forests, the leaves of *Ficus* epiphytes lost turgor at similar relative water contents to those of leaves of the supporting trees, but at higher (less negative) water potentials (Holbrook and Putz, 1996); such differences, which could be attributed to a higher degree of osmotic regulation and the greater wall elasticity of leaf cells of the supporting trees, appear to combine with other leaf characters in the reduction of exposure of the epiphyte to water stress.

In a photosynthesizing leaf, CO_2 molecules must traverse the leaf boundary layer and the stomatal pore before diffusing through the cell walls and plasmalemmas of mesophyll cells to the site of fixation. Because the resistance of the pathway from the substomatal cavity to the site of fixation includes a poorly characterized biochemical component, the term residual resistance is used in preference to mesophyll resistance. In contrast, there is no mesophyll/ residual component in the pathway of water loss by stomatal transpiration. Consequently, the diffusive resistance offered by the leaf to CO_2 uptake is greater than that offered to water loss, and any change in the resistance of the common part of the pathway (stomatal pore and boundary layer) will have a greater influence on transpirational loss of water than upon CO_2 uptake.

This phenomenon is clearly illustrated by the progressive improvement in water use efficiency accompanying stomatal closure (Fig. 4.15). In addition to this universal reversible effect, many species from drier habitats possess permanent morphological features which favour photosynthesis over transpiration by increasing stomatal resistance; these include various methods of extending the effective length of the stomatal pore (pits, antechambers, chimneys). In conifers, stomatal resistance is increased by occlusion of the pores with loosely packed plugs of wax. Adaptations of this kind, of course, also

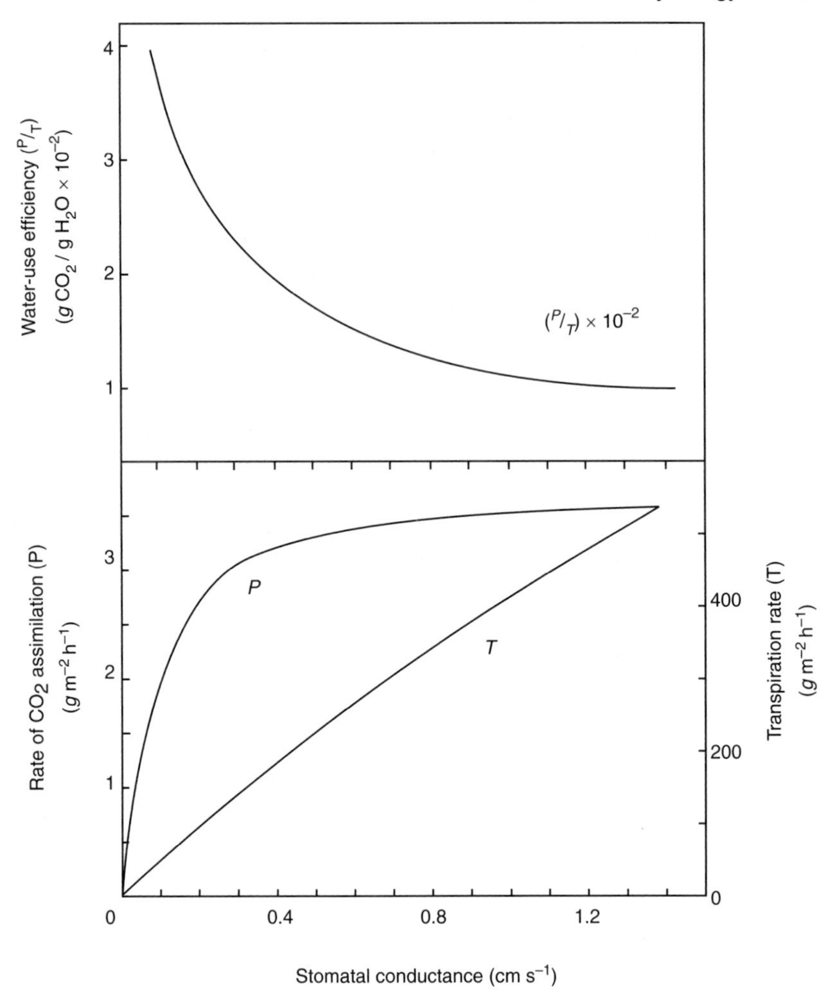

Figure 4.15

Calculated effects of variation in stomatal conductance on photosynthesis, transpiration and water use efficiency of leaves of *Xanthium strumarium* (conditions of original measurements: VPD 1kPa, air temperature 20°C, *c*1900 $\mu mol\,m^{-2}s^{-1}$ total radiation, boundary layer conductance 10 $cm\,s^{-1}$, CO_2 concentration 350 $cm^3\,m^{-3}$) (Adapted from Raschke K (1996) How stomata resolve the dilemma of opposing priorities, Philosophical Transactions of the Royal Society of London **B273**, 551–560. ©1976, with permission of the Royal Society of London.)

reduce the influx of CO_2, and any improvement in water-use efficiency is gained at the expense of net photosynthesis and dry matter production.

Evaluation of the adaptive significance of *stomatal frequency* is complicated by the fact that it is sensitive to a number of environmental factors (principally irradiance, temperature, and CO_2 concentration) during leaf development. Furthermore, if the mean cell size of any species (mesophyte or xerophyte) is reduced by loss of turgor during expansion, then the resulting leaves will tend to have a *higher* stomatal frequency than those growing in the absence of stress because the number of potential guard cells will be unaffected. As illustrated by Peat and Fitter (1994b), stomatal frequency tends to be broadly related to the degree of shade in the natural habitat of a species (Fig. 4.7), but there are some differences in stomatal frequency between groups of plants which appear to be adaptive in relation to water supply; most notably, CAM plants, from the most arid of environments, have frequencies that are approximately one-tenth of those for C_3 and C_4 plants (Osmond *et al.*, 1982; see p. 176).

Once stomatal closure is complete, CO_2 uptake and stomatal transpiration cease but water loss continues, at a lower rate, through the cuticle. Consequently, for plants growing or surviving on limited water supplies, it is important for water loss through the cuticle to be kept to a minimum. Thus many xerophytes have cuticular resistances in the range 60–400 $s\,cm^{-1}$ compared with typical values of 20–60 $s\,cm^{-1}$ for mesophytes; without such high resistances, attributable to thick layers of cutin and additional layers of waxes, most desert species could not survive prolonged droughts lasting from months to years.

4. Other xeromorphic characteristics

Features which are characteristic of certain groups of plants from hot, dry environments include the following.

Leaf pubescence. Depending on their location, frequency, dimensions and colour, leaf hairs can affect the water and energy relations of plant tissues in different, and sometimes conflicting, ways. For example, as explained in Chapter 5 (p. 235), the observed correlation between the density of leaf pubescence and the severity of the habitat (temperature, water deficit) in the native range of perennial shrubs of the genus *Encelia* can be interpreted in terms of the fraction of incident solar radiation which is reflected by the leaf surface. However, increase in pubescence is also associated with increased boundary layer resistance to the loss of water and heat from the leaf, and to the uptake of CO_2 by the leaf. Protection of the leaf from photochemical and thermal damage by increased reflection of solar radiation can, therefore, be at a cost in terms of rate of net photosynthesis; in this context, it is important to note that the highly pubescent white leaves of *E. farinosa* are replaced by green, less-pubescent leaves under wet conditions (Ehleringer, 1980). The clustering of hairs round stomatal pores can increase effective stomatal resistance with less influence on the overall energy relations of a leaf, but it should be noted that, in some species, pubescence may have evolved as a protection against herbivory.

Adjustment of leaf canopy properties. Examples include leaf rolling, which tends to reduce water loss and interception of radiation by one surface of the leaf. Thus a combination of leaf rolling and thick epicuticular waxes, rather than more effective osmoregulation, is responsible for the dominance of *Andropogon halii* over *A. gerardii* in drier areas of the bluestem complex of the Nebraska sandhills (Barnes, 1985). The facultative or obligate adjustment of leaf angle, which determines the fraction of solar radiation that is intercepted (see p. 234), can also have implications for the water economy of the plant. For example, experimental variation of the orientation of the prairie compass plant *Silphium lacinatum* showed that the highest water use efficiency was achieved when the leaf laminae faced east and west, as in nature (Jurik *et al.*, 1990).

Storage of water. Water is stored in a wide variety of tissues: bulbs, tubers, swollen roots, tree trunks (e.g. baobabs), and the stems and leaves of succulents (Plate 9). For example, many herbaceous dicotyledons of savanna zones survive prolonged dry seasons as leafless underground tubers, whereas the growing points of grasses survive above ground, protected by senescent leaf tissues. Storage of water is, therefore, not a universal feature of plants of arid zones; in a study of three perennial species at the same desert site in California, the available stored water could, potentially, support maximum rates of transpiration of the CAM succulent *Agave deserti* for 16 h, compared with only 7 min for the C_3 shrub *Encelia farinosa* and 4 min for the C_4 grass *Hilaria rigida*. However, the difference between *A. deserti* and the other species is at least partly the consequence of a much lower rate of transpiration (Nobel and Jordan, 1983).

5. Alternative pathways of CO_2 assimilation

Because of the ability of their leaves to concentrate CO_2 in bundle sheath cells, plants using the C_4 pathway of photosynthetic CO_2 fixation can maintain lower intercellular concentrations, and steeper gradients of CO_2 concentration between bulk air and substomatal cavity than can C_3 plants (see Chapter 2 for a full account of the alternative pathways). Consequently, optimum fluxes of CO_2 into the leaf can be maintained at higher stomatal resistances (smaller stomatal apertures) than in C_3 plants, leading to higher intrinsic efficiency of water use by C_4 plants (see Fig. 4.15). Furthermore, temperature optima for C_4 photosynthesis tend to be higher, and the photosynthetic apparatus of many C_4 species is not saturated at full midday irradiance (e.g. over 2500 $\mu mol\, m^{-2} s^{-1}$ total solar radiation; Fig. 2.16), whereas many C_3 species, including most crop plants, are saturated within the range 900–1350 $\mu mol\, m^{-2} s^{-1}$. These characteristics suggest that C_4 plants in general should be well adapted to hot dry habitats receiving high levels of irradiance – a hypothesis supported by the observation that the classic C_4 grasses are native to the dry tropics and subtropics.

However, the possession of a higher water-use efficiency can serve only to extend the period of availability of water, and may not be an advantage in competition with more profligate users of water. Unless a species has evolved to *avoid* water deficit, its success in a dry environment involves the survival of low tissue water potentials; possession of the C_4 pathway *alone* does not confer an overwhelming advantage on tissues exposed to severe water stress, and it must be associated with appropriate life cycles, morphologies etc. Furthermore, high optimal temperatures are of value only if the entire photosynthetic apparatus is adapted to high temperatures over prolonged periods; for example, Table 5.5 indicates that photosystem II may be more susceptible than the apparatus as a whole. Finally, examination of a wide range of species has revealed C_4 species whose photosynthesis saturates at one tenth of full sunlight, and C_3 plants which do not saturate within the terrestrial range of irradiance values (e.g. Osmond *et al.*, 1982; Pearcy and Ehleringer, 1984; Smith *et al.*, 1997).

Reservations about the adaptive value of C_4 photosynthesis in dry habitats (e.g. Ehleringer and Monson, 1993) have been reinforced by detailed studies of the relative distribution of C_3 and C_4 plants. The progressively-decreasing proportion of C_4 species in N. American grass floras from the arid south-west to cool moist areas (Fig. 4.16a), or in E. Africa with increasing altitude, does suggest that possession of the C_4 pathway confers an advantage as the climate becomes hotter and drier. Nevertheless, multiple regression analysis has revealed that the distribution shown in Fig. 4.16a is determined primarily by night temperature, since most C_4 plants are chilling-sensitive (see p. 206). Furthermore, although dicotyledons show a similar trend (Fig. 4.16b), C_4 plants represent a very small proportion of dicotyledons throughout N. America (for example there are few C_4 woody plants); and when the total number of all plant species is evaluated for many dry areas of the world, C_3 species tend to predominate (Smith *et al.*, 1997; Ehleringer *et al.*, 1999). By contrast, the C_4 pathway has appeared unexpectedly in warm shaded habitats, such as the understorey of tropical rainforests (e.g. Osmond *et al.*, 1982; Skillman *et al.*, 1999).

It can, therefore, be concluded that possession of the C_4 pathway is by no means indispensable for success in, or restricted to plants native to, arid environments (for example, compare the C_3 and C_4 species of the Sonoran Desert, p. 183). On the other hand, the CAM pathway (strictly the stomatal CAM pathway; Cockburn, 1985) of photosynthetic CO_2 fixation (see p. 59) is virtually restricted to plants native to hot, dry environments (desert succulents) or dry habitats/seasons within otherwise mesic environments (e.g. the moist tropics, Fig. 1.9; temperate succulents, such as *Umbilicus rupestris* growing on cliffs or walls with very restricted rooting volume, Daniel *et al.*, 1985).

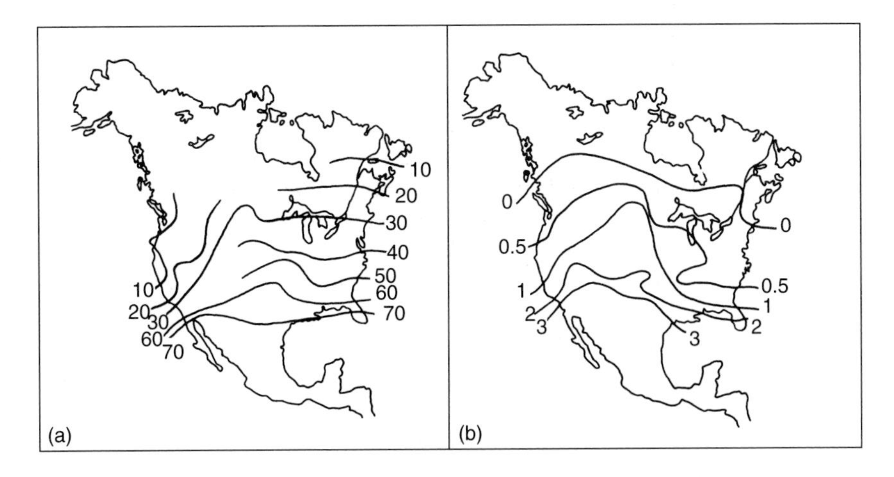

Figure 4.16

'Contour' lines indicating (a) the percentage of grass taxa, and (b) the percentage of dicotyledon taxa, which are C_4 plants, in different parts of North America. Constructed from the regional floral data of Teeri, J.A and Stowe, L.G. (*Oecologia* **23**, 1–12 (1976) and *American Naturalist* **112**, 609–623 (1978))

CAM is also, with few exceptions, associated with a range of other adaptations that facilitate the continuation of photosynthetic CO_2 fixation under drought (Ehleringer and Monson, 1993): a degree of succulence (providing storage for water and organic ions, and thermal buffering); nocturnal opening of stomata (admitting CO_2 at lower bulk air VPD than during the day, conserving tissue water); very high cuticular resistance to water loss; and relatively high critical leaf water potentials (*c*−1.0 MPa) for stomatal closure. Furthermore, a high proportion of above-ground tissue is photosynthetic, counteracting to a certain extent, low rates of net photosynthesis (typically 5% of the rate of C_4 plants per unit leaf area under the same conditions). On the other hand, since their root : shoot ratios tend to be low, desert CAM plants are generally unable to exploit groundwater, and must rely on current precipitation or water stored within their tissues (see Graham and Nobel, 1999; Ehleringer *et al.*, 1999). This, and the low rates of convective cooling of their bulky tissues, is presumably the reason for their absence from the most arid habitats within deserts.

There are few obligate CAM species; most are facultative, fixing CO_2 by the C_3 pathway under favourable conditions but switching to CAM within a few days of the onset of water stress (caused by drought or salinity) or in response to a critical photoperiod or temperature signal which marks the start of dry conditions in a zone of regular wet and dry seasons (e.g. *Sedum telephium;* Borland, 1996). Some species rely more heavily on CAM as ontogeny proceeds (Cockburn, 1985).

Although some CAM plants can be managed to give high rates of dry matter production (e.g. Nobel *et al.*, 1992), the life cycles of plants which rely heavily on the CAM pathway for net photosynthesis are essentially conservative, making use of the available water for survival, protecting the photosynthetic apparatus, and growing very slowly; successful reproduction by seed may occur only after a period of unusually heavy rainfall. On the other hand, the switch from C_3 to CAM at the end of a wet season can extend the growing season sufficiently to permit the completion of the annual life-cycle. In some investigations, the induction of CAM is accompanied by very low rates of net fixation of CO_2, although the plant appears to be recycling CO_2 released by respiration: such 'idling' or 'cycling' (depending on the degree of stomatal opening) has been interpreted as a means of conserving carbon and water, while at the same time protecting the photosynthetic apparatus from photoinhibition (Cockburn, 1985; Daniel *et al.*, 1985; Borland, 1996). In summary, the possession of alternative pathways confers a biochemical flexibility, which can be invaluable in irregularly fluctuating environments; the biochemical details of this flexibility are now being revealed with the use of molecular biological tools (e.g. Cushman and Bohnert, 1999).

6. Ontogeny

The ability of plants to use water efficiently and to avoid the damaging effects of water stress tends to vary with ontogeny. In particular, most species are much more sensitive to drought at the start of the reproductive phase of development than during vegetative development. The reasons for this phenomenon include the large leaf areas carried by plants at the end of canopy development, the diversion of assimilate from roots to developing fruit, and, in monocotyledonous species, the severe disruption of the stem vascular system associated with internode elongation. Thus Potvin and Werner (1983) found that for co-occurring *Solidago* species, successful adaptation to dry habitats was associated with a prolonged vegetative phase and a rosette habit. The very rapid development and maturation of desert ephemerals can be interpreted as adaptations favouring the completion of reproduction before the water supply has been exhausted; many desert perennials (e.g. *Agave deserti*) demonstrate the opposite approach, remaining vegetative throughout most of the life cycle (see p. 184).

7. Stable isotope discrimination and water use efficiency

The water economy of plants, stands and communities has traditionally been measured and expressed in terms of long-term water use per unit of dry matter produced (e.g. Fig. 4.13) or short-term CO_2 fixation per unit of water transpired (water use efficiency; e.g. Fig. 4.15) (Ehleringer *et al.*, 1993). However, stable carbon isotope discrimination by plants can also be a powerful tool in assessing the water economy of certain groups of plants.

Of the two naturally-occurring stable isotopes of carbon (^{12}C, ^{13}C), the lighter is by far the predominant in atmospheric CO_2 (98.9%). This predominance is, in general, even greater in plant tissues because of a discrimination against the ^{13}C isotope by the plant. In particular, discrimination in favour of the ^{12}C isotope occurs during diffusion of CO_2 into the leaf and during photosynthetic fixation of CO_2 by Rubisco (Chapter 2), although discrimination is weaker during the fixation of CO_2 by PEP carboxylase (compared with Rubisco) (Farquhar *et al.*, 1989). Consequently, expressing discrimination against ^{13}C by plant tissue by:

$$\delta^{13}C = f(\text{isotope abundance in plant/isotope abundance in air}) \qquad (4.13)$$

most C_4 species give low values (around 14×10^{-3} or 14‰, with little variation among species and sites) compared with those for C_3 plants (commonly in the range 20–30‰). $\delta^{13}C$ values for CAM plants vary according to the extent to which they rely upon Rubisco or PEP carboxylase for the primary fixation of CO_2 (O'Leary, 1988; Borland *et al.*, 1994).

Since

$$\delta^{13}C = a + (b - a)p_i/p_a \qquad (4.14)$$

where a is discrimination against ^{13}C owing to diffusion; b is the net discrimination owing to carboxylation; and p_a, p_i are the partial pressures of CO_2 in the bulk air and intercellular spaces respectively (Farquhar *et al.*, 1989), $\delta^{13}C$ is also an index of intercellular concentration of CO_2 in the leaf. It can, therefore, be considered as an index of the potential water-use efficiency of the plant since, as p_i falls, a given flux of CO_2 can be maintained at a smaller stomatal aperture (Fig. 4.15). This explains the observation that, whereas there is little variation in $\delta^{13}C$ amongst C_4 plants (which generally maintain very low levels of p_i), by no means all C_3 plants give values within the above-expected range; populations and species adapted to dry habitats commonly give values of 10–20‰ (O'Leary, 1988; Fig. 4.17), which can be explained by unusually low intercellular CO_2 concentrations (Farquhar *et al.*, 1989). Values in this range are common for *communities* (e.g. 12.9 for xerophytic woody vegetation; 14.2 for C_3 plants in semideserts; 18.0 for paddy rice; and 19.6 for cool and cold deciduous forest; Lloyd and Farquhar, 1994). The scope of isotope discrimination in evaluating plant water economy has yet to be fully explored, but it is already clear that within-population differences in $\delta^{13}C$ of as little as 2‰ can affect the competitiveness and survival of perennial species in arid environments (e.g. *Encelia farinosa*, a C_3 perennial shrub in the Sonoran Desert; Ehleringer, 1993b). Carbon isotope discrimination is also widely used to evaluate the water-use efficiency of crop plants (e.g. Knight *et al.*, 1994; Henderson *et al.*, 1998).

3. Tolerance of desiccation

In many environments, adaptations such as those outlined in the preceding two sections can be sufficient to prevent native species being exposed to more than moderate stress, and relatively modest losses of tissue water. In contrast, species

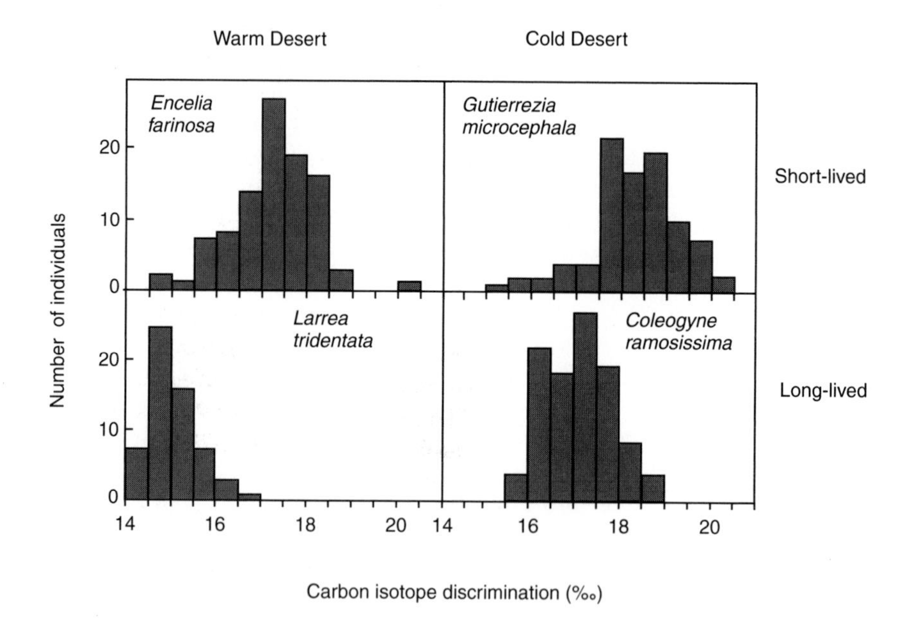

Figure 4.17

Frequency distribution of carbon isotope discrimination values of populations of C_3 perennial species native to the Colorado Plateau (cold desert) and Sonoran Desert (warm desert). These values of $\delta^{13}C$ from dry environments are generally lower than for C_3 mesophytes; there is substantial intraspecific variation in $\delta^{13}C$; and longer-lived species tend to have lower values of $\delta^{13}C$, possibly reflecting the benefits of superior water-use efficiency (Ehleringer, 1993a). (Reproduced from Smith, J.A.C. and Griffiths, H. (Eds) (1993) Water Deficits, with permission of BIOS Scientific Publishers, Oxford)

Plate 1

Four individual *Pinus sylvestris* from the Caledonian Pinewood, Loch Maree, Scotland, showing the diversity of form generated by the interactions of genotype and environment.

Plate 2

(a) Large sunfleck under the canopy of a *Sequoia sempervirens* forest in California. (b) and (c) Temperate deciduous forest in lowland England before and after the expansion of the leaf canopy.

Plate 3

(a) Aerial view of FACE (Free Air CO_2 Enrichment) rings in a *Liquidambar styraciflua* forest at Oak Ridge, USA. (b) Closer view of the FACE rings showing the vents from which CO_2 is released. Photographs courtesy of Dr R Norby.

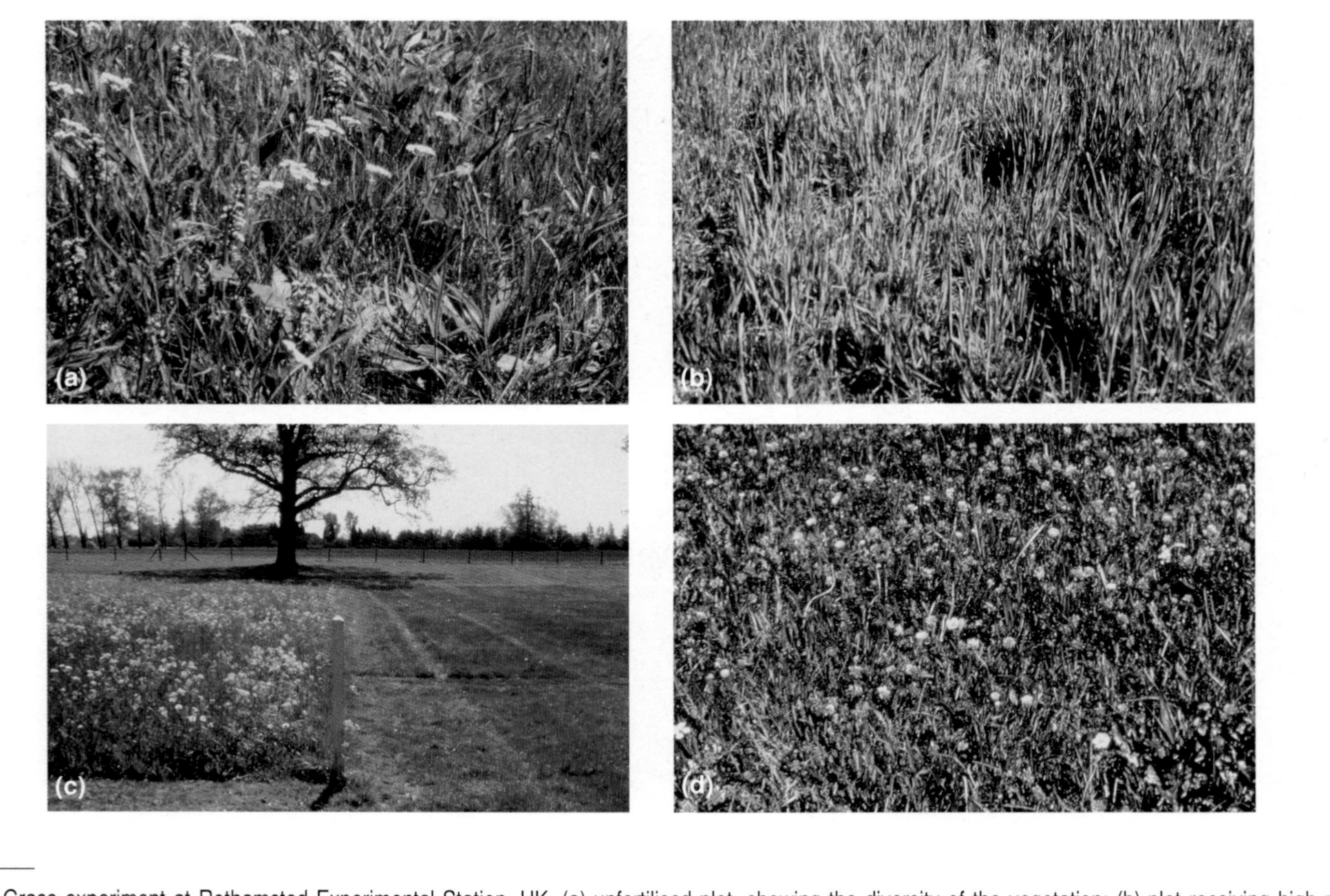

Plate 4

The Park Grass experiment at Rothamsted Experimental Station, UK. (a) unfertilised plot, showing the diversity of the vegetation; (b) plot receiving high rates of ammonium fertiliser (twice that in (c), right side) and no lime, now very acid with complete dominance by the grass *Anthoxanthum odoratum*; (c) boundary between productive, but still diverse plot that receives complete nutrients with N as nitrate (left), and unproductive, acid plot that receives N as ammonium (right); the right-hand plot is dominated by *Agrostis*. Neither plot is limed; (d) plot receiving high phosphate but no nitrogen, dominated by red clover *Trifolium pratense*. Photographs courtesy of ICAR Rothamsted and Dr Paul Poulton.

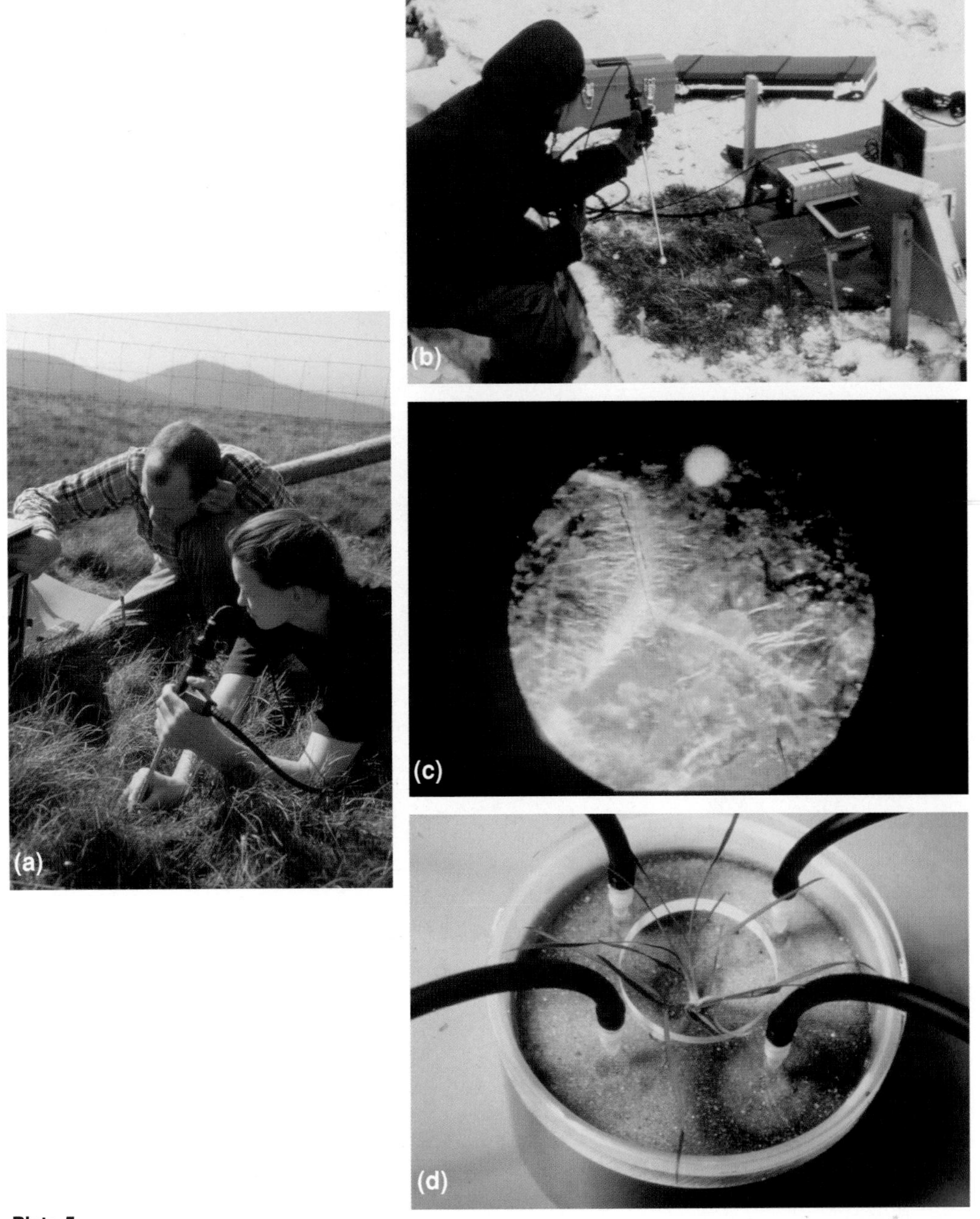

Plate 5

Experiments and techniques for studying roots. (a) and (b) Mini-rhizotron equipment for capturing images of roots in soil allows continuous and long-term recording of root growth under all conditions. (c) Typical image of roots captured by a mini-rhizotron (note the prominent root hairs); image represents *ca* 7 mm diameter view. (d) Design used by Campbell *et al.* (1991) to study root foraging, showing algal growth on the nutrient-rich quadrants. Photographs courtesy of Dr G Self (b), Dr Angela Hodge (c) and Professor Philip Grime (d).

Plate 6

Chemical changes in soils induced by roots. (a) (right hand panel) pH changes around the roots of a maize plant grown with either nitrate or ammonium as the sole nitrogen source, demonstrating strong acidification with ammonium and a marked increase in rhizosphere pH when nitrate is used; (left hand panel) Similar changes observed when sorghum and chickpea take up nitrate from soil. The chickpea acidifies the soil because it is N-fixing and therefore metabolizes ammonium rather than utilizing soil nitrate. (b) Oxidation of the rhizosphere by the roots of *Phragmites australis* in solution culture, as revealed by oxidation of methylene blue (and in soil by the precipitation of iron oxide on the root surface). Photographs courtesy of (a) Dr V Römheld, and (b) Dr J Armstrong and Mr D Holt.

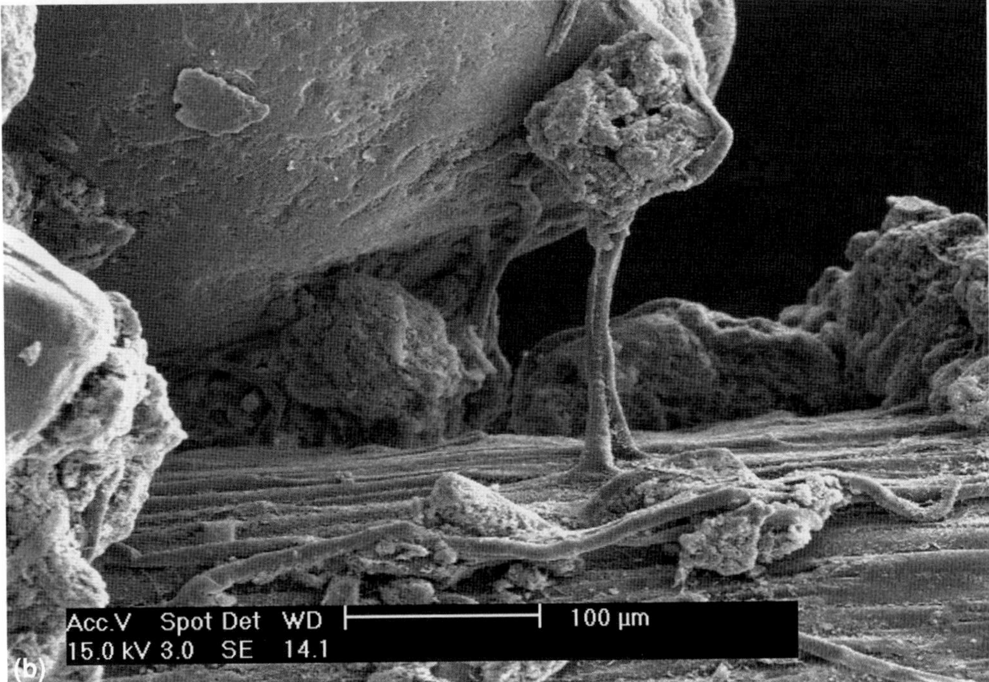

Plate 7

The rhizosphere. SEMs of a barley root in soil, showing (a) soil particles adhering to the root surface, and prominent fungal hyphae; (b) root hairs bridging the gap between the root surface and a soil particle. Photographs courtesy of Dr Colin Campbell.

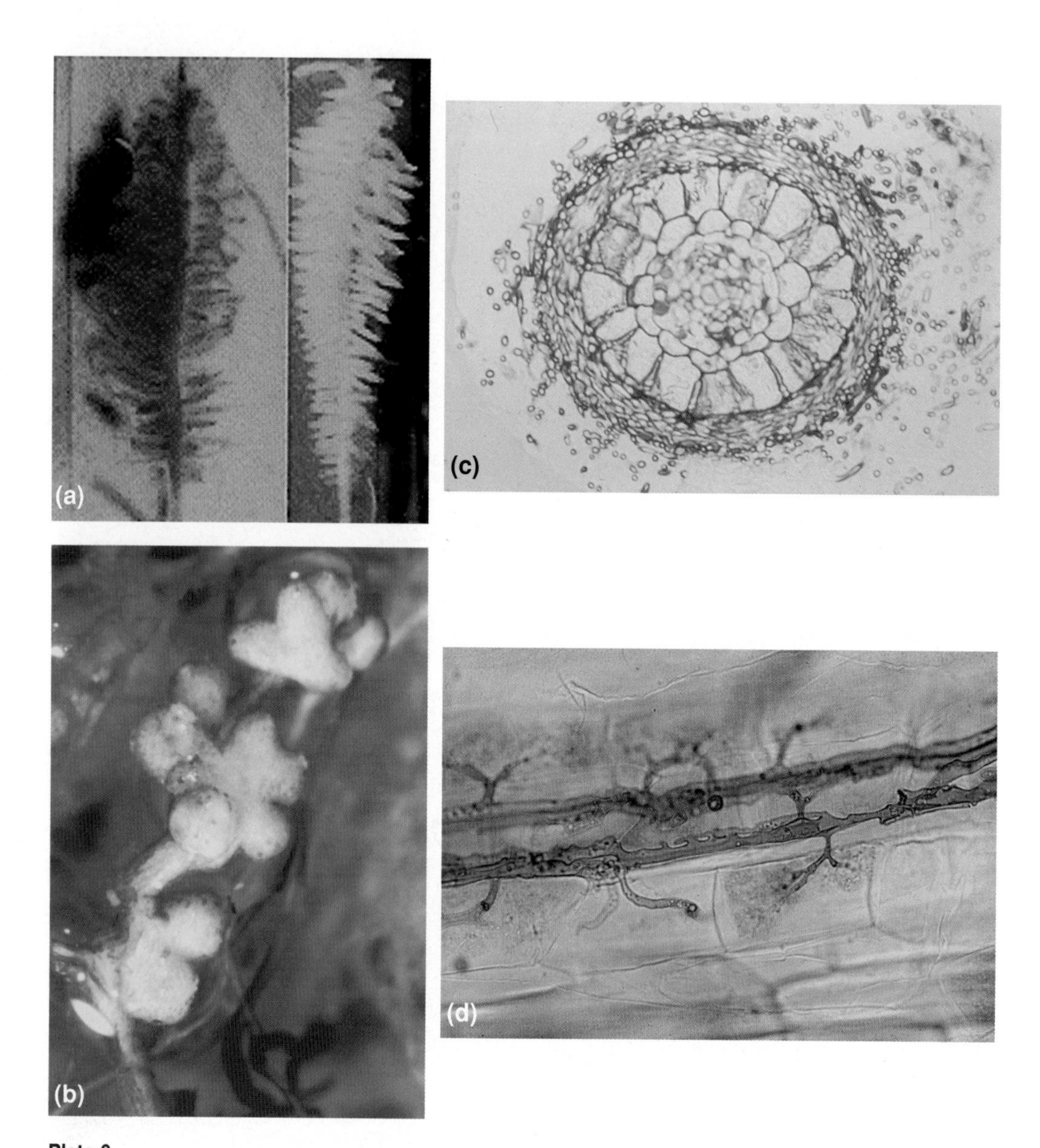

Plate 8

Roots and symbioses. (a) Cluster roots of *Grevillia robusta*; (b) nodules formed by the nitrogen-fixing bacterium *Rhizobium* on the roots of *Robinia pseudoacacia*; (c) cross-section of an ectomycorrhizal root of beech *Fagus sylvatica*; (d) intercellular hyphae and arbuscules of an arbuscular mycorrhizal fungus in a root of bluebell, *Hyacinthoides non-scripta*. Photographs courtesy of (a) Dr Keith Skene, (b) Dr Karyn Ridgway, (c) Dr Angela Hodge, (d) Dr James Merryweather.

Plate 9

Habitats in which water management is important. (a) The cactus *Brachycereus nesioticus* on lava, Isla Fernandina, Galapagos, dependent on water storage; (b) desert in New Mexico after rain, showing the ephemeral and deciduous flora flowering, with surrounding succulents; (c) and (d) giant trees which must raise water to >70 m: c, karri *Eucalyptus diversicolor* in the relatively arid south-west of Australia; d, Douglas fir *Pseudotsuga menziesii* in the temperate moist forest of Vancouver Island, Canada.

Plate 10

Plants of Death Valley, California. (a–c) *Larrea tridentata* ssp *divaricata*: (a) aerial view showing evenly-spaced individuals; (b) single shrub; (c) flowering; (d) *Tidestromia oblongifolia*. Photographs 10(b–d) courtesy of Dr Stanley Smith.

Plate 11

Desiccation resistance. Hydrated and air-dry states of the resurrection plant *Craterostigma wilmsii*. Photographs courtesy of Professor Jill Tarrant.

Plate 12

Regeneration of Australian plants after fire. (a) *Eucalyptus* shoots developing from epicormic buds on the woody stem; (b) vigorous growth of understorey vegetation after fire (note the fire-blackened trunks)

Plate 13

Treeline. (a) Natural treeline in the Canadian Rockies at *ca* 2500 m; (b) Norway Spruce (*Picea abies*) near the treeline after heavy snowfall, Southern Norway; (c) a tree at the treeline on Mount Fuji, Japan, showing the lower skirt that develops under the snowpack, and the flagging effect caused by windblown ice particles.

Plate 14

Alpine plant growth forms. (a) Giant rosettes of *Espeletia lopezii* in the Sierra Nevada de Cocuy, Colombia; (b) *Oxyria digyna* flowering at 600 m in the West Highlands of Scotland; (c) and (d) *Silene acaulis* and *Crepis nana* at *ca* 3000 m in the Canadian Rockies, demonstrating cushion growth form. Photograph 14(a) courtesy of Dr Jason Rauscher

Plate 15

Habitats where toxicity is an important environmental factor. (a) strand-line, Isles of Scilly, England (salinity); (b) Salt flats, Coto Donana, Spain (salinity); (c) acid bog, Ontario, Canada (low pH, waterlogging); (d) heather moor, Cumbria, England (low pH, aluminium).

Plate 16

(a) Copper-tolerant *Mimulus guttatus* flourishing on severely contaminated soil in California. (b) Severely contaminated ground after copper mining at Parys Mountain, Anglesey, Wales. Photograph 16(a) courtesy of Professor Mark Macnair.

adapted to arid zones experience severe water stress as a normal feature of their life cycles, as a consequence of prolonged water loss by cuticular transpiration, or of rapid water loss through stomata held open by osmoregulation. The survival of such plants depends on the ability of their cells to tolerate desiccation.

The desiccation of plants has been less thoroughly studied than other aspects of plant/water relations, primarily because it is not of great significance in agriculture; few crops are grown where there is a consistent risk of plant desiccation and serious loss of yield. However, it is also an intrinsically difficult subject for research since it should involve the simultaneous study of whole plant physiology, cell ultrastructure and function, and a range of metabolic pathways. Furthermore, it can be difficult to establish the primary cause of cell death when so many potentially lethal processes are taking place in the same dehydrated cell: disruption of metabolic pathways; denaturation of macromolecules; failure of membranes and membrane-bound functions; gross damage to organelles; mechanical damage to the protoplast as a whole.

Although it is clear that there are substantial differences among plant species in their resistance to injury by dehydration, little progress has been made in quantifying such differences because of uncertainties in establishing appropriate indices of dehydration 'stress' and the resulting 'strain'. In pioneering studies, Barrs (1968) proposed that the level of dehydration should be expressed by a combination of Ψ_{leaf} and relative water content of the leaves (recognizing that stress-resistant plants can give large depressions of Ψ_{leaf} in response to modest losses of water); and Parker (1972) monitored injury by assessing cell survival using tetrazolium dyes. Subsequent studies have involved assessment of the integrity of membrane-bound systems (e.g. photosystem II using chlorophyll fluorescence; Eickmeier *et al.*, 1993). However, because detailed studies of both stress and strain are rare, it is customary to fall back on minimum recorded leaf water potentials (Ψ_{min}) or the lowest relative humidity which the plant can tolerate, in order to establish relative differences in desiccation resistance between groups of species. For example, Table 4.5 shows that, in general, species from xeric sites are exposed to, and can survive, much lower values of Ψ_{min} than those from more mesic sites.

Table 4.5

Minimum recorded leaf water potential values for plants native to different habitats (from a range of sources)

	Ψ_{min}(MPa)
Resurrection plants	−16.0 (and lower)
Other desert plants	−1.8 to −16.3 (mainly −6 to −10)
Plants from zones with periods of pronounced drought	−3.2 to −7.0
Woody plants from mesic sites	−1.5 to −2.6
Mesophytic herbs	−1.4 to −4.3

Because of these inherent problems, the mechanisms of desiccation tolerance are still poorly understood. In general, younger tissues appear to be more tolerant than mature, presumably because they have a higher proportion of non-vacuolate cells. Modifications to the lipid and protein components of the membrane systems of mitochondria and chloroplasts, and the secretion of compatible organic solutes, appear to provide some protection to bound and soluble enzymes, respectively (e.g. Schwab and Gaff, 1990; Stevanovic *et al.*, 1992; Eickmeier *et al.*, 1993); and there is evidence of a role for abscisic acid (Bartels *et al.*, 1990). Nevertheless, owing to the wide spectrum of gene expression induced by drought, and the similarity of the gene products to generalized stress proteins, molecular genetics has yet to provide precision tools for the elucidation of desiccation tolerance (e.g. Ingram and Bartels, 1996).

The most celebrated and extreme examples of true desiccation-tolerance are provided by the so-called resurrection plants (Plate 11) (over 100 herbaceous and woody perennial species, from a number of genera, originally thought to be restricted to extremely arid regions of South Africa, but subsequently found elsewhere in Africa, Australia, India, S. America and even in Europe; Stevanovic *et al.*, 1992) whose *mature* leaves can tolerate many months of severe desiccation and recover to function normally when rehydrated. In the laboratory, resurrection plants can survive rehydration after prolonged equilibration with dry air.

Bewley and Krochko (1982) proposed that resurrection plants possess three critical abilities:

(i) to limit damage during dehydration;
(ii) to maintain physiological integrity in the dry state (as do the *seeds* of most species);
(iii) to initiate repair processes on rewetting.

Extensive research, notably in South Africa, has established that, in general, the *protection* of cell components during the dehydration of resurrection plants is more important than subsequent *repair* on rehydration. Processes of protection include secretion of compatible solutes, notably sucrose; production of antioxidants (see p 248, 273); subcellular reorganization; and dismantling of the photosynthetic apparatus, with either loss, or protection against photoinhibition, of chlorophyll. There is, however, considerable variation among species in the dynamics of each phase. For example rooted plants of *Xerophyta humilis* (monocot) and *Myrothamnus flabellifolius* (woody shrub) could complete the full programme of protection when reduced to the air-dry state over 200 hours, but uprooted plants dried over 50 hours did not survive rehydration; by contrast, plants of the herbaceous dicot *Craterostigma wilmsii* (Plate 11) could survive both treatments (Farrant *et al.*, 1999). Recovery after rehydration tends to be slower for those species where extensive reconstitution of the chloroplasts and photosynthetic pigments is necessary (Sherwin and Farrant, 1996), and in woody species where the mechanical stresses association with dehydration and rehydration can cause xylem embolism (see p 186) (Sherwin *et al.*, 1998). Few of the mechanisms responsible for such tolerance are common to 'normal' plants.

4. Contrasting life histories in arid environments

Survival and reproduction in a given dry environment do not depend on any one specific adaptation or set of adaptations for the acquisition, conservation and optimum use of water. Some idea of the variety of possible life histories, morphologies and physiological responses can be gained from the range of plants native to the Sonoran Desert, which includes arid, but generally warm (>10 °C) regions of Mexico, California and Arizona (Shreve and Wiggins, 1964; Smith *et al.*, 1997). Most of the species belong to one of four types: evergreen shrubs; drought-deciduous shrubs, some of which continue net photosynthesis by stem tissues; CAM succulents; and annuals/ephemerals. Some thoroughly characterized species, mainly from creosote bush (*Larrea*) scrub, include the following.

Atriplex hymenelytra (C_4 photosynthesis). This is a long-lived, small-leaved evergreen perennial shrub (height 20 cm to 1 m), whose life history is conservative, making use of episodic precipitation for survival rather than annual reproduction. The adaptation of the photosynthetic apparatus to high temperatures and to the seasonal variation in temperature (adjustment of optimum by up to 10 °C; Fig. 5.6) means that assimilation is possible throughout the year if sufficient water is available. However, dry matter production is generally negligible during the cooler, dry autumn and winter months (Fig. 5.16) because of long-term closure of stomata, and the secretion of white salt on the surface of the leaves, which increases the reflection of intercepted solar radiation in much the same way as the seasonal thick white pubescence of *Encelia farinosa* (a drought-deciduous C_3 shrub of the Sonoran Desert; see Fig. 4.17 and pp 175, 180, 285 for other aspects of the stress biology of *Encelia*). Interception of solar radiation can also be reduced by adjustment of leaf angle (Gulmon and Mooney, 1977).

Larrea tridentata (divaricata) (C_3 photosynthesis) (Plate 10). This is an evergreen perennial shrub of up to 3 m, which has a similar life history to *A. hymenelytra* but is less conservative, achieving higher rates of net assimilation throughout the year (Fig. 5.16) as a consequence of biochemical and physiological adaptations (e.g. Fig. 4.17) rather than the morphological adaptations of *A. hymenelytra* (and in spite of employing the C_3 pathway). The photosynthetic apparatus shows considerable capacity to adapt to seasonal temperature changes (optimum varying between 20 and 40 °C, Fig. 5.6; Smith *et al.*, 1997). Leaf turgor, stomatal opening and net photosynthesis can be maintained down to leaf water potentials of −6 MPa, and Mabry *et al.* (1977) quote a Ψ_{min} of −11.5 MPa from New Mexico, with net photosynthesis recorded at a Ψ_{leaf} of −8 MPa. Under extreme drought, *Larrea* can shed leaves, twigs and branches, retaining dormant leaves in buds which are extremely tolerant of desiccation. The reproductive biology of *Larrea* is described as *opportunistic* by Smith *et al.* (1997), with flowering initiated by rainfall, but very high seed mortality.

Tidestromia oblongifolia (C_4 photosynthesis) (Plate 10d). Unlike the two preceding perennials, which can remain active under water stress at a range of temperatures, plants of *T. oblongifolia* (15–30 cm in height) are highly adapted to life at high temperature (Fig. 5.6) and high irradiance but are relatively drought sensitive. Their activity is restricted to the hot summer months (Fig. 5.16) when their deep root systems exploit stored water in moister habitats within the desert (water courses and depressions where run-off water has accumulated and infiltrated). When the stored water is exhausted, individual plants die but the species survives because viable seed is released annually from the end of the first growing season. Gulmon and Mooney (1977) class *T. oblongifolia* as a 'long-lived ephemeral', with a normal life span of approximately five years. The genus *Tidestromia* also includes a true desert ephemeral *T. lanuginosa* (Shreve and Wiggins, 1964).

Cammissonia claviformis (*C_3 photosynthesis*). This is a classic winter annual that completes its life cycle in spring after winter rainfall (Fig. 5.16) by means of very high rates of net photosynthesis and growth (achieving a height of up to 50 cm). As a general rule, the ephemeral plants which flourish under higher temperatures after summer and autumn rainfall employ the C_4 pathway of CO_2 assimilation (Pearcy and Ehleringer, 1984).

Agave spp., Opuntia spp. These and a range of other long-lived and bulky succulents have conservative lifestyles (see p. 222), based upon CAM photosynthesis, that are made possible by their tolerance of high temperatures. They are generally absent from colder, and the most arid, deserts. Clonal reproduction is common (Smith *et al.*, 1997), but reproduction by seed is infrequent. However, in *Agave deserti* the severe high-temperature hazards of seedling establishment can be reduced by the protection afforded by canopies of nurse plants of the grass *Hilaria rigida*. Protection of this kind may be associated with reduced rates of growth because of competition for water and nutrients (Franco and Nobel, 1988; but see Briones *et al.*, 1996).

5. Some special problems in tree/water relations

Because of their height and the structure of their leaf canopies, tall trees face unique difficulties in maintaining water relations that are favourable for leaf growth and function. As a result, tall forest (e.g. tropical and temperate rainforest, northern deciduous and coniferous forest) is the characteristic vegetation of only the most humid areas of the planet.

As a consequence of the normal increase in wind velocity with height, the leaves of trees tend to transpire more rapidly than the leaves of shorter plants, especially in discontinuous stands. Furthermore, the 'sun' leaves at the top of the tree canopy are exposed to greater water and temperature stresses than those lower down or shaded within the canopy. Thus the leaves that require most water are furthest from the point of supply in the soil, and although most forest trees do not exceed 30–60 m in height, mechanisms must exist to supply water to the leaves of the world's tallest trees which, in the recent past, have exceeded 110 m (Plate 9).

A third feature of tree physiology is that although the leaf area indices of forests or plantations are broadly similar to those of stands of productive agricultural crops (e.g. 3 to 8), substantial leaf areas tend to be supported and supplied by a single trunk rather than by a number of smaller stems; thus the water relations of an enormous number of leaves or needles can depend on the health and normal function of the vascular system of a single trunk. The scale of trees must also be taken into account when considering the coordination of leaf, stem and root; for example, Schulze (1991) has calculated that it would take 10 days for a root signal to be carried in the xylem from the roots to the top of the canopy of a mature *Sequoia* tree, compared with the minutes elapsing for herbaceous plants. Tall trees must, therefore, rely on longer-term strategies for water economy.

In addition to these physiological difficulties, trees must do work to raise water against the force of gravity. Although this is true of all plants, the effect is small and can be neglected for short vegetation. When considering the water potential of trees, a gravitational component must be added to equation 4.3 to give:

$$\Psi = \psi_s + \psi_m + \psi_p + \psi_g \quad (4.15)$$

where ψ_g, the gravitational potential, lowers leaf water potential by 0.01 MPa for each metre increase in height. If the cohesion theory of water transport holds, then this implies that for water to move from the root to a transpiring leaf at a height of 50 m, a lowering of leaf water potential of at least 0.5 MPa is required in addition to the lowering required to overcome the hydraulic resistance of the plant (see p. 153).

Much of the more modern evidence in support of the cohesion theory relies on measurements of xylem water tensions using the Scholander–Hammel pressure bomb technique. Zimmermann *et al.* (1994) and Canny (1995, 1998) have challenged the validity of such measurements, and the concept that transpiration alone drives water flow to the canopy. They propose a set of additional mechanisms, notably the compensatory pressure theory (Canny, 1995), by which radially-directed inward pressure on the xylem, exerted by the surrounding tissues, increases with transpiration rate. Under such circumstances, both xylem water tension and leaf water potentials would be less negative (than predicted by the cohesion theory) at a given transpiration rate. These ideas have stimulated intense theoretical and experimental activity, including the application of centrifugal forces, rather than direct suction, to generate high tensions in the xylem (e.g. Holbrook *et al.*, 1995). The finding that xylem contents can sustain tensions lower than −3 MPa without undergoing cavitation (Tyree, 1997; Comstock, 1999; Stiller and Sperry, 1999) has confirmed the cohesion theory as the current consensus to explain the transpiration physiology of tall trees.

Thus, as long as they are taking part in transpiration, the water potential of leaves at 50 m can never be higher, and will normally be much lower, than −0.5 MPa, a value at which leaf turgor, leaf expansion and a range of other processes are reduced significantly in many plant species (Fig. 4.3). Tall trees, therefore, require:

(i) a vascular system of high capacity that can deliver water rapidly and preferentially to those parts of the canopy which are most active in photosynthesis and transpiration; the system must also be highly resistant to environmental and biological stresses, including low temperature, freezing, mechanical stresses (wind) and pathological infection;

(ii) leaves that use water efficiently but also continue to grow and assimilate at low water potentials.

1. Vascular system

As in all higher plants, the movement of water through the trunks and branches of trees takes place in the lumina of dead xylem elements; in trees the lignified xylem is also important in providing the mechanical strength necessary to support the leaf canopy. The structure of the vascular system differs markedly among groups of trees. In ring-porous trees, including many north-temperate deciduous species (*Fraxinus*, *Quercus*, *Ulmus*), water moves predominantly in an outer ring of wide vessel elements (60–400 μm in diameter; Fig. 4.18) arranged to give long continuous pipes (vessels), without cross walls, which can be several metres in length. There is a functional relationship between the development of new xylem and leafing out at the start of the growing season. Older, non-functional vessels can be either gas-filled or plugged with lignin and other substances as part of heartwood formation. In contrast, the functional xylem of diffuse-porous trees (e.g. *Alnus, Betula, Populus*) and conifers is not restricted to the youngest layers but is composed of finer, shorter vessel elements or tracheids, interconnected through pores or pits (Fig. 4.18; Tyree and Ewers, 1991).

According to Poiseuille's Law, the flow of water along a cylindrical pipe (down a constant gradient of hydrostatic pressure between its ends, and at a constant temperature and viscosity) is proportional to the square of the radius of the pipe. Consequently, the conducting elements of ring-porous trees can transport water at much greater velocities (normally in the range 15–45 m h^{-1}; Fig. 4.18) than can the elements of diffuse-porous (1–6 m h^{-1}) or coniferous trees (up to 2 m h^{-1}). (Note that the Poiseuille Law should, strictly, be applied to the radius of the pores between elements in many cases.) Even such large differences do not necessarily mean that ring-porous trees can support higher rates of transpiration, because the rate of flow also depends on the total cross-sectional area of functional xylem. For example, Jordan and Kline (1977) demonstrated that the rate of transpiration of tropical forest trees and of Douglas Fir was highly correlated with the sapwood area of the trunk, irrespective of species and site. Nevertheless, as we shall see later, trees tend to have more functional xylem than they require under normal conditions.

In temperate and boreal zones, the geometry of xylem elements may be even more important for xylem function in spring after severe winters. In ring-porous trees, new xylem elements are laid down around the time of leafing out, and a few wide and long vessels can provide sufficient conduit volume to supply the requirements of the transpiring leaves. However, in a severe winter, after leaf fall, the contents of the xylem will undergo cycles of freezing and thawing, leading to the release of small volumes of dissolved gases; the released gases merge to give air bubbles which lodge at constrictions between elements, thereby breaking the continuity of liquid water (embolism or cavitation). Since the vessel can no longer maintain a water column under tension, it ceases to

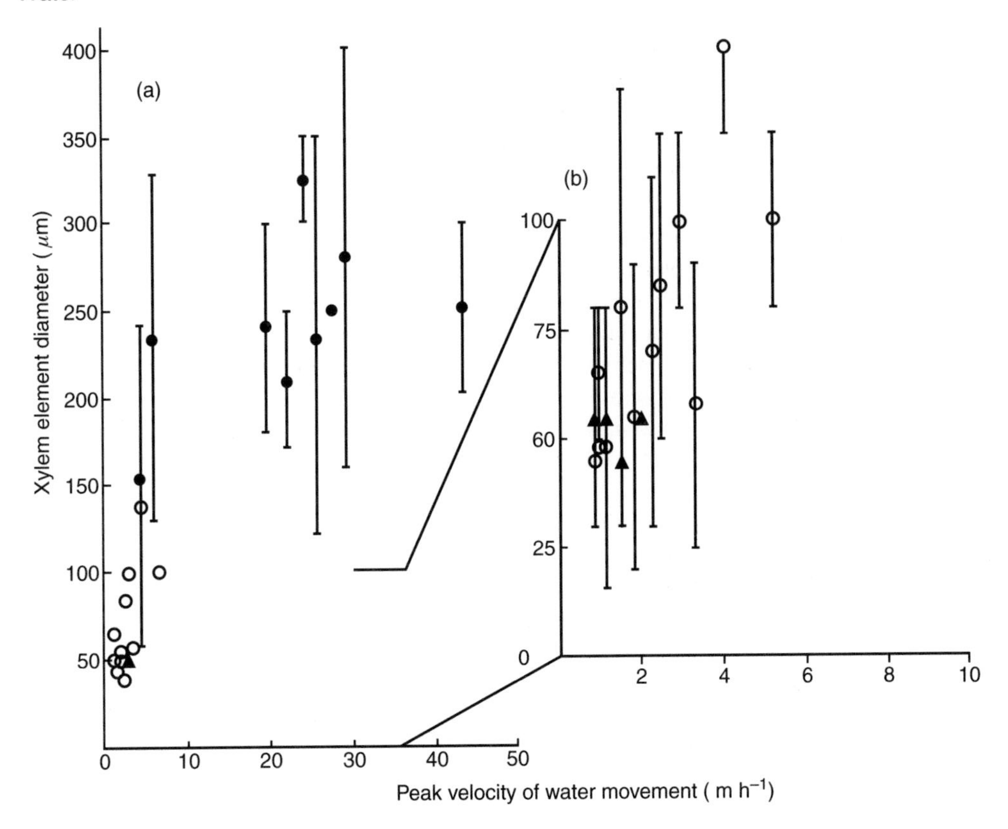

Figure 4.18

Relationships between the diameter of the functional xylem vessel elements (vessels, tracheids) of selected tree species and the midday peak velocities of water movement through them, measured at breast height. ●, indicates ring porous; ○, diffuse porous; ▲, coniferous species (highest values), and the bars indicate the range of values recorded. The lower part of (a) is expanded in (b) to reveal the range of measurements for the diffuse porous species (from data compiled by Zimmermann and Brown, 1971)

function as a water conduit and must be replaced by new water-filled xylem in the spring.

For example, Sperry *et al.* (1988), using 5–8-year-old saplings of sugar maple (ring porous) in Vermont, showed that water stress during the growing season caused the cavitation of 11–31% of the xylem of larger branches and trunks, but that this rose to 69% in winter (84% in small twigs), especially on the south side of the trees which experienced the greatest variation in temperature. However, the extent of embolized xylem decreased to around 20% between March and June *before* the effect of new xylem was experienced because of partial refilling of vessels by positive xylem pressures ('root pressure').

Although they achieve an efficient functional relationship between xylem development and leaf production, ring-porous trees are faced with other serious problems. For example, mechanical damage (e.g. the effect of strong winds) can also cause embolism, and as the spread of Dutch Elm disease has shown, their xylem is peculiarly susceptible to pathogens (Zimmermann and McDonough, 1978).

In general, gymnosperms are more resistant to embolism than angiosperm trees, an extreme example being *Juniperus virginiana*, whose xylem begins to suffer cavitation at a xylem water potential of −4 MPa (a value at which most species have reached zero hydraulic conductance; Fig. 4.19). Gas released in the xylem of conifers tends to be trapped in small volumes within the fine tracheids where it can be redissolved in the xylem water. Indeed, Edwards *et al.* (1994) have shown that reversal of embolism in *Pinus sylvestris* can take place even when the xylem water is under tension. Consequently, only a small proportion of the conducting volume will be lost permanently under drought or during cycles of freezing and thawing. Similarly, since the functional xylem is not restricted to the periphery of the trunk, and since successive tracheids are arranged in a complex, commonly spiral, pattern, the vascular system is less susceptible to mechanical damage and pathogens. These features presumably contribute to the prevalence of coniferous species in more extreme boreal zones, and at treelines (see p. 211).

The hydraulic conductivity of a stem section (water flow per unit gradient of pressure/tension) can be converted into leaf specific conductivity (LSC) by dividing by the area of leaf distal to (and supplied with water by) the section; at a given rate of transpiration, the higher the LSC, the lower will be the gradient in water potential required to maintain the water flow (Tyree and Ewers, 1991).

Measurements on the trunk and branches of a range of tree species have shown that LSC decreases with stem diameter, with the result that the pressure/tension gradient required to maintain the same flow in the finest twigs is of the order of 100 times that required in the main trunk. Thus, because the hydraulic resistances of the trunk and major branches are low, the water potential in these branches is similar throughout the tree, and each small branch can compete for water on an approximately equal basis throughout the canopy. For example, Yang and Tyree (1994) found that, on a leaf area basis, only 15% of the resistance to water flow within young *Acer* trees resided in the trunk, compared with 35% in crown branches and 50% in petioles and leaves. This phenomenon, coupled in some species with constrictions in the xylem elements leading to principal (lower) lateral branches, explains how water can be supplied preferentially to those parts of the canopy which are actively engaged in transpiration.

2. Leaves

In addition to the gradient in water potential necessary to raise water against gravity, reductions in leaf water potential are required to overcome the hydraulic resistance of stem, branch, petiole and leaf (as in herbaceous plants; see p. 153). The combined gradient will normally be of the order of 0.02 MPa m^{-1}, giving leaf water potentials around −1.0 MPa at a height of 50 m, even when the soil is at field capacity and there is no shortage of water. It is, therefore, imperative that the leaves of tall trees use water efficiently so as to minimize further lowering of leaf water potential. Many of the relevant characteristics of tree leaves (sunken stomata with wax plugs; thick cuticle; leaf

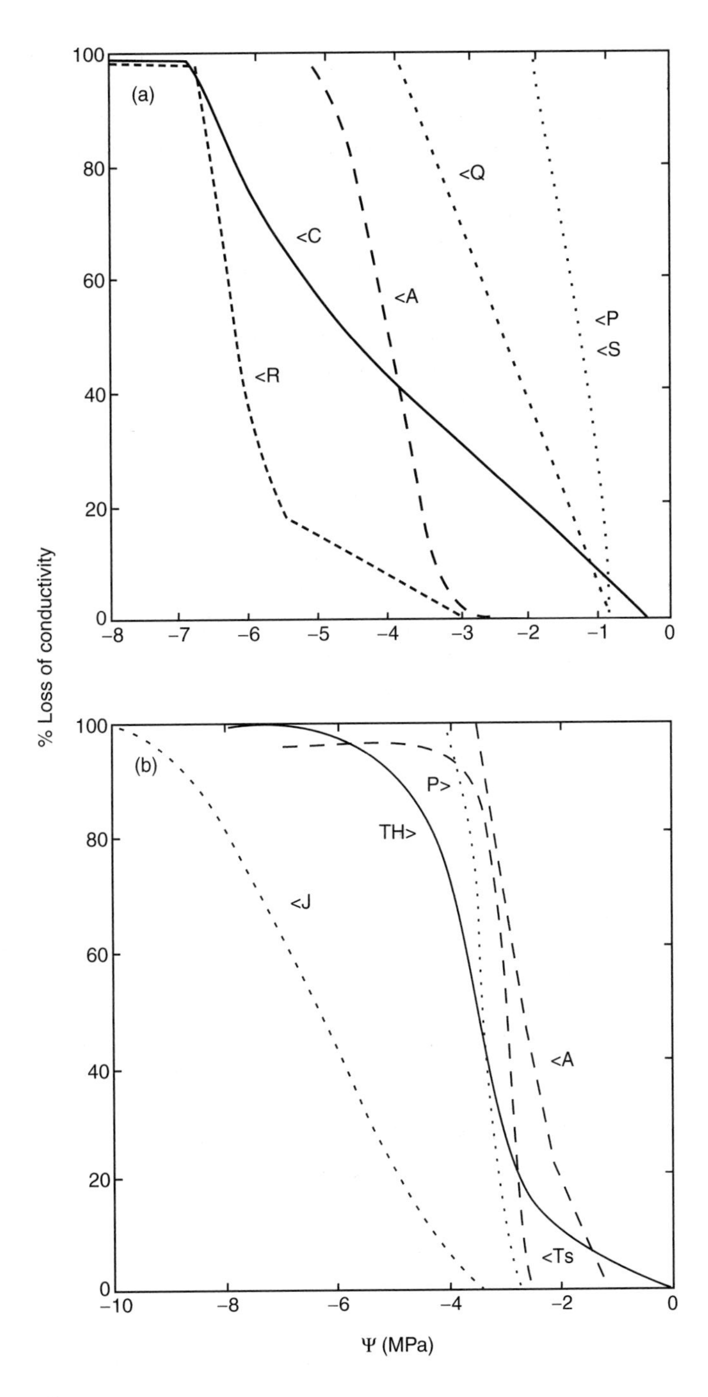

Figure 4.19

The vulnerability of tree species to xylem embolism evaluated by monitoring the hydraulic conductivity of the xylem of intact plants over a range of water potentials imposed by varying the rate of transpiration. (a) Angiosperms: A, *Acer saccharum*; C, *Cassipourea elliptica*; P, *Populus deltoides*; Q, *Quercus rubra*; R, *Rhizophora mangle*; S, *Schefflera morotoni.* (b) Gymnosperms: A, *Abies balsamea*; J, *Juniperus virginiana*; P, *Picea rubens*; Th, *Thuja occidentalis*; Ts, *Tsuga canadensis* (from Tyree and Ewers, 1991)

abscission) have already been considered in detail above. Nevertheless, the leaves of tall trees must be able to grow and function for much of their lives at leaf water potentials substantially lower than −1.0 MPa (i.e. under moderate to severe water stress; Fig. 4.3); cell expansion must take place largely at night, with turgor maintained by osmoregulation (see p. 166).

It is likely that, in cold regions, trees are most at risk from drought during early spring when transpiration, in evergreens, is beginning to increase in response to seasonal increases in irradiance and temperature. At the same time the roots are unable to extract water if the soil is frozen, or the hydraulic conductivity of the root system is depressed as a consequence of low temperatures. The leaves (needles) of most species adapted to such areas are equipped with appropriate xeromorphic features or are seasonally deciduous (e.g. *Larix* spp.). Adaptation to such combinations of environmental factors (water, low- and high-temperature stresses) is considered in greater detail in relation to the physiology of trees at the treeline (see p. 211).

Part II

Responses to Environmental Stress

5

Temperature

1. The temperature relations of plants

Since, unlike homeothermic animals, plants are unable to maintain their cells and tissues at a constant optimum temperature, their metabolism, growth and development are profoundly affected by changes in environmental temperature. Nevertheless, it can be difficult to establish precise relationships between temperature and plant processes, particularly in the field, because of the variability of soil and air temperatures. Plants are commonly 'uncoupled' from the surrounding air, particularly when boundary layer resistances are high (Monteith, 1981): for example, the absorption of solar radiation by tundra plants can raise the temperature of their tissues by 20 °C compared with the bulk air (e.g. Figs. 5.8 and 5.9; see Chapter 2). Care must, therefore, be exercised in measuring appropriate environmental temperatures.

The complexity of the thermal environment of plants is matched by the complexity of their responses to temperature. Nevertheless, in view of current interest in climatic change, there has never been a greater need to understand, model and predict these responses. In this chapter, we first establish some general principles, as a foundation for later consideration of plant responses to temperature, and their interactions with supplies of CO_2, pollutants, mineral nutrients and other resources (in Chapter 7). We then concentrate on those adaptations which enable plants to survive in areas where extremes of temperature are normal features. Although, as elsewhere in this book, the treatment is largely restricted to higher plants, it should be pointed out that lower plants play major ecological roles in extremely cold and hot environments. The chapter concludes with a brief treatment of fire in mediterranean habitats.

1. The Thermal Environment of Plants

The temperature of a leaf at a given time is determined by a range of factors which influence the amount of solar radiation intercepted, and the potential for energy exchange with the environment:

(i) time of day (regular diurnal variation of solar elevation);
(ii) month of the year (regular seasonal variation);
(iii) cloudiness, wind velocity, and origin of the air mass (irregular, short-term variation);

(iv) position in the canopy (e.g. 'sun' or 'shade' leaf);
(v) height above the soil surface;
(vi) canopy characteristics, including leaf shape, dimensions and surface properties.

The net effect of these factors can be assessed by drawing up an energy budget. Thus, a leaf will absorb between 20 and 95% of incident solar radiation, depending on wavelength (Figs 2.1 and 2.2), but only a small percentage of this absorbed energy is used in photosynthesis; the remaining energy is transformed into heat and, unless the leaf can lose this excess heat, its temperature will rise, leading ultimately to death under thermal stress. In a well-watered plant, there are three major processes that can act to dissipate heat from leaves: re-radiation (long wave), convection of heat, and transpiration.

If the leaf is to remain at a constant temperature, then its energy budget must balance:

$$Q_{abs} = Q_{rad} + Q_{conv} + Q_{trans} \tag{5.1}$$

Q_{abs}	Q_{rad}	Q_{conv}	Q_{trans}
Energy absorbed by leaf	Energy lost by radiation	Energy lost by convection of heat	Energy lost by transpiration of water

Therefore if

$$Q_{rad} + Q_{conv} + Q_{trans} > Q_{abs} \tag{5.2}$$

the leaf will be cooled, whereas, if

$$Q_{rad} + Q_{conv} + Q_{trans} < Q_{abs} \tag{5.3}$$

leaf temperature will rise. Equation 5.1 can be expanded (see equation 4.11) to give the form:

$$\begin{aligned} Q_{abs} = {} & \varepsilon\sigma T_l^4 \quad (Q_{rad}) \\ & + k_l(V/D)^{0.5}(T_l - T_a) \quad (Q_{conv}) \\ & + \frac{L.d_l^s(T_l) - \mathrm{RH}.d_a^s(T_a)}{R_l} \quad (Q_{trans}) \end{aligned} \tag{5.4}$$

where

ε is the emissivity of the leaf surface (long wave radiation);
σ, k_l are constants;
T_l, T_a are the temperatures of the leaf and bulk air, respectively;
V is wind velocity;
D is leaf width;

L is the latent heat of vaporization of water (dependent on leaf temperature);
$d_l^s(T_l)$, $d_a^s(T_a)$ are the saturation vapour densities in the leaf and in the bulk air, respectively;
RH is the relative humidity of the bulk air;
R_l is the leaf diffusive resistance (see p. 148).

The temperature of a leaf is, therefore, determined not only by the interplay of a set of environmental variables (radiant flux density; air temperature and humidity; wind velocity), but also by a range of *plant* characteristics, including leaf radiative properties (reflection coefficient; emissivity; colour); dimensions; shape and angle; stomatal responses; and height above the soil surface. Thus a leaf tends to cool down under conditions favouring convection (high wind velocity; small leaves) and/or transpiration (low air humidity; low leaf diffusive resistance – open stomata and thin boundary layer); whereas it tends to warm up with low wind velocity, large leaves, high humidity and high diffusive resistance (closed stomata and thick boundary layer). In temperate zones, the latter set of conditions is the more likely to cause the 'uncoupling' of plant leaves from the ambient air than the former.

The temperatures of organs below the soil surface similarly depend on the seasonal and diurnal variation in energy exchange (factors i–iii above), and the interception of solar radiation by the canopy (vi), which together determine how much energy reaches the soil surface. An energy budget for a root segment would also take account of:

(vii) depth below the soil surface;
(viii) soil properties which influence the energy balance at the soil surface, and the transfer of heat through the soil (e.g. moisture content, bulk density, colour/albedo and the vegetative or litter cover).

In summary, the leaf canopy and the soil profile are complex mosaics of fluctuating temperature, with each group of leaves or root elements responding to a unique pattern of fluctuation. Under high irradiance, the variation can be very substantial: for example, in clear summer weather in the European Alps the diurnal variation in temperature of the soil surface and the lower leaves of rosette plants can be over 40 °C, with differences of up to 20 °C within the canopy and 40 °C in the top 5 cm of soil, at midday, but only a few degrees at sunrise (Körner, 1999; compare with Fig. 5.15). Considerable variation can also be found between the exposed and shaded leaves of forest trees.

2. The Temperature Relations of Plant Processes

As a consequence of the complex variation in environmental and plant temperatures, it is difficult to quantify the temperature relations of plant processes in the field over periods of days or weeks. Plant growth and reproduction (and, ultimately, plant distribution) can depend on one or more of a range of thermal parameters, including: mean, minimum and maximum temperature, and the amount of accumulated temperature (thermal time;

degree days) above a threshold during the whole growing season, or a shorter critical phase such as seed production (e.g. Fig. 5.1).

Furthermore, in a single species or population, physiological processes, and stages of ontogeny, can have different temperature optima. For example, the temperature optima for the various stages of reproductive development in *Tulipa* species vary from 8 °C to 23 °C, the values being correlated with seasonal temperature trends in their native range (Pisek *et al.*, 1973). The development of some species is controlled by night, rather than day,

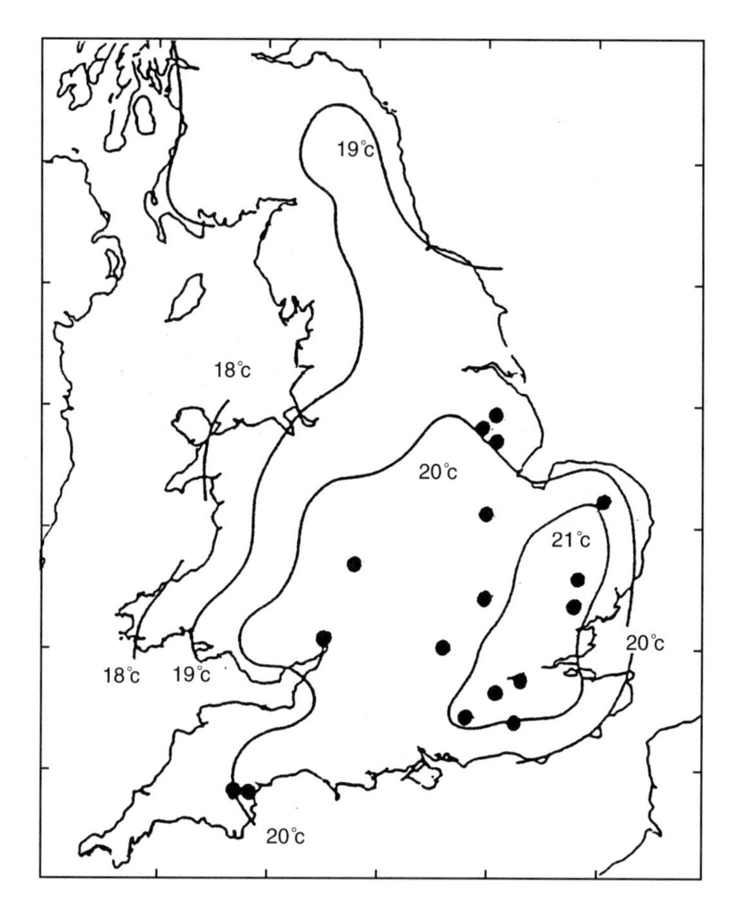

Figure 5.1

The distribution of fertile fruit production by the lime tree *Tilia cordata* in the British Isles (•), in relation to the mean maximum air temperature in August (1901–30, estimated sea level equivalent) (simplified from Pigott and Huntley, 1981). Laboratory studies indicate that a threshold temperature of 20 °C is necessary for successful pollen tube extension, fertilization and viable seed set. The native *T. cordata* trees in cooler areas to the north and west of this critical isotherm (down to 18 °C) are relics of a period when temperatures were higher. In general they are infertile but can produce viable seed in unusually warm years such as 1976.

temperatures; and seed germination can be enhanced by temperature fluctuation (e.g. in *Sorghum halepense*; Ghersa *et al.*, 1992). It can also be difficult to establish the relative importance of soil and air temperatures for plant processes; this is particularly true during the vegetative and early reproductive development of grasses and sedges, when apical development and leaf extension take place near the soil surface, whereas the mature photosynthesizing leaves are subject to air temperature (Hay and Kirby, 1991).

In view of this complexity, it is not surprising to find that most characterizations of plant response to temperature have been carried out under controlled conditions with root and shoot at the same constant temperature or, more rarely, with root and shoot at different, but still constant, temperatures. Some experiments have involved different day and night temperatures but few have attempted to simulate even a regular diurnal variation. Extrapolation of such data to the field must be done with caution, and only since the 1980s have concerted attempts been made to study plant response to temperature in the field, notably in investigations of the *development* of crop plants.

Plant responses to temperature can be expressed in terms of three cardinal temperatures: the *minimum* and *maximum* temperatures at which the process ceases entirely, and the *optimum range* of temperature over which the highest rate can be maintained, assuming that temperature is the limiting factor. In practice, the optimum range normally covers the range of temperatures over which the rate is within (say) 10% of the highest rate (e.g. Fig. 5.4). The cardinal temperatures of higher plant processes vary widely within the range −10 to 60 °C, and are normally related to the temperature regime in the native range of the species (e.g. Figs 5.5, 5.6 and 5.11).

3. Plant Development

As noted in Chapter 1, it is necessary to distinguish between *growth* (increase in dry weight) and *development* (increase in the number and/or dimensions of organs by cell division and/or expansion: leaves, branches, spikelets, florets, root apices etc., including those present in seed embryos). Development also includes the *scheduled* death of organs that have not been killed prematurely by grazing, disease or the effects of stress; thus leaves have a genetically determined life span, which can vary from days in fast-growing crops, through weeks in slow-growing herbaceous perennials in stressful environments, to years in the case of the needles of coniferous trees. The rates of plant developmental processes tend to be controlled primarily by temperature, and to be less sensitive to other environmental factors; thus, the magnitudes of the plastochron (the time interval between the *initiation* of successive leaves, or of spikelets, at a shoot meristem) and the phyllochron (interval between the *unfolding* or *appearance* of successive leaves) in shoots of graminoid species tend to be unaffected by variation in irradiance, mineral nutrition, water supply, and plant population density, unless the plants are exposed to severe stress (e.g. Longnecker and Robson, 1994). This accounts for the familiar observation that, at maturity (i.e. completion of the cycle of development, driven primarily by temperature), individual plants of a given species can vary widely in dry weight (since growth rate is primarily determined by interception of solar radiation).

Figure 5.2 illustrates commonly observed patterns of response for developmental processes: linear increase in rate from a minimum (or threshold) cardinal temperature up to a relatively distinct optimum temperature, followed by a linear (germination; Fig. 5.2a), or less regular (leaf expansion; Fig. 5.2b), decrease at superoptimal temperatures. Few developmental processes, other than germination, have been studied in sufficient detail to give reliable estimates of maximum cardinal temperatures; the available evidence suggests that the values for temperate species are similar to those for germination (e.g. *c*40 °C for the near-bilinear response of leaf extension in *Lolium perenne;* Pollock, 1990).

One important consequence of the close linear relationships between development and temperature, below the optimum range, is that the temperature responses of developmental processes under *fluctuating* temperature in the field can commonly be described by simple linear relationships between number (or size) of organs and accumulated temperature/thermal time above the temperature threshold for the process (e.g. generation, unfolding and death of leaves; Fig. 5.3). Little is known about the factors which determine the responsiveness of plant development to temperature (e.g. the magnitude of the increment in leaf length, or in number of organs, per degree day), although photoperiod may play a role in certain circumstances (Hay and Kirby, 1991). Relationships such as those in Fig. 5.3 break down if temperatures exceed the optimum for prolonged periods.

The ultimate size of a developing organ depends on the duration as well as the rate of expansion. The available information for leaves (Hay and Walker, 1989) suggests that duration is relatively unaffected by factors other than temperature, and can normally be expressed in terms of degree days, whereas the rate of expansion is also strongly affected by factors such as leaf turgor and cell wall properties (see the Lockhart Equation, p. 136). Thus, in comparisons at the same level of insertion, the final area of leaves growing under water stress will be smaller than that of unstressed leaves (Fig. 4.4), even though the duration of expansion is unaffected except under severe stress. In general, thermal regime during ontogeny, and its interactions with irradiance, photoperiod and nutrient supply, can influence the morphology and dimensions of the resulting plant parts, as well as the partitioning of dry matter within the plant (e.g. root diameter and branching pattern, the size and shape of leaves, and root:shoot ratio) (see Chapters 2 and 3).

Finally, temperature can also play a part in controlling the *pattern* and *timing* of plant development. For example, many species native to cool temperate zones (e.g. most perennial grasses; species overwintering as bulbs or rosettes) are unable to initiate reproductive development until they have been *vernalized* by a period of low temperature. This requirement improves the probability of successful reproduction by ensuring that flowering and seed production do not begin until the beginning of the next period favourable to growth. Similarly, the seed dormancy possessed by a wide range of temperate plants, which prevents premature germination in autumn, can be broken by low-temperature stratification, which simulates the low temperatures of winter.

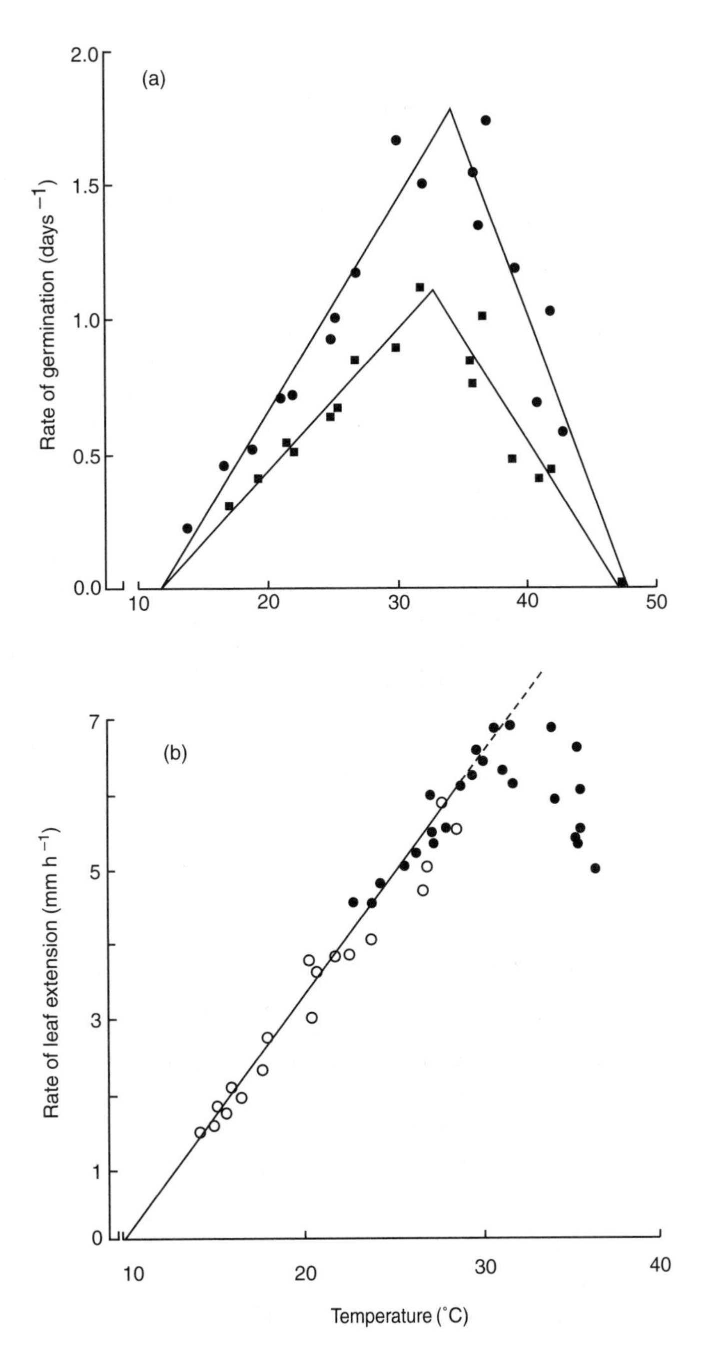

Figure 5.2

The influence of variation in temperature on (a) the rate of germination of seeds, and (b) the rate of extension of leaf 9 of plants, of pearl millet (*Pennisetum typhoides*). In (a), the rate of germination is expressed in terms of the reciprocal of the time to 10% (●) or 60% (■) germination, under constant temperature. In (b), the rate of leaf extension was measured using auxanometers, and different meristem temperatures (plotted) were achieved at daily mean temperatures of 19 °C (○) and 31 °C (●) (from Garcia-Hudibro *et al.*, 1982; and Ong, 1983, with permission from Oxford University Press).

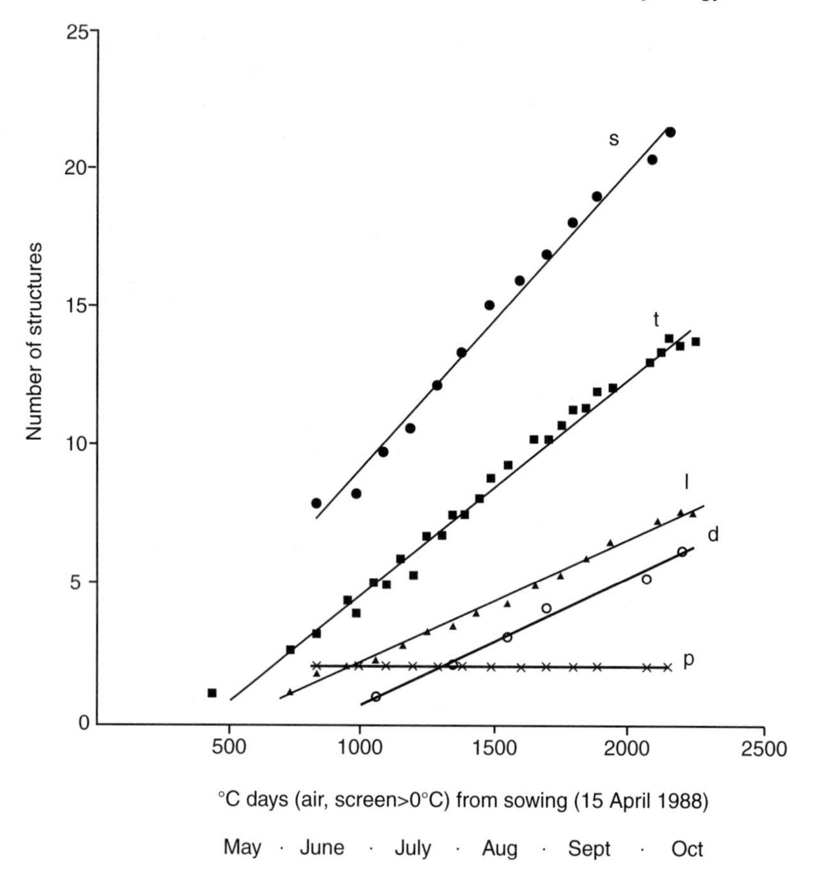

Figure 5.3

Thermal time courses (threshold = 0 °C) of (s) the cumulative production of primordia/total structures, (t) the appearance/unfolding of leaf tips, (l) the appearance of leaf ligules, (d) the death/senescence of leaves, and (p) the number of unexpanded primordia on the shoot meristem, of leek plants grown in the field in the South-West of Scotland, 1988. The divergence of lines (s) and (t), and of (t) and (l) indicates that the appearance of leaf tips and ligules was progressively delayed by the extension of successive leaf sheaths (from Hay and Kemp, 1992).

4. Plant Growth and Metabolism

The response of plant growth rate to a wide range of (constant) temperatures (all other factors being held constant) can commonly be represented by an asymmetric bell-shaped curve; this differs from the response of developmental processes mainly in the sigmoid shape of the suboptimal response (compare Figs 5.2 and 5.4; although linear suboptimal relationships may, in many cases, be obtained by neglecting the early, minor, curvilinear portion of the relationship). This characteristic response of plant growth to temperature arises because increase in temperature affects biochemical processes in two mutually antagonistic ways. As the temperature of a plant cell rises, the velocity of movement (vibrational, rotational and translational) of the reacting molecules increases, leading to more frequent intermolecular collisions and more rapid reaction rates; this effect is common to most chemical reactions. On the other hand, virtually all of the reactions contributing to growth are catalysed by enzymes whose activity depends on their precise, three-dimensional, tertiary structures, to which the reacting molecules must bind exactly for each reaction to proceed. As the temperature rises, increased intra- and intermolecular motion tends to damage tertiary structures, leading to reduced enzyme activity and reaction rates. The asymmetry of response curves such as Fig. 5.4 is the net result of an exponential increase in rate, caused by increased collision frequency, increasingly modified by the thermal denaturation of macromolecules.

The component processes of plant growth do not all respond to temperature in the same way. For example, in most temperate crop species, gross photosynthesis ceases at temperatures just below 0 °C (minimum) and above 40 °C (maximum), with the highest rates being achieved in the range 20–35 °C (Fig. 5.4b). In contrast, rates of respiration tend to be low below 20 °C but, owing to the thermal disruption of metabolic controls and compartmentation at

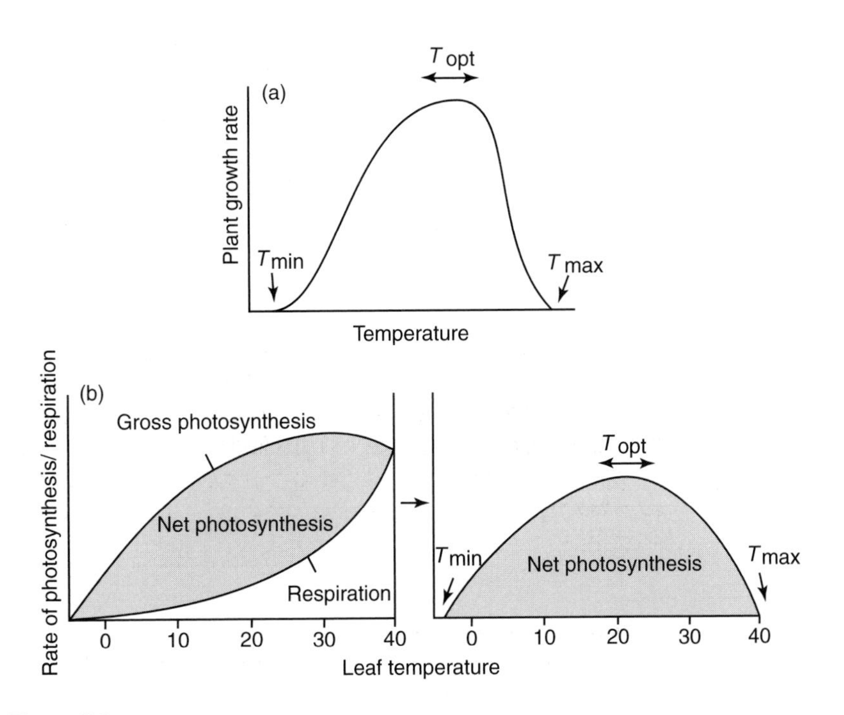

Figure 5.4

Schematic representations of plant responses to temperature. (a) Generalized diagram of the response of plant growth rate, illustrating the three cardinal temperatures: the minimum (T_{min}) and maximum (T_{max}) temperatures, and the optimum range (T_{opt}). (b) The influence of temperature on gross photosynthesis (net of photorespiration) and respiration in a representative plant (adapted from Pisek *et al.*, 1973).

higher temperatures, they rise sharply up to the compensation temperature, at which the rate of respiration equals the rate of gross photosynthesis, and there can be no net photosynthesis (Fig. 5.4b). In consequence, the shape of the response of *net* photosynthesis is broadly similar to that of growth (Fig. 5.4), although the cardinal temperatures for a given species or population will depend on the thermal environment of its native range (e.g. Fig. 5.5). That there is considerable scope for phenotypic plasticity of the thermal relations of metabolism is shown by Figs 5.6 and 5.11.

Quantitative comparisons of the temperature relations of different plant processes have tended to concentrate on the exponential or near-exponential sections of response curves below the optimum range. Most commonly, Q_{10} or Q_5 have been used, where

$$Q_{10} = \frac{\text{rate at temperature } T + 10\ ^\circ\text{C}}{\text{rate at temperature } T} \quad \text{(similarly for } Q_5\text{)} \qquad (5.5)$$

Because Q_{10} values for chemical reactions *in vitro* are usually around 2 (i.e. a doubling of the rate for each increment of 10 °C), it is normally assumed that measured values of 2 or more indicate that a plant process is under metabolic control, whereas values lower than 2 are taken to indicate that the rate of the process under study is limited by a purely physical step, such as diffusion or a

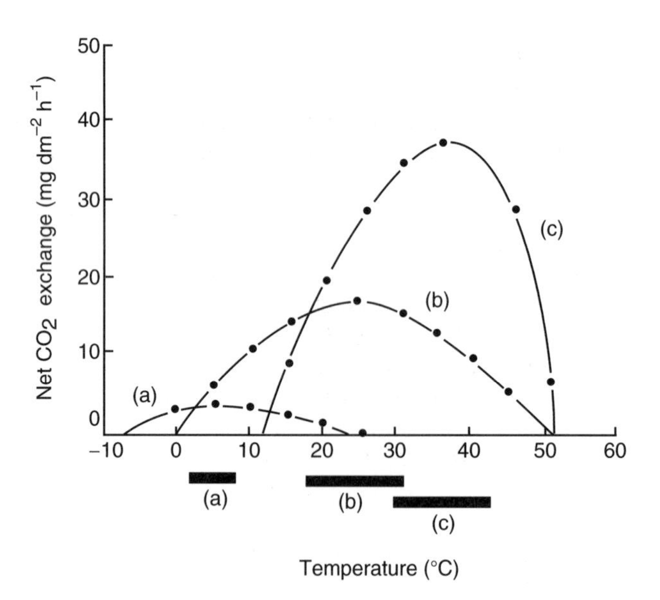

Figure 5.5

The temperature relations of net photosynthesis in three species of the Gramineae from contrasting environments: (a) *Chionochloa* spp. tussock grass (alpine C_3 photosynthesis), (b) wheat (temperate crop, C_3) and (c) maize (subtropical crop, C_4). The horizontal bars indicate the optimum ranges (adapted from Wardlaw, 1979).

photochemical reaction (e.g. the light reactions of photosynthesis; Table 5.1). In general, Q values for plant processes should be treated with caution; in particular, since temperature responses are not strictly exponential, principally owing to the thermal deactivation of enzymes, Q values are themselves temperature-dependent.

An alternative approach has been to apply the Arrhenius equation (from chemical kinetics) to plant processes:

$$k = A \exp(-E_a/RT) \tag{5.6}$$

where k is the rate constant and E_a the activation energy for the process, A is a constant, R is the gas constant, and T (temperature) is expressed on the absolute temperature scale. Plotting $\ln k$ against $1/T$ (for the exponential section of the temperature response), therefore, gives a straight line whose gradient is the characteristic Arrhenius constant (E_a/R) for the process. Arrhenius constants can be useful in biochemical comparisons between species (e.g. Criddle *et al.*, 1994), and in the analysis of plant membrane changes under cooling and freezing.

5. Responses to Changes in the Thermal Environment

Plants can adapt to changes in the temperature regime through the evolution of genotypes with more appropriate morphologies, life histories, or physiological

Table 5.1

Representative values of temperature coefficients (Q_{10}) for selected plant processes measured at varying intervals within the range 0–30 °C. (Compiled from several sources, including Pollock *et al.*, 1993.)

Process	Q_{10}
Diffusion of small molecules in water	1.2–1.5
Water flow through seed coat	1.3–1.6
Water flow into germinating seeds	1.5–1.8
Hydrolysis reactions catalysed by enzymes	1.5–2.3
Photosynthesis (light reactions)	~1
(dark reactions)	2–3
Phosphate ion uptake into storage tissue	0.8[a]–3[b]
Potassium ion uptake into seedlings	2–5
Root axis extension	2.3
Grass leaf extension	3.2
Relative growth rate	7.2

[a] At high external concentrations (50 mM) where uptake is largely by passive diffusion.
[b] Active uptake at low external concentrations (0.1 mM) (see p. 98).

and biochemical characteristics; or by plasticity. Figure 5.5 gives an indication of the potential for evolution of the photosynthetic apparatus; beginning from a common ancestor grass, species have evolved which are adapted to alpine, temperate and subtropical environments. Where changes are rapid (on a geological/evolutionary time scale), and especially where they occur over periods shorter than the life span, there is insufficient time for evolution; survival depends on plasticity, or migration of the plant community (principally by the movement of seeds and other propagules) to a more favourable zone (see p. 229). Plants also adapt to changing temperatures *within* a growing season by plastic responses.

Much can be learned from a series of reciprocal transplant experiments in California, between a cool coastal site (annual mean temperature 13–18 °C), and the floor of Death Valley (18–46 °C) (Björkman, 1980). The experimental plants were fertilized and irrigated to ensure that the principal differences between the sites were in temperature (and radiant flux density). On the basis of their response to transplantation, each of a range of species could be placed in one of three classes:

(i) those which were incapable of surviving the hot desert summer but survived and were summer-active at the coast. This class included species native to the coast, e.g. *Atriplex glabriuscula* (C_3 photosynthesis) and *A. sabulosa* (C_4);
(ii) those which could survive the desert summer but were winter-active in the desert and summer-active at the coast. This class included desert evergreens of both the Old and New Worlds, e.g. *A. hymenelytra* (C_4), *Larrea tridentata* (C_3) and *Nerium oleander* (C_3) (Fig. 4.17) (Plate 10);
(iii) a class of only one species, *Tidestromia oblongifolia* (C_4) (Fig. 4.17) which is both native to, and summer-active in, Death Valley, but is incapable of surviving at the coast (Plate 10).

These patterns of growth and survival were reflected in the temperature relations of photosynthesis for each species (Fig. 5.6, at the appropriate growing temperature: 'cool' for *A. glabriuscula* and *A. sabulosa*, and 'hot' for the others). For example, the optimum temperature for *T. oblongifolia* was 42 °C, compared with 25 °C for *A. glabriuscula*. Figure 5.6 also shows that alteration of the growing temperature can have a profound influence on temperature responses. Thus, growing *T. oblongifolia* at 17 °C rather than at 42 °C lowered the temperature optimum of net photosynthesis to below 30 °C; however, because of the associated severe depression in the *rate* of net photosynthesis, presumably owing to increases in the rate of respiration, its survival at low temperature was endangered. The opposite trend was observed for coastal species; raising the growing temperature did raise the temperature optimum by a few degrees, but again there was a serious reduction in the rate of net photosynthesis. By contrast, the survival of the three evergreen species (class ii) at both sites in the transplant experiment, albeit with lower growth rates than the more extreme species, can be attributed to the plasticity of their photosynthetic systems; changes in optimum temperatures by up to 15 °C were associated with modest penalties in terms of the rate of net photosynthesis (Fig. 5.6). Phenotypic

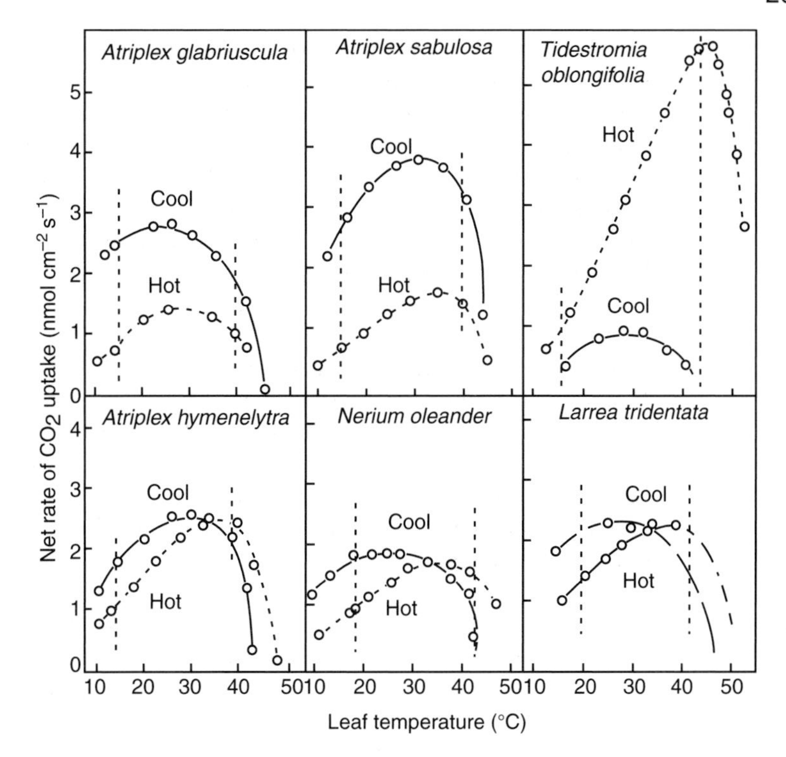

Figure 5.6

The effects of growing temperature on the temperature relations of light-saturated photosynthesis of six species native to contrasting habitats in California. The vertical broken lines indicate the daytime temperatures of the 'cool' and 'hot' growing regimes for each species (from Björkman, 1980).

plasticity of this kind is common to species adapted to both hot and cold environments. For example, Billings *et al.* (1971) showed that the optimum temperature for net photosynthesis in alpine populations of *Oxyria digyna* fell by 1 °C for each reduction of 3 °C in growing temperature.

2. Plant adaptation and resistance to low temperature

1. The Influence of Low Temperature on Plants

Cooling plants to below their optimum temperature range causes a slowing of rates of metabolism, growth and development (Figs 5.2, 5.4–5.6), with far-reaching consequences for plant survival, including reduced availability of surplus assimilate for storage. The duration of the annual life cycle (which includes the acclimation of tissues that will subsequently be exposed to freezing or drought) increases as the climate becomes cooler, and there may be a critical mean temperature below which reproduction cannot be completed within the season. This is clearly illustrated by the characteristic altitudinal zonation of certain herbaceous perennials in temperate zones: the upper limit of *fertile* plants is relatively sharply delimited (at the appropriate isotherm) but vegetative or non-fertile individuals (originating in an unusually warm season, or as a result of long-distance seed dispersal) can be irregularly distributed well above this limit (e.g. *Juncus squarrosus*; Pearsall, 1968). Figure 5.1 presents an analogous distribution, by latitude. For woody species, the immaturity of tissues at the end of the growing season can be a factor in generating the sharp species limits observed at some treelines (see p. 211).

Exposure of temperate plants to low but positive temperatures triggers the expression of suites of up to 20 low temperature-induced (LTI) or low temperature-responsive (LTR) genes, many of which are also triggered by other stresses such as drought, or treatment with abscisic acid (Hughes and Dunn, 1996; Atherley and Jenkins, 1997; Thomashow, 1999). Although the functions of the associated gene products are still largely unknown, the net effects are: (i) enhanced levels of soluble components in the cytoplasm (carbohydrates, proteins, amino acids and organic acids); (ii) increased degrees of unsaturation of membrane lipids; and (iii) the hardening or acclimation of the tissues against subsequent frost damage. In temperate cereals, these genes appear to act in association with those bringing about vernalization. The various solutes may play a cryoprotective role, as the cytoplasm is dehydrated by ice formation, resulting in subcellular changes similar to those caused by water stress (see below and p. 142, 180). The increased proportion of unsaturated fatty acids in the plasma membrane has for long been interpreted in terms of greater membrane stability (resistance to 'freezing' of the membrane), and continued enzyme function, at subzero temperatures (Clarkson *et al.*, 1980; Browse and Somerville, 1991; Thomashow, 1999).

By contrast, cooling tropical and subtropical plants down to temperatures in the range 0–10 °C tends to cause a precipitate decrease in the activity of metabolic processes, notably respiration, which can be fatal within a few hours or days. 'Chilling injury' of this kind is generally associated with a phase change ('liquid' to 'solid') in membrane lipids; this leads to the deactivation of membrane-bound enzymes, including the respiratory enzymes of the mitochondria, and disruption of the water and ion-uptake activities of the root system.

In general, temperate plants are not susceptible to such chilling injury, and tend to show signs of serious damage only after ice has formed within their tissues. For example, Neilson *et al.* (1972) showed that photosynthesis of Sitka spruce needles did not cease completely until extracellular ice had formed (−3 to −5 °C), and even then the initial cessation of activity was probably a purely physical phenomenon: the blocking of diffusion of CO_2 by ice. Under relatively low rates of cooling (<1 °C h^{-1}), ice tends to form preferentially in the apoplast of plant tissues (i.e. extracellular rather than intracellular ice) because of the higher solute concentrations in cytoplasm and vacuole. As long as such periods of freezing are not prolonged, and the rate of thawing is not too rapid, the formation of extracellular ice may not cause permanent damage to hardened plants.

However, if extracellular ice persists, the low water vapour pressure in the apoplast will cause water to migrate out of the cells and into the apoplast, where it freezes, thereby increasing the amount of ice in the tissue. As well as causing mechanical damage, this process results in the progressive dehydration of the cell contents and an increase in the concentration of the cell sap. Consequently, the biochemistry of the cytoplasm is seriously disrupted: proteins, including enzymes, are denatured; various components are precipitated; compartmented molecules such as hydrolytic enzymes are released into the cytoplasm; the buffer system may be insufficient to control cell pH; and there will be a tendency for macromolecules to condense when forced into close contact by the dehydration of the cytoplasm. Under most circumstances, such effects lead inevitably to cell death, and rapid thawing can be equally lethal, owing to further disruption of cell integrity; these effects are analogous to the results of post-anoxic stress (see p. 273). Prolonged 'encasement' of plants in extracellular ice can result in a particularly hazardous combination of freezing and anaerobic stresses (Andrews, 1996).

Hardened ('acclimated') plants of many species can survive prolonged periods of very low temperature and high degrees of desiccation, owing to the resistant nature of their cytoplasm. However, it appears that the cells of even the most resistant plants cannot survive intracellular freezing caused by very rapid rates of cooling (Guy, 1990). What constitutes a rapid rate of cooling in this context is difficult to establish because there have been few relevant field measurements of rates of plant cooling or warming; the rate of 10 °C per hour, observed in tropical alpine regions of South America (Squeo *et al.*, 1991), is probably near to the upper limit in the field. Many of the earlier reports of frost resistance are unreliable because of (i) the emphasis on intracellular ice formation, which is uncommon outside the laboratory (as opposed to severe cell desiccation); and (ii) difficulties in discriminating between damage caused by inter- and intracellular freezing in the laboratory.

2. Characteristic Features of Cold Climates: Arctic and Alpine Environments, Temperate Winters

In subsequent sections, we shall examine plant adaptations favouring survival in cold environments, especially arctic and alpine areas, but also temperate regions during the winter months. However, it should be emphasized that low temperature is only one of a number of potentially unfavourable factors in such areas, and that plant growth, survival and distribution may not be determined primarily or even directly by temperature.

Although it defies precise definition, the term *arctic* is generally used in ecology to describe regions stretching from the limit of tree growth (the treeline or timberline) into higher latitudes. It, therefore, includes tundra areas in both the Arctic and the Antarctic. Many of the characteristic features of the arctic environment (with particular reference to higher plants) can be summarized as follows:

(i) *Temperature*: (air and soil): very low temperatures during the winter; low, and commonly below-zero, temperatures in summer (Table 5.2).
(ii) *Solar radiation*: very long photoperiod in summer (continuous for several weeks) although much of the radiation is received at relatively low radiant flux densities. The amount of photosynthetically active radiation can be

Table 5.2

Environmental conditions in three contrasting regions of North America (adapted from Billings and Mooney, 1968).

	Arctic tundra[a]	Alpine tundra[b]	Temperate forest[c]
Solar radiation			
Mean July flux density (W m^{-2})	209	391	405
Quality	Low in short wavelengths, particularly short UV	High in short UV	—
Maximum photoperiod	84 days	15 h	15 h
Air temperature (1 m, °C)			
Annual mean	−12.4	−3.3	8.3
January mean	−26.7	−12.8	−1.7
July mean	3.9	8.3	20.6
Absolute min.	−48.9	−36.6	−33.8
Soil temperature (15 cm, °C)			
Annual mean	−6.2	−1.7	8.3
Absolute min.	−15.5	−20.0	−10.0
Precipitation (mm)			
Annual mean	107	63.4	53.3
Wind (km h^{-1})			
Annual mean	19.3	29.6	10.3
Air composition			
CO_2 (mg 1^{-1})	0.57	0.36	0.44
O_2 (partial pressure, mm)	160	100	122
Depth of soil thaw	20–100 cm	>30 cm	—

[a] Barrow, Alaska (altitude 7 m, latitude 71 °20′ N).
[b] Niwot Ridge, Colorado (altitude 3749 m, latitude 40 ° N).
[c] Bummer's Gulch, Colorado (altitude 2195 m, latitude 40 ° N).

reduced by snow cover, even at midsummer. (i) and (ii) combine to give a very short arctic growing season (typically 6–8 weeks).

(iii) *Water relations*: in spite of the presence of very large quantities of snow and ice, arctic plants are exposed to drought in summer because of frozen soil moisture; low precipitation; high rates of transpiration caused by high wind velocities; and the direct sublimation of snow into dry air. Since the processes of pedogenesis are slow, mineral soils tend to be unstructured sands and gravels, with low water-holding capacities, although arctic peat soils generally have more favourable water relations. Substantial areas of the Arctic are classed as polar deserts or semi-deserts, but the hydrology of coastal sites means that long-term flooding of tundra soils is also not uncommon.

(iv) *Wind*: in the arctic environment, high winds can be particularly hazardous owing to abrasion of plant tissues by sand particles and by angular snow and ice crystals, especially in winter. Frozen plant tissues are also more brittle and susceptible to mechanical damage.

(v) *Inorganic nutrition*: as a result of the inhibition of soil microbial activity (mineralization of organic nitrogen; nitrogen fixation; etc., Atkin (1996)) by low temperature and water stress, and the virtual absence of legumes in the tundra flora, the supply of inorganic nitrogen by arctic soils is poor. Phosphorus nutrition may be limited as a consequence of inadequate root development.

(vi) *Mechanical effects*: frost heaving of soils during freeze/thaw cycles can cause the uprooting of plants, especially seedlings during the early stages of establishment.

(vii) *Reproductive problems*: because of a scarcity of insects, and the relatively sparse vegetative cover in many tundra areas, the probability of successful cross-pollination may be low. Low population densities of vertebrates rule out several important seed-dispersal mechanisms.

In summary, it can be seen that arctic habitats pose serious problems for plants, not least in terms of their variability (Ives and Barry, 1974). Representative data from an arctic tundra site are presented in Table 5.2.

The *alpine* zone in mountainous regions is normally defined as those areas above the natural treeline. The altitude of the treeline boundary in this definition varies considerably, depending on latitude, exposure and distance from the sea; extreme examples include the Arctic, where the treeline is at sea level, and parts of the Andes where forest can be found up to altitudes of nearly 5000 m. In many parts of the world it can be difficult to determine the altitude of the treeline accurately because of early forest clearance by humans.

The fact that alpine conditions are broadly similar to those in arctic environments is demonstrated by the number of plant species that are common to both zones (e.g. the circumpolar species: *Silene acaulis*, *Trisetum spicatum*, *Oxyria digyna*; Fig. 5.7; Plate 14b). Nevertheless, there are several important differences (Table 5.2):

(i) *Solar radiation*: irradiance and photoperiod depend on latitude, season and snow cover. Radiant flux density can be very high in clear summer weather, and can exceed the solar constant owing to reflection by snow, leading to very high soil temperatures. The supply of ultraviolet radiation increases with altitude.

(ii) *Composition of the atmosphere*: as a result of decreasing air pressure, the partial pressures of O_2 and CO_2 in the atmosphere fall with altitude.

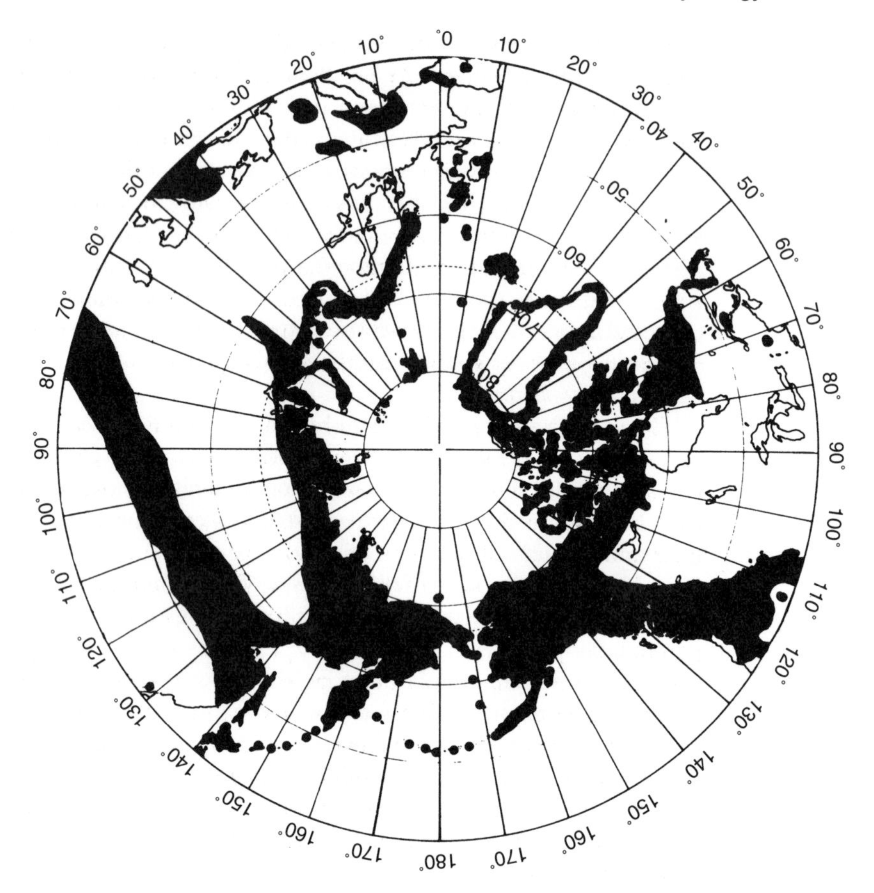

Figure 5.7

Continuous circumpolar and alpine distribution of the mountain sorrell *Oxyria digyna* southwards to 40 °N (simplified from Hultén, 1962; and Billings, 1974).

(iii) *Variability*: the alpine environment is exceptionally variable. For example, there are large differences in radiant flux density between north- and south-facing slopes at the same altitude. Tranquillini (1964) reported an astonishing difference of 57 °C in soil surface temperature, as a result of slope aspect effects; such variation is matched by variation with altitude in a range of factors. There is corresponding variation in the length of the growing season and in water supply.

Nevertheless, the tissues of adapted plants growing in alpine environments tend to have higher nitrogen contents than corresponding lowland plants (Körner, 1989; Morecroft and Woodward, 1996), suggesting that growth may be limited by factors other than nutrient availability; and water stress is more likely in winter and early spring than in summer (Körner, 1999b).

These zones are very extensive; Good (1964) estimated that, together, arctic and alpine regions occupy 23.6 million km^2 in the Northern Hemisphere and 1.3 million km^2 in the Southern; and Körner (1999b) has calculated than alpine vegetation covers 3% of the surface of the Earth (4 million km^2).

Temperate winter climates vary according to altitude and the warming influence of the sea. For example, in the (maritime) lowlands of the British Isles, soil and air temperatures vary in frequent but irregular cycles between −5 °C and 10 °C, according to whether the prevailing air masses originated in the Arctic, the Atlantic or the subtropics. By contrast, the winter climates of continental North America and Europe tend to be much more severe, with subzero temperatures persisting through several months (i.e. approaching an arctic climate; Table 5.2).

The most important differences between arctic/alpine and temperate zones are, therefore, the length and quality of the growing season. In temperate climates, the season normally exceeds six months per year, and native plants can use this period of favourable conditions to grow vigorously and complete their annual cycle of development; most species can avoid the hazards of the following winter in a state of dormancy (seeds or inactive hardened plants – see below), although 'opportunist' species continue to grow whenever conditions permit (e.g. graminoids, whose stem apices near the soil surface are insulated from extremes of air temperature). By contrast, the tundra growing season (i.e. the snow-free period) is short and severe weather can occur at any stage of plant development. To survive under such conditions, plants must be adapted to grow under unfavourable conditions, and to recover quickly from exposure to stress; as in arid zones, these characteristics tend to be associated with conservative life cycles, with reproduction taking place over several seasons (i.e. annual species are rare).

3. Adaptations Favouring Plant Growth and Development In Arctic and Alpine Regions

1. Limiting Conditions at the Treeline (Plate 13)

Surveys of the treeline in several mountain ranges, in the mid-latitudes of both hemispheres, have revealed that even though the boundary varies considerably in altitude, the air temperature in mid-growing season is within the range 10–12 °C; at other times, especially in winter, it diverges sharply (e.g. Grace, 1977). A more extensive study, including tropical alpine areas, has now suggested that the common factor is a mean growing season temperature of 5–6 °C (Körner, 1999b). Although they differ in detail, each of these findings indicates that a critical duration of the growing season, in degree days, must be exceeded for tree growth, development and survival around the treeline (e.g. 100 days above 5 °C; Körner, 1999b). This conclusion is supported by the observation that treelines tend to migrate with climatic change (Kullman, 1988), albeit after a substantial time lag in many areas (Paulsen *et al.*, 2000). Nevertheless, the lack of success of trees above the treeline need not be a *direct* effect of low temperature.

Woody plants at the treeline are threatened in two principal ways (damage and slowing of development). First, in comparison with shorter vegetation, trees are subject to: greater wind damage; associated abrasion by wind-borne particles (ice and soil); winter desiccation (high evaporative demand combined with low root permeability and frozen or highly viscous soil moisture); winter photoinhibition of photosynthesis of exposed tissues; as well as gross physical damage caused by the weight of snow on their branches (Plate 13b). Secondly, unlike herbaceous plants, trees make an annual investment of (surplus) assimilate in woody tissues and supportive root systems, which does not contribute to future productivity, and cannot be remobilized for use in subsequent seasons (note the cycling of carbon in herbaceous perennials; see p. 222). Since the upright tree canopy is well-coupled to the atmosphere (Fig. 5.8), the depression of rates of net photosynthesis by low temperature around the treeline makes it difficult for trees to generate this surplus (the 'carbon balance' hypothesis; Stevens and Fox, 1991) and/or to complete the full cycle of development within the available growing season (the 'season length' hypothesis).

These effects are, of course, not independent. Since evergreen plants rely on long-lived leaves of relatively low photosynthetic potential (see p. 113, 222), the commonly observed winterkill of one-year old needles of conifers at the treeline (by wind and/or particle abrasion, winter desiccation or photoinhibition) compromises the ability of the tree to generate the necessary surplus of assimilate in the following year (James *et al.*, 1994). Furthermore, from the evidence assembled by Körner (1999b), it appears that low root temperatures under tree stands at high altitude act to inhibit the above-ground partitioning of assimilate to new tissues. On the other hand, if the development of new photosynthetic tissues is incomplete by the end of the growing season (e.g. incomplete maturation of the leaf cuticle; Tranquillini, 1979), then these tissues will be vulnerable to severe conditions in winter.

The physiology of woody plants is complex, and less explored than that of herbaceous plants; for example, interpretation of tree physiology at the treeline is complicated by the fact that performance in a given year depends on conditions in previous seasons (James *et al.*, 1994). Nevertheless, a consensus is developing that a range of, largely interdependent, factors is at work at the treeline, and that the relative importance of these factors varies between sites. This is consistent with the observation that treelines can vary from sharp boundaries to extensive *krummholz* zones, where the trees are distorted and prostrate (Plate 13).

2. Dwarfing

In response to abrasion and drought in winter, and low temperatures in summer, most arctic and alpine species have evolved a dwarfed habit, either as a permanent feature of the genotype or as a plastic response. Consequently, tundra plants are normally protected by layers of snow during winter except on exposed ridges. Evergreen and deciduous shrubs, the characteristic plants of exposed, drier sites, are rarely taller than 15 cm, but individual plants of *Betula* and *Salix* spp. can exceed this height when growing in the shelter of large boulders. Tundra herbs occur as rosettes or cushions (Plate 14), or short tufted grasses and sedges whose apical meristems are close to the soil surface. Other morphological features which protect young tissues from abrasion and desiccation include the clustering of twigs round the buds of deciduous shrubs,

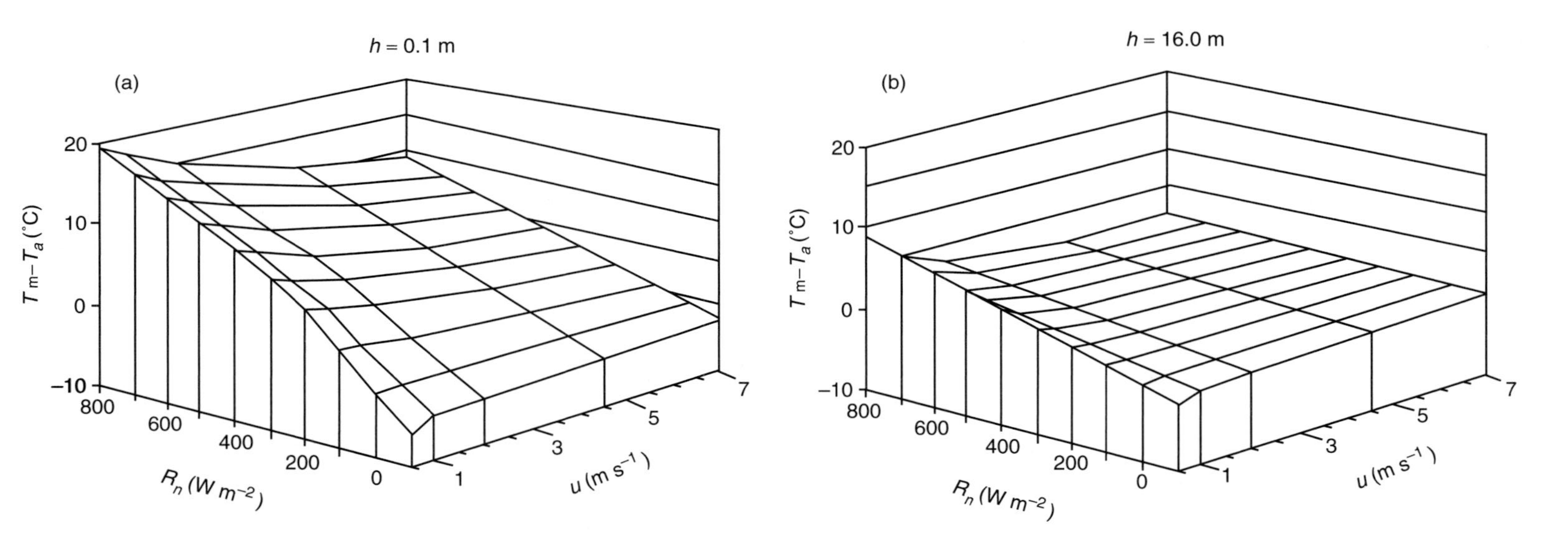

Figure 5.8

The influence of wind velocity (u) and net irradiance (R_n) on the difference in temperature between bud meristems and the bulk air ($T_m - T_a$), for (a) a dwarf shrub (height 10 cm) and (b) a tall tree (height = 16 m), as calculated by the simulation model of Grace *et al.* (1989). The data used in the model originated from (a) above the treeline (*Arctostaphylos uva-ursi* and *Loiseleuria procumbens* at 650–850 m a.s.l.), and (b) in the forest below the treeline (*Pinus sylvestris* at 450 m) in the Cairngorm Mountains of Scotland. (Reprinted from Grace, J., Allen, S. and Wilson, C. (1989). Climate and meristem temperatures of plant communities near the tree-line. *Oecologia* **79**, 198–204. ©1989, with permission from Springer Verlag, Berlin.)

the formation of dense grass tussocks, and the accumulation of insulating layers of dead tissues. For example, dead material accounted for 94% of the above-ground mass of a stand of *Luzula confusa* in the Canadian high arctic (Addison and Bliss, 1984); because of the low level of grazing and microbial activity, windbreaks of this type can persist for decades.

Dwarfing means that, during the growing season, growth and metabolism take place in the warmest zone near the soil surface. When associated with canopy characteristics favouring thicker boundary layers (see p. 152, 170), this can result in daily maximum temperatures within stands of tundra plants which are 10–20 °C higher than values recorded at standard meteorological screen height (1.22 m) (e.g. Fig. 5.9). As shown by Fig. 5.8, the meristem temperatures achieved depend on the interaction of wind velocity and supply of radiation (see equation 5.4). High boundary layer resistances can also play a significant role in the water economy of plants adapted to polar deserts.

The giant (herbaceous) rosette plants of tropical and subtropical alpine areas (in the Andes, Africa, New Guinea, Hawaii and Tenerife), which can grow to a height of 3 m, appear to be notable exceptions to the rule of dwarfing above the treeline (Fig. 5.15; Plate 14a). However, they grow in a unique environment where the daily range of temperature is as great as the seasonal range in most areas: frequent night frosts are associated with high radiant flux densities during the day, and the combination of cold soils and high evaporative demand means that there is a high risk of water stress. The giant rosette form (an example of convergent evolution: unrelated species adopting similar morphologies in widely-separated geographical areas) appears to be successful in such environments because of the thick insulating layer of dead

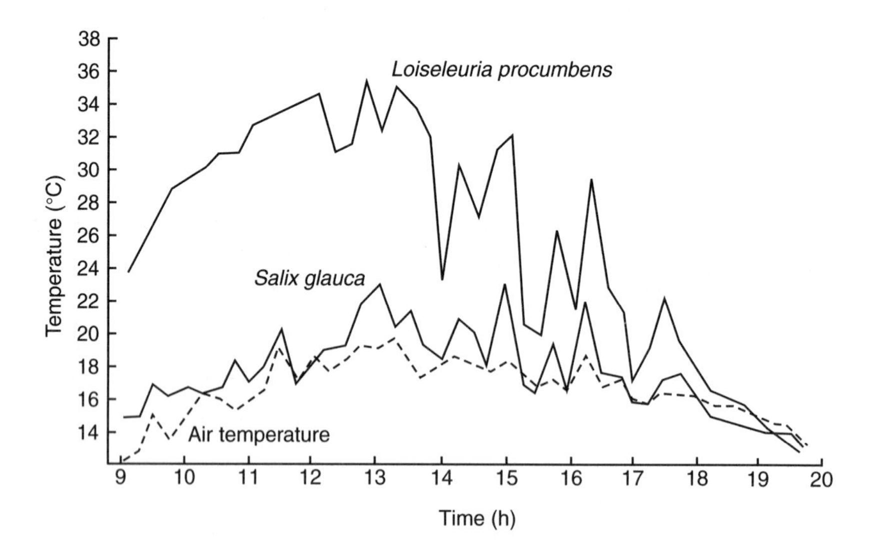

Figure 5.9

Leaf temperatures in stands of *Loiseleuria procumbens* (cushion) and *Salix glauca* (single stem, erect) at Kongsvoll, Norway, 900 m a.s.l. on 17 July 1979. Air temperatures were measured in a screen at 2 m above the ground surface (adapted from Gauslaa, 1984).

leaves which protects the living tissues, and especially the central pith with its reservoir of liquid water, against extremes of temperature (Smith and Young, 1987); giant rosettes also display a range of adaptations which minimize the damage caused to the photosynthetic apparatus by extracellular freezing (e.g. Bodner and Beck, 1987) (see p. 229). Similar adaptations are found at lower altitudes in some dry areas.

3. Canopy Characteristics

Plants adapted to high altitudes tend to have smaller, thicker leaves than those growing in surrounding lowlands. Transplantation studies and comparisons of closely related alpine and lowland species indicate that such differences are brought about by a combination of genotypic effects and plastic responses (changes in number and size of cells; e.g. Table 5.3) (Körner *et al.*, 1989). Although individual leaf area is also small in many arctic plants, exposure of some species, notably grasses, to the long photoperiods characteristic of high latitudes results in larger, thinner leaf blades and longer leaf sheaths (Hay, 1990).

Normally, reduction in leaf dimensions tends to lead to improved coupling of the leaf with the ambient atmosphere, owing to *cooling* by convective heat loss (equation 5.4); however, as a consequence of the growth habit of most tundra plants (dwarf rosettes, cushions, tussocks etc.; Plate 14), individual leaf boundary layers overlap, giving a thicker overall boundary layer for the canopy. For many species, this effect, combined with high levels of anthocyanin pigments (increased absorption and reduced transmission of solar radiation) and the trapping of substantial volumes of warmed air within the canopy (damping of temperature variation), can give very high tissue temperatures; these effects are most strikingly shown by cushion plants (e.g. Fig. 5.9). Short-term variation in temperature within a given canopy will, of course, depend on

Table 5.3

Comparison of leaf characters of alpine plants grown in their natural habitat (2000–3200 m in the Austrian Alps) and in a transplant garden at 600 m in Austria (from Körner *et al.*, 1989)

	Geum reptans		*Oxyria digyna*	
	600 m	2000–3200 m	600 m	2000–3200 m
Individual leaf area (cm^2)	16.5	3.7 (***)	4.8	1.7 (ns)
Specific leaf area (dm^2/g)	1.6	1.4 (ns)	3.1	3.9 (**)
Leaf thickness (μm)	157	309 (***)	414	454 (ns)
Palisade layer thickness (μm)	60	148 (***)	217	205 (ns)
Palisade cell length (μm)	59.5	63.0 (ns)	175.4	144.0 (*)
breadth (μm)	16.9	22.5 (**)	51.1	50.8 (ns)

Where ***, **, * and ns indicate that the paired values are significantly different at $P < 0.001$, $P < 0.01$, $P < 0.05$ or not significantly different.

the interaction between wind velocity and irradiance (e.g. Fig. 5.8). A more extreme example of tissue heating is provided by lichens growing on bare rock surfaces; for example, Coxson and Kershaw (1983) recorded tissue temperatures of 30–50 °C in the Canadian high Arctic.

Because of these elevated temperatures, the growing season (expressed in degree days) is extended considerably, and rates of development and dry-matter production of tundra plants are much higher than would be predicted from bulk air temperatures. For highly pigmented plants, it may also prove possible for net photosynthesis to begin before snow cover has disappeared completely at the start of the season, if the plant has access to liquid meltwater (Körner, 1999b; Lee, 1999a). In different species, the level of pigmentation appears to be genetically determined, or induced by the environment; but the accumulation of anthocyanins is a common response to plant stress *per se* (drought, mineral deficiency etc.) rather than to low temperature in particular, and may, in alpine plants, be related to exposure to UV radiation (see below).

The degree of pigmentation of the corolla can also be of importance in the thermal economy of flowering plants of the tundra, with dark flowers tending to be several degrees warmer than the surrounding air on clear days. This effect, which serves to accelerate the rate of reproductive development, has reached an advanced stage in the many species (e.g. *Dryas integrifolia, Papaver radicatum, Ranunculus acris, R. adoneus*) whose flowers track the sun across the sky, apparently in response to blue light (Corbett *et al.*, 1992; Stanton and Galen, 1993). The resulting increases in temperature of up to 10 °C in still air are of direct benefit to the plant (e.g. earlier ripening, increased seed weight), but they may also improve reproduction indirectly by encouraging pollinating insects to bask in the warm flowers for prolonged periods (thereby advancing the development of the insects, although Totland (1996) was unable to confirm this effect for *R. acris*). The 'woolly' pubescence on arctic willow catkins also serves to raise tissue temperatures.

Up to this point we have concentrated on the potential benefits of deep pigmentation for tundra plants. Nevertheless, in addition to improving the absorption of intercepted solar radiation, dark leaves are also more effective at emitting long-wave radiation (high emissivity, ε; equation 5.4). Consequently, the accumulation of pigment molecules in leaves and flowers may lead to very low temperatures and the risk of freezing damage during long alpine nights in the growing season. The risk will tend to be less under the short nights of the arctic summer.

On the other hand, there is increasing evidence that cytoplasmic flavonoids (whose synthesis is induced by exposure of plants to UV), can intercept the 5% of incoming UV radiation which normally passes through the leaf epidermis, thereby protecting the leaf tissues from UV-induced photoinhibition and other damage (Bornman and Sundby-Emanuelsson, 1995; Veit *et al.*, 1996). Alpine plants are generally more resistant to UV damage than lowland or arctic plants (possibly owing to effective nucleic acid repair mechanisms), and resistance may increase with altitide (Körner, 1999a), but the development, morphology and physiology of plants from a range of tundra environments can be influenced by the supply of UV radiation. For example, exposure to additional supplies of UV-B radiation caused reductions in internode length and increases in leaf number in two species of columbine, but the decrease in stem height was greater for the alpine species (*Aquilegia caerula*) whereas the increase in leaf number was greater for the lowland *A. canadensis* (Larson *et al.*, 1990). The experimental enhancement of UV-B supply, to levels predicted under O_3-depletion, caused reductions in the growth rate of several subarctic shrubs, but the morphogenic effects varied from leaf blade thickening in evergreen species to thinning in deciduous species (Johanson *et al.*, 1995).

4. Metabolism

Many studies have tested the hypothesis that the photosynthetic apparatus of arctic and alpine plants is specifically adapted to operate at low temperatures (e.g. Tieszen and Wieland, 1975). This hypothesis has raised several important questions. First, has sufficient time elapsed for the appropriate adaptation of arctic plants to take place: did most species migrate polewards from temperate zones over the last few thousand years, or were there extensive ice-free refugia at high latitudes for adapted plants during the most recent glaciations (e.g. Sonesson and Callaghan, 1991)? It now seems likely that both scenarios held, although rapid migration appears to have been the more important (Huntley, 1991). In any case, the rate of adaptation to high latitudes can be very rapid: for example, within a century of the introduction of the pasture grass timothy (*Phleum pratense*) to Scandinavia from North America, populations had evolved with distinct physiological responses to very long days (Hay, 1990).

Secondly, although tundra plants do show some distinctive features (e.g. high levels of Rubisco; see Chapter 2), the optimum temperatures for light-saturated photosynthesis of arctic and alpine plants do not differ greatly from those of temperate species (commonly within 10 °C of the values for lowland temperate plants, when grown at the same temperature, although minimum values tend to be lower) (Körner and Larcher, 1988; but see below for the effects of temperature pretreatment). However, the results of alpine *field* studies, where the plants were operating at a disadvantage owing to lower partial pressures of CO_2, do suggest that the potential rate of net photosynthesis in alpine plants is generally higher than that of corresponding lowland species and populations (evidence reviewed by Körner, 1999a). This finding is considered to be consistent with the observed reduction in discrimination against ^{13}C discrimination with altitude (see p. 179).

Thirdly, as discussed on p. 214, several characteristics of the canopies of tundra plants lead to tissue warming, commonly by 10–20 °C above bulk air temperature on clear still days. These findings led Körner (1982) and Chapin (1983) to propose that, in the absence of water- and nutrient-stress, the overall rate of net photosynthesis is limited by the *length* rather than the *quality* of the growing season: arctic and alpine plants tend to be adapted to keep warm rather than to operate at low temperatures.

In many, but by no means all, comparisons, the rate of (dark) respiration at a given temperature has been found to be higher for tundra plants than for those adapted to warmer habitats – an effect which has been explained in terms of higher populations of mitochondria (Tieszen and Wieland, 1975; Berry and Raison, 1981; but see Atkin and Day, 1990; Collier, 1996). Larigauderie and Körner (1995) have confirmed that this effect is not universal for alpine species; of the eight comparisons in Fig. 5.10, the rate of respiration was higher for the alpine population (grown and measured at 10 °C) in only three cases (*Ranunculus acris, Taraxacum* and *Cirsium*) but lower in two (*Poa* and *Anthoxanthum*). More importantly, when the plants were grown and measured at 20 °C, most of the alpine species/populations showed very much higher rates of dark respiration, giving "Q_{10}" values of 2.4 or higher for 7 out of the 11 species studied; the rate of respiration of plants of *Saxifraga biflora* at 20 °C was 5.5 times that at 10 °C. Perhaps the most surprising finding was that the respiration of at least two of the lowland species was very unresponsive to changes in temperatures.

The advantages of such an adaptation (or, rather, a lack of acclimation to higher temperatures) are clear: a small increase in temperature, at near-freezing temperatures, at the start of the growing season, could lead to a doubling of the rate of respiration and a substantial growth response (e.g. when growing new leaves from stored reserves, see below). On the other hand, such an increase at higher temperatures leads to a damaging depletion of reserves (Crawford and Palin, 1981), particularly under long nights/short days. Thus, as originally proposed by Dahl (1951), some alpine species may be restricted to higher altitudes because of the adaptation of their respiratory apparatus to low temperatures.

The discussion so far has tended to concentrate on comparisons between species grown under the same, normally constant, temperatures, but growing

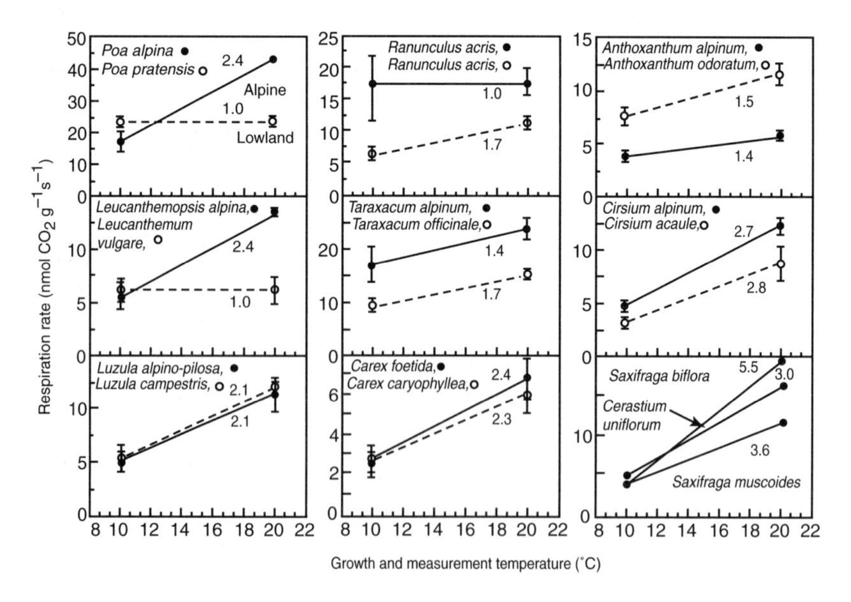

Figure 5.10

The effect of growth temperature on leaf dark respiration rate (per unit dry weight), measured in the laboratory, of a range of alpine (●) and lowland (O) species. The bars indicate s.e.m. and the figures indicate the "Q_{10}" for each species (i.e. applying equation 5.5 to values for plants *grown* as well as measured at different temperatures) (from Larigauderie and Körner, 1995).

seasons in the tundra are extremely variable, with tissue temperatures ranging irregularly and unpredictably over 30 or 40 °C (e.g. Fig. 5.9). The high degree of plasticity of cell metabolism required for plants to survive and reproduce in such an environment is well illustrated by *Saxifraga cernua*, a perennial herb that is widely distributed in the Canadian arctic (Mawson *et al.*, 1986; McNulty and Cummins, 1987).

S. cernua plants raised at 10 °C showed the patterns of response to temperature shown in Fig. 5.11. The optimum for net photosynthesis was around 5 °C, with near-optimum rates maintained around 0 °C (Fig. 5.11a), whereas dark respiration increased linearly from 0 °C (Fig. 5.11b). This corresponded to an optimum temperature for *gross* photosynthesis of 9 °C (Fig. 5.11c). Ten days after these plants had been transferred to 20 °C, the

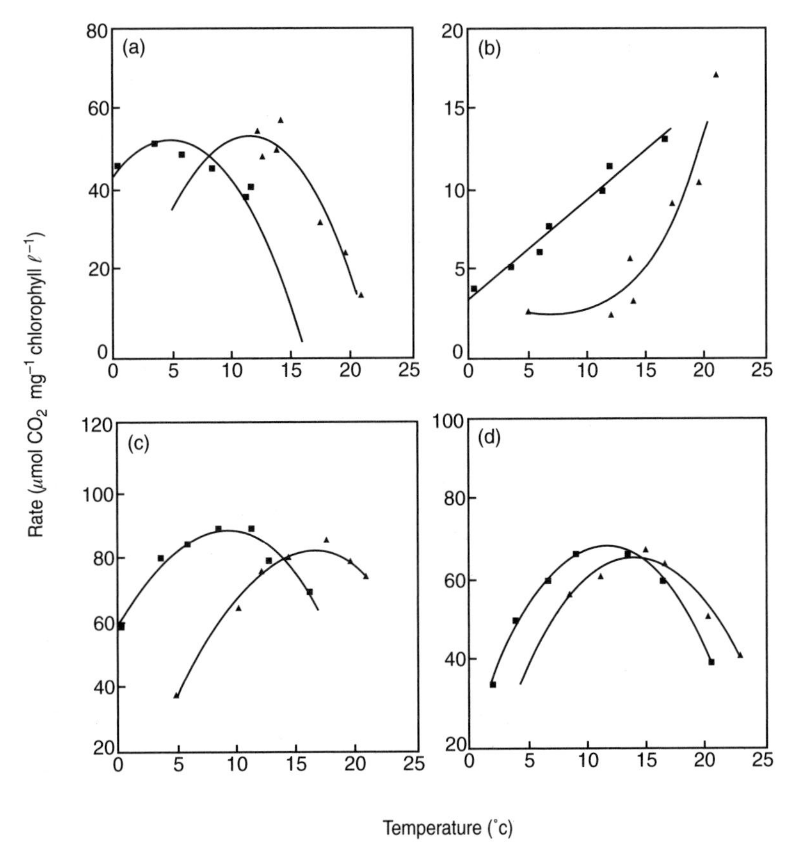

Figure 5.11

Temperature relations of the photosynthesis and dark respiration of plants of *Saxifraga cernua* from the Canadian Arctic, grown and pretreated at different temperatures, but otherwise identical conditions in the laboratory. (a) Net photosynthesis, (b) dark respiration; (c) gross photosynthesis (net photosynthesis + dark respiration) of plants grown at 10 °C (■) or grown at 10 °C followed by 10 days at 20 °C (▲); (d) net photosynthesis of plants grown at 20 °C (▲) or grown at 20 °C followed by 13 days at 5 °C (■) (from Mawson *et al.*, 1986).

optima for net and gross photosynthesis had risen to 12 and 17 °C, respectively (Fig. 5.11a, c), and the temperature response of dark respiration had changed to a curve more typical of temperate plants, with very low rates below 10 °C, rising sharply above (Fig. 5.11b). However, when plants that had been acclimatized to 20 °C were transferred to 5 °C, the shift in photosynthetic optima was less marked (e.g. 3–4 °C for net photosynthesis; Fig. 5.11d). In neither of these transfers was there a significant change in the highest *rate* of net photosynthesis (compare with Fig. 5.6). Such a pattern of response is appropriate for tundra plants, which need to exploit the seasonal rise in temperature without suffering long-term inhibition by short periods of very low temperature.

Although pronounced metabolic plasticity of this kind is not uncommon among arctic and alpine plants, its effectiveness does vary between species and populations (see Fig. 5.10). For example, in a classic study of 17 ecotypes of *Oxyria digyna* from throughout the Northern Hemisphere (Fig. 5.7; Plate 14b), Billings *et al.* (1971) showed that alteration in the mean growing temperature from 26.5 to 8 °C reduced the temperature optimum for net photosynthesis by amounts varying from 0 °C (Silver Gate, Montana) to 12.5 °C (Tarmachan Crags, Scotland), but that this was associated with an increase in the maximum rate of net photosynthesis in nine out of the 17 ecotypes (all of the arctic ecotypes and 5 out of 13 alpine ecotypes).

Arctic and alpine plants, therefore, appear to be well-equipped to maintain relatively high rates of CO_2 assimilation, per unit leaf area, during cold and variable growing seasons. Nevertheless, annual dry matter production per unit area is generally low, and tundra plants tend to remain small in size. Biomass production is limited primarily by season length, but it has been suggested that there are other, intrinsic, features of the physiology of plants adapted to these environments which limit the exploitation of photosynthetic activity in terms of plant development and dry matter accumulation (e.g. intrinsically-low growth rates for alpine *Poa*; Atkin *et al.*, 1996). For example, Körner and Woodward (1987) found that leaf extension (a component of plant *development*) in *Poa* species adapted to different altitudes in the European Alps showed distinctly different (*in situ*) patterns of response to temperature: the threshold/minimum temperature for extension fell from 7 °C around 1000 m to 0 °C at 3000 m, but the slope of the response also decreased with altitude (Fig. 5.12). Thus, in generating leaf area, plants at high altitude had an advantage over those in lower-lying areas at temperatures below 10–15 °C; above this temperature range, the reduced responsiveness to temperature overcame any advantage conferred by a lower threshold. Since there were no pronounced differences in the functioning of the photosynthetic apparatus of these plants, and relatively high tissue temperatures are common during the day at high altitudes in summer, the plants adapted to high altitude were less efficient at deploying photosynthate in the generation of leaf area.

'Disparities' of this kind between carbon assimilation and carbon investment have been recognized in a number of investigations, and suggestions, yet to be tested extensively, have been made about how plants, adapted to the tundra, use or dissipate 'surplus' fixed carbon (inefficient respiration; root growth; development of reserves) (Körner and Woodward, 1987; Atkin and Day, 1990;

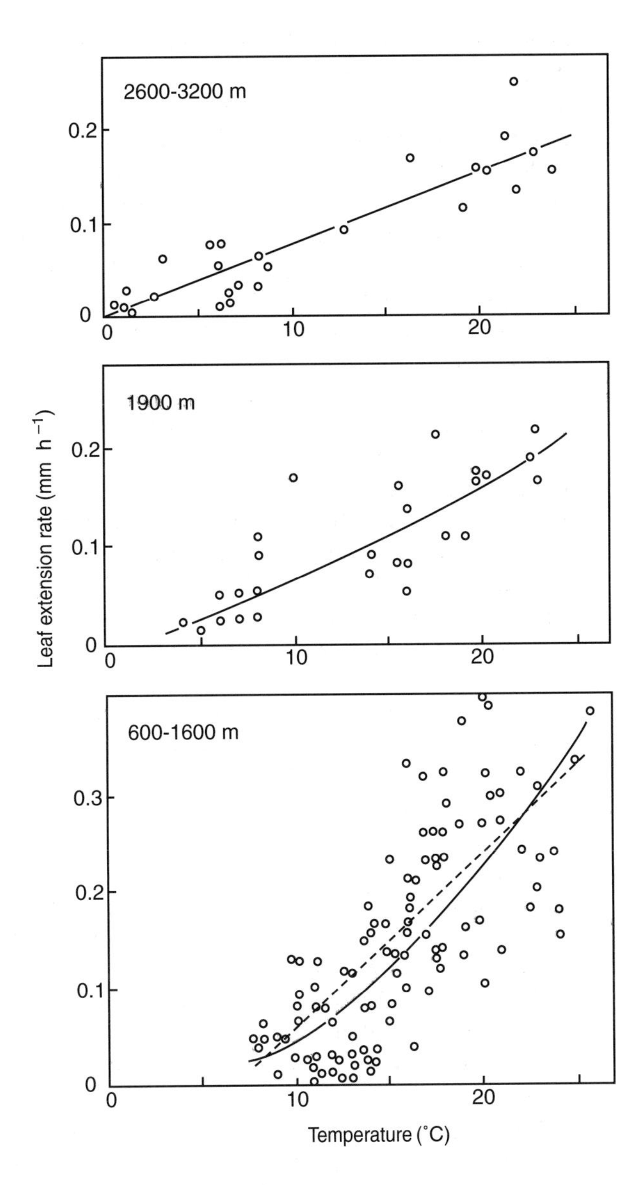

Figure 5.12

Temperature relations of leaf extension in *Poa* spp. measured *in situ* at different altitudes in the Austrian Alps ($r = 0.74$ (linear) and 0.67 (exponential) at 600 m; 0.82 at 1900 m; and 0.90 at 2600 m). Temperatures were measured using thermocouples within the plant stand, and leaf length was monitored using auxanometers. (Reprinted from Körner, C. and Woodward, F. I. (1987). The dynamics of leaf extension in plants with diverse altitudinal ranges. 2. Field studies on *Poa* species between 600 and 3200 m altitude. *Oecologia* **72**, 279–283. ©1987, with permission from Springer Verlag, Berlin.)

Atkin *et al.*, 1996). This phenomenon may prove to be an important component of the conservative life cycle, as discussed in the following section.

5. Life Histories and Resource Allocation

The morphological features of arctic and alpine plants described in earlier sections are commonly associated with characteristic life-histories and patterns of resource allocation (Johnson and Tieszen, 1976). For example, many herbaceous perennials have substantial organs for the storage of carbohydrate below the soil surface (swollen roots, rhizomes, corms, bulbs) which can represent a large proportion of plant biomass. In other species, lipids are stored above ground in older leaves and stems.

Investigation of arctic/alpine species such as *Oxyria digyna* (Fig. 5.7; Plate 14b) has shown that such reserves play a crucial role in enabling the plant to make the best use of short growing seasons. The translocation of stored carbohydrate from below ground to the shoot at the beginning of the season supports the generation of new leaf and stem tissue at a time when the rate of net photosynthesis is low, but respiration is less affected by low temperatures (see above and Figs. 5.4b, 5.11b). Thus, for a few weeks, the dry weight of the plant declines, but, as conditions improve, net assimilation becomes positive and surplus carbohydrate can be partitioned between developing reproductive organs and recharging storage organs (Billings, 1974). This pattern of carbohydrate cycling, coupled with preformed flower buds (generated one or more seasons before, e.g. in the arctic/alpine snowbed grass *Phippsia algida*; Heide, 1992), increases the probability of survival and successful reproduction. Nevertheless, as demonstrated by Chapin and Shaver (1989), it is by no means restricted to the tundra, but occurs in a wide range of environments in which there is a need for perennial plants to generate the leaf canopy and flower rapidly at the start of the growing season (e.g. the temperate woodland bluebell, *Hyacinthoides non-scriptus*, which flowers early in the year before the tree canopy forms; see p. 43).

An alternative life history, which ensures that the interception of solar radiation occurs from the beginning of the short growing season, is shown by evergreen shrubs (e.g. *Ledum palustre, Vaccinium vitis-idaea*) and certain grass species with long-lived, overwintering leaves (Johnson and Tieszen, 1976; Robertson and Woolhouse, 1984). As is the case for evergreens in general (Schulze *et al.*, 1977), the photosynthetic potential of their leaves tends to be substantially lower than that of deciduous plants, but this must be set against increased leaf area duration, and a lower annual investment of carbon in new leaf tissue.

In general, as the environment becomes more severe, the life histories of tundra species tend to become more conservative, geared to the survival of the individual plant rather than to regular annual reproduction. Thus, it is not uncommon for high-arctic plants to produce no more than two leaves per season, or for grasses to grow for four to seven years before flowering (Addison and Bliss, 1984). Grulke and Bliss (1988) have documented an extreme case: individual plants of the grass *Puccinellia vaginata* which, in the high arctic of Canada, can be up to 48 years of age before they reach sexual maturity. Similarly conservative life-cycles are found in other stressful environments (e.g.

p. 184). In spite of such a range of adaptations, it is common for the growing season not to be long enough for the production of mature viable seed, and there may be difficulties in ensuring pollination. Furthermore, seedlings attempting to establish in such hostile environments cannot fall back on substantial carbohydrate reserves, and they are at risk from frost-heaving. Consequently, reproduction by seed declines sharply with increasing latitude and altitude, accounting for the scarcity of tundra annuals, especially in the Arctic. For some perennials, these difficulties have been overcome by vegetative reproduction (by rhizomes, layering of branches, vivipary etc.), and, in some cases, this may confer an advantage in dispersal since large propagules can more easily be transported by wind.

6. Other Environmental Factors

The recognition that plants in many tundra areas are not, in fact, exposed to low temperature stress during much of the growing season has stimulated research into other potentially limiting factors. An important clue in this search has been the observation, particularly on arctic coasts, that vegetation proliferates at sites of accumulation of mineral nutrients in the form of bird excreta. A series of nutrient addition experiments has shown a range of responses, to N mainly but also P, from no effect, through increased biomass, to significant effects on development and phenology and extension of the growing season (e.g. Henry *et al.*, 1986; Shaver *et al.*, 1986; Havström *et al.*, 1993; Lee, 1999a). However, in many cases it is not possible to identify the limiting factors operating in undisturbed tundra vegetation because, in addition to affecting productivity, nutrient enrichment can cause changes in species composition (e.g. by selective nitrogen toxicity (Morecroft and Woodward, 1996); increase in the proportion of graminoids, Fig. 5.13) or a shift to another limiting factor such as irradiance (Körner, 1989a).

There is now sufficient evidence to confirm that the survival and growth of plants growing in drier habitats within arctic tundra zones can be influenced by water relations. For example, Grulke and Bliss (1988) showed that grasses adapted to the high arctic in Canada experienced very low leaf water potentials (−3 to −5 MPa) and loss of turgor during several days per growing season, in spite of some degree of osmotic adjustment. Nevertheless, the effect of drought need not be direct, as shown by the restriction of nitrogen fixation by low soil moisture levels (Chapin *et al.*, 1991). By contrast, Körner (1999a) concluded that, although variation in altitude was associated with changes in stomatal density, stomatal conductance, leaf thickness, and specific leaf area, water stress is uncommon in alpine plants, possibly owing to generally good water supplies and high boundary layer resistances. Thus the stomata in stands of *Loiseleuria procumbens* have been observed to remain fully-open throughout the growing season, while retaining the ability to respond rapidly to rapid changes in humidity.

This section has emphasized the variability in time and space of arctic and alpine environments, and in the various limiting factors operating within different habitats. In their adaptation to such hostile habitats, plant species display a high degree of 'niche specialization' (Chapin and Shaver, 1985); this has important implications for the influence of climatic change on arctic and alpine vegetation (see p. 229).

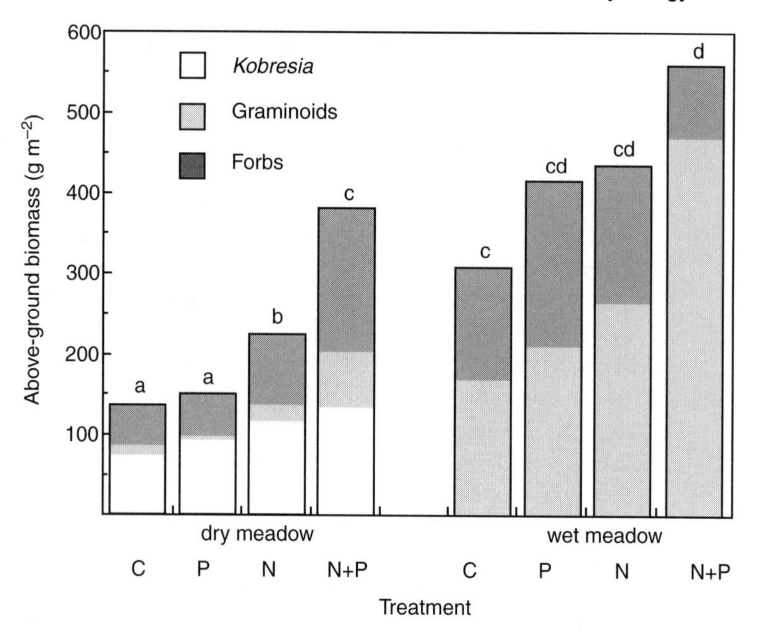

Figure 5.13

Above-ground biomass production in two alpine tundra communities in 1990 on Niwot Ridge, Colorado (see Table 5.2 for environmental conditions). The fertilizer treatments were: C, control; P, 40 kgP/ha; N, 40 kgN/ha as urea; N + P, 40 kg/ha each of N and P). Columns without a common letter are significantly different at $P = 0.05$. The application of N + P caused a significant rise in the biomass of graminoids and forbs in the dry meadow; and an increase in the biomass and proportion of graminoids in the wet meadow (from Bowman *et al.*, 1993).

4. Adaptations Favouring Survival of Cold Winters – Dormancy

As noted earlier, temperate plant species face different problems from those of tundra species. Because climatic conditions are favourable for vegetative and reproductive development for at least six months in each year, temperate plants are normally able to complete their annual reproductive cycles successfully; however, in order to make use of the next growing season, they must be able to survive the intervening winter without sustaining physical damage, or severe respiratory losses. This problem, which is particularly acute in north-temperate and boreal regions, has commonly been overcome by selection for winter dormancy. For example, the seeds of many temperate species, both annual and perennial, are dormant when shed in the autumn and will germinate only after they have experienced a period of low temperature (stratification). For many species, the temperature need not fall below 1–8 °C, and the cold period required for the breaking of dormancy can vary from a few days to several months (Pisek *et al.*, 1973). A clear example of this adaptation is provided by the series of classic experiments carried out by Black and Wareing (1955) on *Betula pubescens*, a north-temperate deciduous tree species. When the seeds are shed in autumn (short photoperiod), they are dormant and can be induced to germinate only when exposed to unseasonable long days. However, after a period of chilling (simulating winter conditions), dormancy is broken and the seeds will germinate whenever temperature and water supply permit. Thus, under natural conditions, autumn germination is suppressed by daylength but, once winter is over, birch seeds can germinate and establish early in spring.

In the same way, the meristems of perennials can cease activity and pass into a resting phase in autumn. For example, the shoot growth of most temperate woody plants ceases in autumn, and dormant terminal buds are formed under the influence of shortening days; as with seed dormancy, bud dormancy is broken naturally by winter chilling. In addition to developing dormant buds, many broad-leaved woody species avoid frost damage to the canopy by autumn leaf abscission, which is triggered by a combination of decreasing daylength and temperature. Winter dormant buds are also found in the bulbs, corms, rhizomes and tubers of herbaceous plants.

Arctic and alpine plants also have to survive very severe winters, although snow cover can provide very effective insulation (Körner, 1999b). However, because of the short duration and irregularity of the tundra growing season, it is essential that growth resumes at the earliest possible date and continues to the end of the period of favourable conditions (compare with temperate perennials which develop dormant buds in early autumn before the radiation and temperature climates have deteriorated seriously). Consequently innate dormancy is less common, and growth generally resumes whenever temperature permits. For example, of 60 alpine species examined by Amen (1966), only 19 possessed seed dormancy, and in eight of these it was caused by a hard, impermeable, seed coat. This, presumably, improves the probability of seedling establishment by spreading germination over several years in soils where freeze/thaw cycles cause abrasion of seed surfaces (see p. 157).

In tundra perennials, the onset of winter, bringing reduced temperature and photoperiod, causes a progressive slowing of metabolism, the *hardening* of plant tissues (see below) and the *enforcement* of dormancy in leaves and perennating buds. Since this dormancy appears to be relieved in spring whenever temperatures rise to around 0 °C, it is important that such dormant buds be able to respond rapidly to changes in temperature once any protective layers of snow have melted. The limited evidence which is available does, indeed, suggest that, unlike their temperate counterparts, the buds of arctic perennials are not enclosed in heavy bud scales which would insulate their contents against fluctuation in air temperature. Whether innate or enforced, dormancy is clearly a widely adopted way of avoiding severe winter conditions. Nevertheless, the survival of plant tissues, even in the form of seeds or dormant buds, depends ultimately on the ability of their cells to avoid freezing injuries.

5. Adaptation Favouring Survival of Cold Winters – Plant Resistance to Freezing Injury

Much of the existing literature on frost resistance deals with temperate species of economic importance, in particular field and horticultural crops, and timber trees. However, where wild species have been studied, the degree of frost resistance of their tissues has been found to be broadly related to the temperature regime in their natural range. For example, Alexander *et al.* (1984) found that the winter 'killing temperature' of twigs of white ash (*Fraxinus americana*) was linearly related to the latitude of origin of the population, falling from −30 °C for saplings from Mississippi (30.9 °N) to −42 °C (Michigan; 46.6 °N). Exceptions to this rule include the tolerance of very low subzero temperatures (−30 to −50 °C, after hardening) in *Salix* species from the lowland tropics (Sakai, 1970).

The survival of plants exposed to freezing temperatures depends primarily on the prevention of the formation of intracellular ice, although cell desiccation and physical damage caused by the accumulation of *extracellular* ice must also be taken into account. Survival can be analysed in terms of a series of successive 'lines of resistance', involving both avoidance and tolerance mechanisms. In general, as the winter climate becomes more severe, the possession of more of these lines becomes necessary for survival. The first line of defence is simply the depression of the freezing point of the cell water as a consequence of its solute content (i.e. avoidance of freezing). Thus, even without hardening (see below), the tissues of most temperate plants can be cooled to a few degrees below zero (typically −1 to −5 °C) before ice forms, and halophytes can have freezing point depressions as much as 14 °C below zero. This colligative property of the cell water can confer complete protection from frost damage in warm areas, such as the Mediterranean and California, where there is a low incidence of frost.

Protection by this mechanism tends to be enhanced for many temperate species if they are 'hardened' for several days at low (positive) temperatures, a process which, simulating autumn, tends to lower the potential for growth (Pollock, 1990). For example, Eagles *et al.* (1993) showed that the lowest temperature tolerated by whole plants of *Lolium perenne* (but, surprisingly, not by the mainstem apices) was strongly dependent on hardening temperature (Fig. 5.14), and that protection down to around −20 °C could be achieved by the appropriate hardening of grasses adapted to high latitudes. It has been suggested that this enhancement arises out of a further depression of the cell freezing point owing to the accumulation of low-molecular-weight organic solutes (sugars, organic acids, amino acids, as well as larger protein molecules) which is a common feature of hardened plants (Pollock *et al.*, 1993). However, such an additional depression could not be more than a few degrees, and it is more likely that the accumulating solutes (whose synthesis is associated with the expression of low-temperature response (LTR/LTI) genes) play a role in protecting cytoplasmic molecules and membranes from the effects of cell desiccation (Hughes and Dunn, 1996).

The second line of defence, protecting critically-important tissues of *woody* plants (dormant buds and xylem ray parenchyma) from freezing at temperatures down to about −40 °C, comes into operation after growth has ceased, dormancy has been established, and the tissues have been hardened by exposure to temperatures below 0 °C for several days (i.e. naturally induced by autumn conditions, before the onset of winter). In spite of a great deal of biochemical work, the processes involved are poorly understood, although it is clear that membrane properties are substantially modified (Sakai and Larcher,

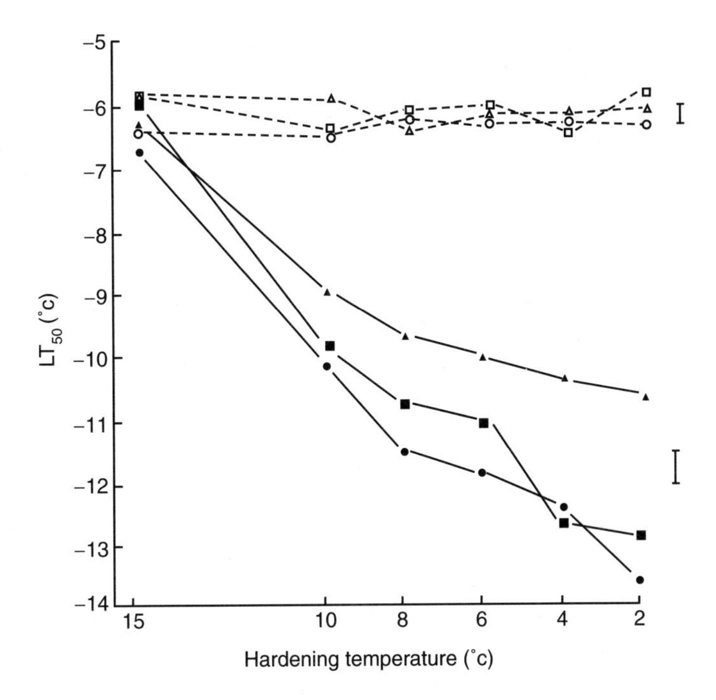

Figure 5.14

The effect of hardening temperature on the freezing tolerance (LT_{50}: the temperature at which 50% of the population is killed) of the main shoot apex (....) and the whole plant (—) of three cultivars of perennial ryegrass, *Lolium perenne.* Where shoot apices were killed but the plant survived, further development and growth took place *via* tiller apices. The bars indicate LSD at $P = 0.05$ (from Eagles *et al.*, 1993).

1987; Pollock *et al.*, 1993). Tissues conditioned in this way tend to behave as if their cells contain ultra-pure water, lacking any nucleating sites where ice formation can initiate; they can thus *avoid* freezing, undergoing *deep supercooling* down to about −38 °C (the spontaneous nucleation temperature of water). This effect has been observed in twigs of several north temperate/boreal tree species (intracellular freezing typically in the range −35 to −50 °C, although in other cases supercooling provides protection only to −20 to −30 °C; e.g. George *et al.* (1977), Gusta *et al.* (1983)). However, although these critical tissues are protected by supercooling, the remaining tissues do suffer freezing injury, or survive by accommodating intercellular ice and tolerating cell dehydration (Ishikawa and Sakai, 1981).

It is clear that this second line of defence will be of little value (and, in fact, positively harmful) to trees growing in areas where winter temperatures lower than −40 °C are common. However, there is a third line of defence, required only in the most severe climates, for example near treelines (see above), by which hardened tissues can be cooled slowly to even lower temperatures, and survive. Thus, Sakai (1970) found that dormant twigs of *Salix* spp. from a number of provenances could be cooled to the temperature of liquid nitrogen

(−196 °C) without seriously impairing subsequent growth. In such cases, it may be that the formation of intracellular ice is prevented because all of the 'freezable' water in the cell has been withdrawn into the apoplast, leaving thin protective layers of water molecules (and accumulated *cryoprotectant* solutes) surrounding macromolecules, membranes and organelles. There is no evidence that tissues protected in this way can survive the formation of ice *within* cells.

This rather simple view of frost resistance as a series of lines of defence must be modified in a number of ways. In particular, rapid rates of cooling (several degrees per hour) tend to cause intracellular freezing and cell death even in hardened plants, at least in the laboratory. This is, presumably, because the hydraulic conductivity of the cell membranes is insufficient to permit the rate of water efflux necessary for extracellular ice formation. The rate of warming is also critically important since time is required not only for the readjustment of cell/apoplast water relations, but also for molecular repair processes. Thus, the rapid changes in temperature and the frequent cycles of freezing and thawing which are characteristic of alpine zones in spring, are considered to be a significant factor in determining the altitude of the treeline (see p. 211).

The degree of protection against frost damage should not be seen as an invariable property even in midwinter. In the severe continental climate of Finland, Scots Pine trees experiencing the natural autumn trends in temperature and day- (or more correctly night-) length, enter winter with the ability to survive temperatures as low as −70 °C (i.e. well below the spontaneous nucleation temperature) (Leinonen *et al.*, 1997). This degree of protection normally endures for at least 100 days, and does not begin to decline until the seasonal rise in temperature in April. Nevertheless, if the trees are exposed to more than a few days of unusually high temperatures, the degree of protection can decline rapidly; this response will be more marked in early spring as there is an interaction between temperature and daylength. Evidence is also accumulating that exposure of evergreen tree species to air pollutants, including acid mists, interferes with frost hardening and lowers frost resistance (Sheppard, 1994). Resistance is not uniform among the tissues and organs of a given species; for example, below-ground organs, which are relatively insulated from the extremes of temperature experienced by shoot tissues, tend to be more sensitive to freezing stress than leaves or stems (Table 5.4).

A similar pattern of cold-hardening occurs in herbaceous tundra perennials, permitting them to survive winter temperatures lower than −20 °C at exposed sites or under snow cover (Körner and Larcher, 1988). However, compared with temperate plants, which deharden rapidly on resumption of growth in spring, there is evidence than dormancy and hardening are not as tightly linked in tundra plants. For example, Robberecht and Junttila (1992) found that plants of an arctic population of *Saxifraga caespitosa*, growing actively at 12 °C, survived exposure to −15 °C by supercooling (compared with −25 °C for plants pretreated at 3 °C); a similar degree of freezing avoidance during the growing season was shown by plants of *Silene acaulis* sampled in the field (Junttila and Robberecht, 1993). Responses of this kind are probably essential for survival in highly variable tundra climates where subzero temperatures can be a feature of the growing season.

Table 5.4

Highest temperature at which freezing injury was detected in the organs of plant species native to the Venezuelan páramos, 4200 m above sea level. Injury was detected by a tissue-staining technique after cooling from 10 °C at a rate of 10 °C h^{-1} in the laboratory (from Squeo *et al.*, 1991).

Species	Injury Temperature (°C)		
	Leaf	Stem	Root
Giant rosette (see Fig. 5.15)			
Espeletia spicata	−11.3	−5.0	−5.0
Espeletia timotensis	−11.9	−6.5	−4.0
Perennial herb			
Senecio formosus	−9.3	−7.9	−3.7
Short rosette			
Draba chionophila	−14.8	−12.0	−14.0
Cushion			
Azorella julianii	−10.6	−9.2	−4.0
Lucilla venezualensis	−14.3	−11.7	−3.5

There is an even greater need for an uncoupling of dormancy and freezing resistance in tropical alpine plants, notably giant rosette plants, which are exposed to daily ranges in temperature that can be greater than the full seasonal range experienced by temperate plants (see p. 215). Thus photosynthetic exploitation of high levels of solar radiation during the day is possible only if the plants can survive subzero air temperatures (typically down to −10 °C in Africa and −5 °C in the Andes) at night (e.g. Fig. 5.15). Protection against frost injury is partly by the temperature-buffering afforded by the high water content of the pith, and partly by insulation, principally by the retention of dead leaves, and, for many species, by the folding of living rosette leaves round the apical meristem at night to give so-called 'nyctinastic buds'. Nevertheless, the survival of the actively-assimilating exposed leaves depends on the supercooling of cell contents to temperatures as low as −10 to −20 °C, with or without extracellular ice formation (Rada *et al.*, 1985; Table 5.4).

6. Life in a Warmer World: The Case of the Arctic Tundra

If the relentless rise in the CO_2 concentration in the atmosphere continues, there will be a mean increase in global temperature of around 2.5 °C by 2050, with higher temperatures predicted by the end of the century. The impacts, on individual plants, populations and communities, of global climate change (involving increases in CO_2 concentration, changes in temperature, alteration in the supply of UV radiation, as well as pollution by acidic and potentially fertilizing nitrogen species) cannot be treated in full in this book, although some aspects are considered in Chapters 2, 3 and 7. Here we concentrate on

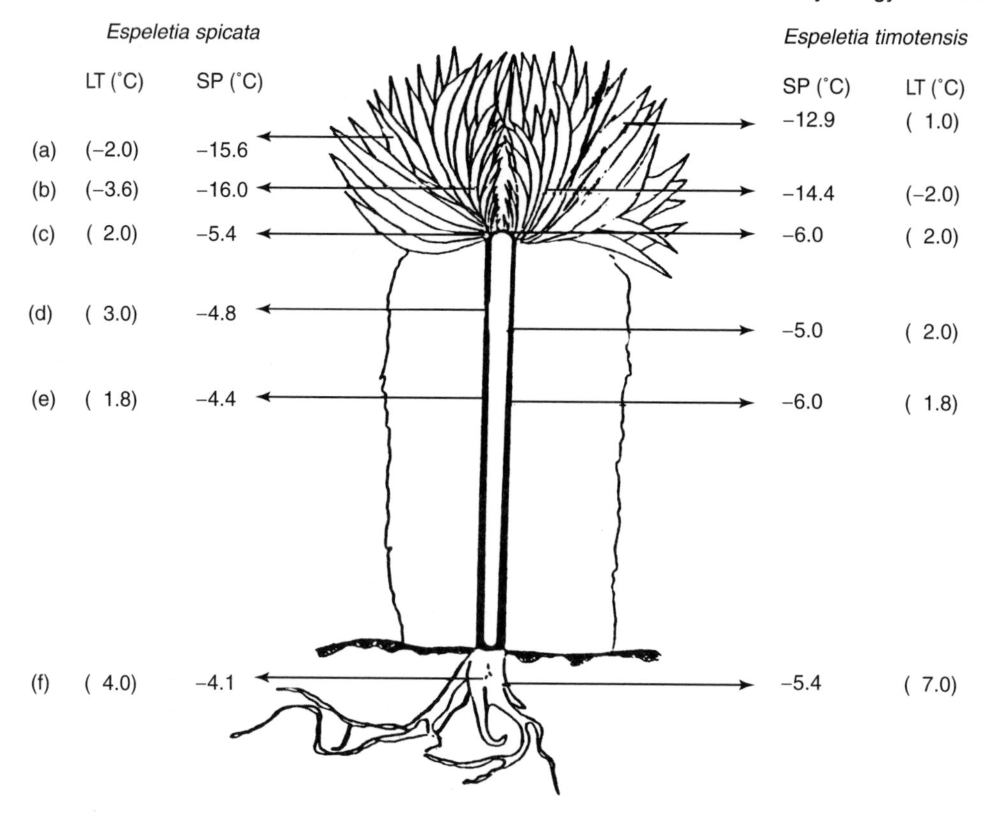

Figure 5.15

The lowest temperatures recorded in the field (LT) and the temperature down to which the tissue can be supercooled without freezing (SP) for (a) fully expanded leaves, (b) expanding leaves, (c) the apical bud, (d) stem pith, (e) phloem and periderm tissue, and (f) roots at 10 cm depth, of plants of two giant rosette species *Espeletia spicata* and *E. timotensis* growing at 4000 m in the Piedras Blancas Páramo, Venezuela. The plants are represented schematically, showing the accumulation of a cylinder of dead tissues surrounding the living stem (from Rada *et al.*, 1985). In this habitat, where tissue cooling was modest, supercooling provided a wide margin of safety.

temperature, considering the very challenging future faced by plants adapted to arctic zones.

Although climatic projections are still far from precise, it is clear that the mean warming of 2.5 °C conceals considerable geographic variation; effects will probably be greater in the mid-latitudes of continental masses, but the earliest impacts of climate change are expected to be seen at high latitudes. It is predicted that temperatures will rise by up to twice the global mean (i.e. by 5 °C), but, paradoxically, this will be associated with increased winter precipitation, resulting in longer snow lie and increased availability of water in summer (Jones *et al.*, 1998; Vaganov *et al.*, 1999). The melting of glaciers and icecaps will lead to the flooding of low-lying areas. One of the more serious findings of 'experimentally induced climate change' in the arctic is that the tundra soil is already a net *source* of CO_2 for the atmosphere rather than a sink; under higher temperatures, the soil microflora will mine the substantial reserves of soil organic matter, generating a powerful feed-forward effect on climate change.

Thus, the principal changes influencing tundra plants will include: a shorter but warmer growing season; increased water supply; higher atmospheric concentrations of CO_2. Experimental modification of the tundra environment to simulate such changes has yielded conflicting results, with increased CO_2 supply generally having little effect; for example, Wookey *et al.* (1993) found that reproduction was stimulated by increased temperature at a polar semidesert site, but by increased supply of water and nutrients (and not by temperature) in a dwarf shrub heath. Reviewing two decades of such experiments, Lee (1999a) has concluded that responses in the Arctic will be determined principally by soil nutrient levels (i.e. by the inputs of mineral nitrogen from the atmosphere). As in all studies of the effect of climate change, the challenge is to scale up from the plant and small plot (Ehleringer and Field, 1993).

If temperature *per se* (or, more correctly, the length of the growing season in degree days) proves to be a decisive factor, the current rate of change in temperature is too rapid for evolution to play a part in adaptation. Consequently, those species, populations or communities, whose plasticity is insufficient to permit them to compete with invading plants adapted to warmer habitats, will need to migrate or face extinction (Huntley, 1991). The picture is reasonable clear for tundra plants: treelines are largely determined by temperature, and warming will tend to drive them to higher altitudes and latitudes, possibly with a significant time-lag (Grabherr *et al.*, 1995). According to the studies reviewed by Chapin *et al.* (1996), tundra communities appear to be equipped to undergo the necessary migration to higher latitudes, as arctic zones warm, but problems will arise if migration is not possible. Thus communities towards the summits of isolated low-altitude mountains, which commonly include endemic species, will disappear (this, of course, will be a more acute problem in certain lower-latitude alpine habitats). Similarly, the advance of the forest into the high Arctic, coupled with the melting of sea ice, will reduce the scope for migration of tundra plants which may, as in periods of glaciation, be forced to rely on isolated refugia.

3. The survival of plants exposed to high temperatures

Plant tissues dissipate heat by three main processes (emission of long-wave radiation, convection of heat, and transpiration of water; see equations 5.1–5.4), of which transpiration tends to be the most effective in many natural situations. High plant temperatures (>40 °C) are, therefore, almost invariably associated with the cessation of transpirational cooling, following stomatal closure in response to drought. Consequently, the vegetation of the drier tropics, subtropics and mediterranean zones commonly experiences a *combination* of water, thermal and photochemical stresses, as do plants in xeric habitats within temperate zones (e.g. sand dunes, shallow soils), and at high altitudes during periods of high irradiance. The effects of high temperatures *per se* are considered here, but the interrelationships amongst temperature, radiation and water relations are considered in Chapters 2 and 4.

Because of the close association between drought and high temperatures, it can be difficult to disentangle the effects of each stress on plants growing in the field; and adaptations to arid environments can be effective only if they lead to avoidance or tolerance of both stresses. Nevertheless, in experiments under controlled conditions, with plants well supplied with water, exposure to temperatures within a relatively narrow range (45–55 °C) for as little as 30 min can cause severe damage to the leaves of plants from most climatic zones. The actual temperature and duration necessary for thermal injury or death will depend on the degree of prehardening by exposure to non-damaging high temperatures. Exceptions to this rule include shade-adapted and aquatic species which are injured by temperatures lower than 45 °C, and some desert plants which can tolerate significantly higher temperatures (see below). This contrasts with the very wide range of temperatures associated with low-temperature damage (+5 to −40 °C or lower, see above).

In general, the *primary* effect of high temperature is on photosynthesis, and, in particular, on the electron transport activities of photosystem II, whose normal function depends on maintaining the correct conformation within the thylakoid membranes of chloroplasts (Weiss and Berry, 1988). This is consistent with the observation *in vitro* that chloroplast lipids undergo a phase transition, involving an abrupt reduction in viscosity, at temperatures within the same narrow range (45–55 °C) (Raison *et al.*, 1980).

The sensitivity of photosystem II can be illustrated by detailed studies of the temperature responses of the species, adapted to different thermal environments, which were included in the Death Valley transplant experiment described earlier (Fig. 5.6). Whole-leaf studies established that photosynthesis was more temperature-sensitive than other processes (respiration, ion uptake; Table 5.5), although the precise temperature threshold for damage varied with species and growing temperature (Björkman, 1980). At the subcellular level, most membrane-related functions and enzyme activities were *less* sensitive than whole-leaf photosynthesis, with the exception of photosystem II which was consistently inhibited by temperatures similar to those affecting photosynthesis (Table 5.5; Björkman, 1980 for other species). For practical reasons, experiments of this kind concentrate on acute thermal stress imposed for periods of as short as 10 min, but lower temperatures (35–40 °C) may prove damaging over longer periods of exposure. Plants can also experience *chronic* exposure to elevated temperatures which does not induce dramatic injuries but which can influence plant performance and survival; for example, plant reserves can be seriously depleted by high rates of dark respiration during long warm nights (see p. 217 for alpine species).

Much of the work on acute thermal stress during the 1980s and 1990s focused on the role of heat shock proteins, which are a feature of a wide range of living organisms. These proteins, which are not synthesized in non-stressful environments, are expressed within minutes of exposure of the plant to high temperature, although their lifetime is measured in hours, and re-exposure to thermal stress does not necessarily evoke further expression (Howarth and

Table 5.5

Minimum temperatures (°C, 10 min exposure) for the inhibition of whole-leaf functions, membrane reactions and enzymes of *Atriplex sabulosa* (C_4 photosynthesis) and *Tidestromia oblongifolia* (C_4) (from Berry and Raison, 1981).

	A. sabulosa	*T. oblongifolia*
Whole leaf functions		
Net photosynthesis	43	51
Respiration	50	55
Retention of ions	52	56
Membrane reactions		
Photosystem I	>55	>55
Photosystem II	42	49
Soluble enzymes		
Rubisco	49	56
PEP carboxylase	48	54
NAD malate dehydrogenase	54	56
NAD glyceraldehyde-3-P dehydrogenase	51	56
NADP reductase	55	55
3-PGA kinase	51	51
Adenylate kinase	47	49
FBP aldolase	49	55
Phosphoglucomutase	51	53
Phosphohexose isomerase	52	55
Light-activated enzymes		
Ru5P kinase	44	52
NADP-glyceraldehyde-3-P dehydrogenase	42	51
NADP-malate dehydrogenase	—	51

Ougham, 1993). Detailed biochemical investigation has indicated a role for heat shock proteins in the stabilizing and repair of tertiary structures of macromolecules, which tend to be damaged by increased inter- and intramolecular motion at high temperature (Fig. 5.4). Nevertheless, it is becoming clear that heat shock proteins are not specific to heat injury, but are similar to other stress proteins (for example the LTI/LTR proteins; see p. 226); it is not yet possible to establish whether their protective role has major significance outside the laboratory, where plants are subject to prolonged and repeated episodes of thermal, and other, stresses.

The very wide range of morphologies and phenologies shown by desert plants (from small ephemerals to enormous, bulky cacti) indicates that survival in hot, dry environments can be achieved in a variety of ways, by combinations of adaptations. Many species have evolved life histories which permit them to avoid the hottest period of the year. This can be achieved by leaf abscission, leaving heat-resistant buds (*Larrea tridentata*, see p. 183; Plate 10), or in desert annuals such as *Camissonia claviformis*, by completing the entire reproductive cycle during the cooler months (Fig. 5.16).

Other species show apparently *inappropriate* adaptations, such as diaheliotropic sun tracking (maximizing the interception of solar radiation, by maintaining the leaves at right angles to the solar beam; Ehleringer and Forseth, 1980; Chapter 2) or high rates of transpiration (Matsumoto *et al.*, 2000), which permit them to grow very rapidly within short periods of favourable conditions (e.g. after episodic rainfall). Alternatively, individual leaves, or the canopy as a whole, may possess characteristics which reduce the interception of radiation or promote leaf cooling, with the result that lethal temperatures are avoided, and the plant can continue to function throughout the year, albeit at a low level of activity (e.g. *Atriplex hymenelytra*; Fig. 5.16). Control of leaf angle and leaf folding has reached a high degree of precision in those species whose leaves are held parallel to the direct rays of the sun (paraheliotropic sun tracking), thereby minimizing the interception of radiation

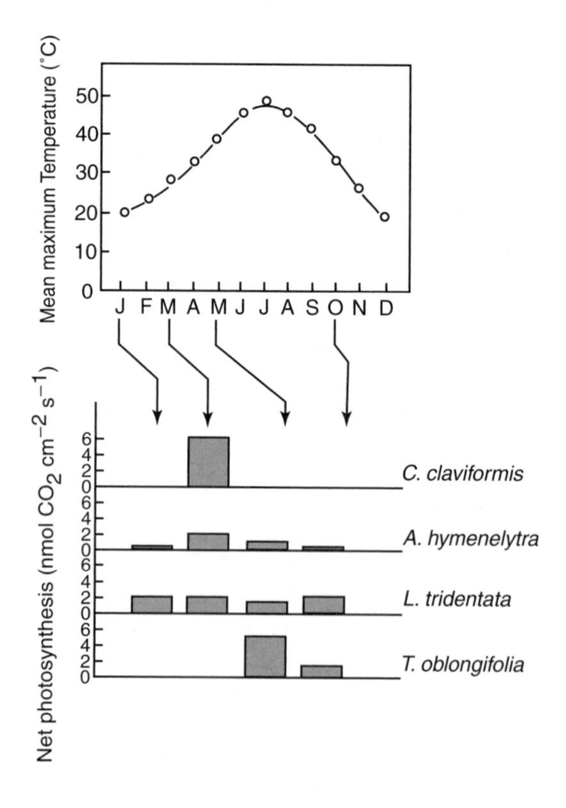

Figure 5.16

The seasonal pattern of photosynthetic potential, and corresponding mean maximum air temperatures, of four species native to Death Valley, California: *Camissonia claviformis*, winter annual; *Atriplex hymenelytra* and *Larrea tridentata*, evergreen perennials; and *Tidestromia oblongifolia*, summer-active herbaceous perennial (from Mooney, 1980).

(Ehleringer and Forseth, 1980). This phenomenon, which can involve leaf-cupping as well as leaf movement, is common to a number of species with compound leaves (e.g. *Lotus, Lupinus, Stylosanthes* spp.). Clearly, both dia- and para-heliotropic leaf movements will be most effective under conditions of clear skies and clean air, where the contribution of diffuse radiation is small, and where the canopy is not unduly affected by wind.

If leaves do intercept excessive amounts of solar radiation under conditions of water shortage, survival will depend on convection of excess heat from the leaf surface or reflection of incident radiation before it can be absorbed. As discussed fully in Chapter 4 (p. 152), the former process is enhanced by morphological features (e.g. small, non-overlapping leaves) which minimize leaf boundary layer resistance. Thus small leaflets of length less than 1.5 cm are a feature of the common evergreen shrubs of North American deserts (e.g. *Larrea tridentata*; Fig. 5.16; Plate 10).

In general, as the climate becomes more hot and arid, the reflection coefficient of the leaves of native species increases. A particularly clear example of this trend is shown by the various perennial shrubs of the genus *Encelia* which are found in California and adjacent states (Ehleringer, 1980). The green, non-pubescent *E. californica* (absorbing 85% of intercepted *PAR*) grows at the coast, under warm moist conditions, but gives way successively, as the climate becomes more severe, to the lightly pubescent *E. asperifolia* and *E. virginensis* (76%) and the highly pubescent *E. palmeri* and *E. farinosa* (<60%). Furthermore, *E. farinosa* exhibits seasonal leaf dimorphism, with white pubescent leaves in summer (giving protection from lethal temperatures but restricting photosynthesis by high boundary layer resistances to CO_2 uptake) being succeeded under moister conditions by greener, less hairy leaves which favour photosynthesis. Where the trend towards greater pubescence is not observed (e.g. the coexistence of *E. farinosa* with the green species *E. frutescens*; Ehleringer, 1988; Fig. 5.17), this may be explained by examining the niches occupied by each species (e.g. *E. frutescens* occupies 'wash' microhabitats where it can maintain sublethal leaf temperatures by more profligate water use). Under the most severe stresses, these shrubs are deciduous.

Such morphological and phenological adaptations are commonly associated with biochemical adaptations favouring net photosynthesis at high temperatures (in particular C_4 and CAM photosynthetic pathways), although C_3 plants are common in desert floras. The comparative physiology of these pathways is considered in detail in Chapters 2 and 4.

Nevertheless, there are species which rely on tolerance rather than avoidance of high temperatures. These include the obligate thermophile *Tidestromia oblongifolia* (Fig. 5.16; Plate 10), which is active in summer in Death Valley at temperatures above 50 °C but cannot survive the moderate climate of the Californian coast (see p. 204). They also include various succulent species whose bulky shapes (low rates of convective heat loss) and CAM photosynthesis (no transpirational cooling during the day) give little scope for leaf cooling, although their high water contents confer a degree of buffering against temperature change owing the the high specific heat of water (p. 131). In fact the highest recorded temperatures of living plant tissues (65–70 °C; with indications that tolerance can be extended above 70 °C by appropriate

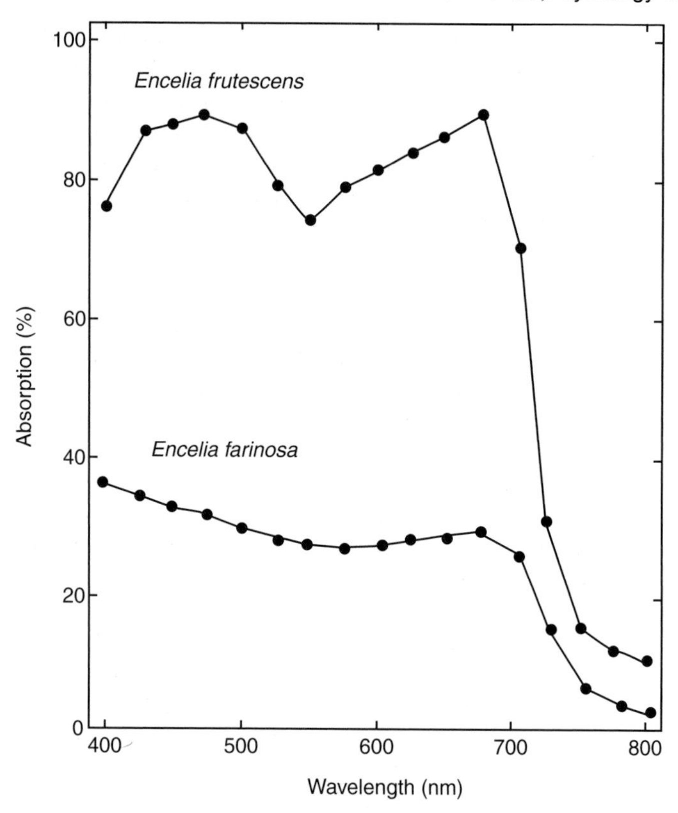

Figure 5.17

The absorption of light of different wavelengths by the leaves of plants of *Encelia farinosa* and *E. frutescens*, measured *in situ* in Death Valley, California. A high proportion of the incident light in the range 400–700 nm was reflected by the white pubescence on the surfaces of leaves of *E. farinosa*. (Reprinted from Ehleringer, J. R. (1988). Comparative ecophysiology of *Encelia farinosa* and *Encelia frutescens*. 1. Energy balance considerations. *Oecologia* **76**, 553–561. ©1988 with permission from Springer Verlag, Berlin.)

hardening) have come from cacti (*Opuntia, Ariocarpus* spp.) and other succulents (*Agave, Lithops, Hawarthia* spp.) (Nobel, 1988, 1989). Tolerance of such high temperatures, which may be partly a consequence of more highly saturated membrane lipids, is particularly important for seedlings and adult plant tissues in contact with bare soil.

4. Fire

1. The Influence of Fire on Plants and Communities

Fire provides the ultimate high temperature stress, and wildfires, caused by lightning strikes or even by volcanic activity (Ogden *et al.*, 1998), are a natural feature of all productive zones with pronounced dry seasons, or in unusually dry years. Spectacular forest fires are common in drought years in temperate and boreal zones, but wildfire is most characteristic of mediterranean zones with cool, wet (and productive) winters followed by warm dry summers. Typical fire-adapted communities include the *kwongan* of Western Australia, the *fynbos* of South Africa, and the Californian *chaparral*, and it plays a significant role in the communities of the Mediterranean Basin itself (*maquis, garigue* etc.).

Where humans intervene, the effects of fire can be devastating. On the one hand, in many parts of the world, prevention of fire in national parks has led to the accumulation of combustible litter, and sporadic very hot and destructive fires. On the other hand, thousands of years of deliberate firing has meant that virtually all plant communities in Australia have become adapted to fire, with the extinction of many fire-sensitive species (Bowman, 1998).

Soil and canopy temperatures during fires depend on the heat released per unit area; this, in turn, depends primarily on the quantity of combustible plant material (litter and living tissues) per unit area, but also on the heat of combustion of the fuel (energy released per unit of dry matter) and other factors affecting the convection of heat away from the fire (notably wind, which also affects the rate of spread of the fire front). Heat of combustion varies among the components of plants (typically 17 kJ/g for carbohydrate; 23.5 kJ/g for protein; and 39.5 kJ/g for lipids) but dry plant materials generally fall within the range 19–22 kJ/g (Allen, 1989). The concentration of volatile oils (e.g. thymol, 37.5 kJ/g) in the shoots of mediterranean labiates may result in hotter fires, but the temperatures achieved probably depend more on the water content of the litter and the proportion of living tissues, since energy is consumed in supplying latent heat of vaporization. Soil temperature is similarly buffered by the evaporation of soil water (Campbell *et al.*, 1995). Thus fire at the end of a cool, wet growing season will tend to be considerably less damaging than during the dry season. Fire temperatures at the soil surface and within the canopy will almost invariably be above the maximum for plant survival (60–70 °C; see above), and values well above 100 °C are common (e.g. up to 150 °C at the soil surface in experimental bushfires in Australia; Bradstock and Auld, 1995).

Apart from killing living tissues (opening the plant canopy), removing surface litter and reducing soil organic matter, fire can play a number of other roles. For example, nutrients which are non-volatile at these temperatures (e.g. K, P) can be more rapidly cycled in the ecosystem, being concentrated in surface ash rather than released by microbial decomposition; if the ash has a high calcium content, then soil surface pH can be raised, improving the environment for germination and seedling development. However, in many fire-tolerant communities, nutrients are withdrawn very effectively from senescing tissues, and enrichment of the surface layers of soil occurs only if living tissues are burned (Pate, 1993). Of course, most of the nitrogen in the plants, litter and surface soil will be lost to the atmosphere during fires.

High soil temperature will tend to kill most of the seeds in the superficial layers of the soil, but at critical depths, dormant seeds of certain fire-adapted species can either receive a high temperature cue for germination, or sustain thermal damage to the seed coat, resulting in increased hydraulic conductivity (e.g. in *Acacia suaveolens*; Bradstock and Auld, 1995). In some communities and species, seed dormancy can also be broken by exposure to smoke, aqueous extracts of smoke, or charcoal (e.g. amongst *chaparral* plants, Keeley, 1991; in the Restionaceae of the *fynbos*, Brown *et al.*, 1994); it is likely that fire-generated ethylene and nitrogen oxides play a part in at least some of these effects (Keeley and Fotheringham, 1997).

Many of the species of fire-adapted plant communities store viable seed for several seasons within the canopy in protective, normally woody, fruit. Release is triggered by the mechanical effects on the fruit of high temperature, or fire damage; dormancy is released; and the (serotinous) seeds are shed on a bed of ashes where competition from established plants is at a minimum. For example, fire-resistant communities in Western Australia store up to 1100 seeds per m^2 in the canopy (Bell *et al.*, 1993), and large woody fruits can be a conspicuous feature of the plant stand (e.g. *Banksia* spp.). In some species, seed release is triggered by wetting fire-affected fruit, thereby ensuring that the seedbed is moist (e.g. Cowling and Lamont, 1985), and the temperature optimum of germination tends to increase the probability of establishment in the most appropriate season (see Chapter 4, p. 158).

Fire can also play a role in altering the relative competitiveness of resistant species. For example, Hodgkinson (1992) found that, in Eastern Australia, there was a complex relationship amongst fire incidence, fire temperature, and

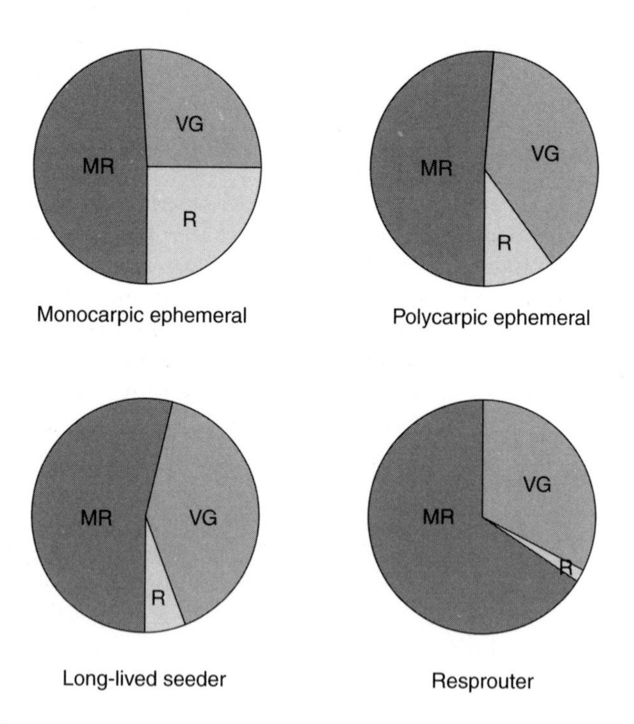

Figure 5.18

The relative partitioning of annual biomass production into maintenance respiration, vegetative growth and reproduction, in four classes of plant species adapted to the kwongan of Western Australia (from data of Bell *et al.*, 1993).

pre-fire plant height, in determining the growth rate of shrubs; this was interpreted in terms of varying availability of soil water and inorganic nutrients.

2. Life Histories in the *Kwongan*: Ephemerals, Obligate Seeders and Resprouters

The *kwongan* of Western Australia, on deep infertile sands subject to seasonal drought and regular firing has been the subject of intense investigation in the 1980s and 1990s (Pate, 1993). Most of the native plant species can be classified as ephemerals, obligate seeders or resprouters, although there is some degree of overlap between classes (Bell *et al.*, 1993). Here we concentrate on adaptations to fire, but kwongan species also display a range of notable adaptations to low supplies of phosphate and nitrogen (e.g. the proteid root mats of the *Proteaceae* and some legumes; Pate, 1993; Chapter 3).

Since the probability of successful germination and establishment is generally low in such dry infertile soils under a relatively continuous mature canopy, the proliferation of annual plants (monocarpic ephemerals) tends to be closely associated with hot fires which remove competitors for water and radiation, and concentrate nutrients at the soil surface. These species, released from seed dormancy by fire or blown in from surrounding areas, show very high relative growth rates, high harvest indices (Fig. 5.18; Plate 12b), and effective partitioning of mineral nutrients to the seeds (Pate *et al.*, 1985). Polycarpic ephemerals, which flower first in the year after a fire and continue to set seed for a further two or three years, can be seen as an intermediate class between monocarpic ephemerals and obligate seeders.

Plants classed as *obligate seeders*, including herbaceous perennials, shrubs and trees, and representing around one-third of kwongan species, are killed by bushfires and must re-establish themselves from seed stored either in the soil or in the canopy. The remaining two-thirds of the vegetation is largely made up of *resprouters*, whose photosynthetic tissues are destroyed by fire, but which can regrow from epicormic buds protected under thick bark (Plate 12a), or from underground lignotubers, corms, bulbs, rhizomes or roots. Resprouters, ranging from small monocots to forest eucalypts (in wetter areas), do also set seed but less prolifically and successfully than obligate seeders (Bell *et al.*, 1993).

These two groups have fundamentally different approaches to the reconstruction of the leaf canopy. Obligate seeders show very much higher rates of photosynthesis, growth and leaf expansion than resprouters in the first three or four years from germination (i.e. after a fire). Growth tends to be vertical at first, followed by spreading of the upper canopy to maximize interception of radiation. Reproduction by seed begins early in the life history, in some species in the second season after establishment. On the other hand, resprouters tend to partition a higher proportion of current assimilate to storage.

Figure 5.18 shows that these differences are clearly reflected in the partitioning of annual dry matter production. Resprouters restart photosynthesis from a substantial plant mass which makes higher maintenance demands than is the case for seedlings, which devote more resources to vegetative growth and early reproduction. Ephemerals, of course, devote a much higher proportion of photosynthate to seed production.

These metabolic differences are matched by other adaptations. For example, in a study of the kwongan *Restionaceae* (rhizomatous monocots), resprouter rhizomes (which survive fires) were larger, represented a higher proportion of plant biomass, contained higher concentrations of soluble carbohydrates, and maintained perennating buds deeper in the soil than obligate seeders (whose below-ground tissues tend to be killed by fire) (Pate *et al.*, 1991). A similar range of differences is shown by kwongan dicot species (Pate *et al.*, 1990).

There is, thus, a contrast between the conservative life styles of the resprouters (recreating the leaf canopy slowly from buds released from dormancy, making modest demands on soil resources, and relying more on the long life of each individual than on reproductive propagation) and the obligate seeders (which, by relying on propagation by seed and rapid growth, risk losses under unfavourable conditions at germination, and seed predation (Enright and Lamont, 1989), and make heavy demands on soil nutrients and water in the first few years of life). The relative success of each group, and their representation within the canopy depend, therefore, on a range of factors in addition to the intensity and frequency of fire (Fox and Fox, 1986).

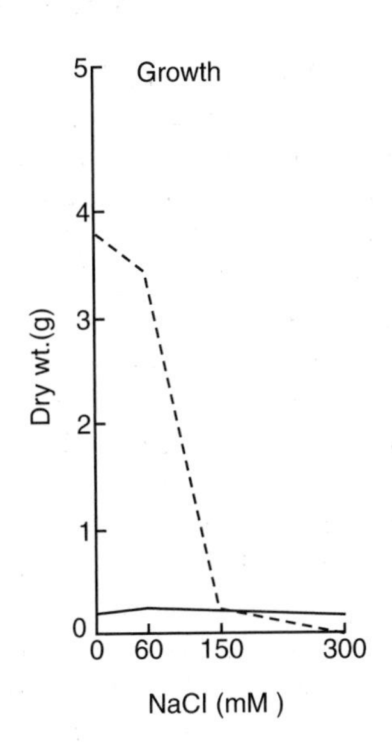

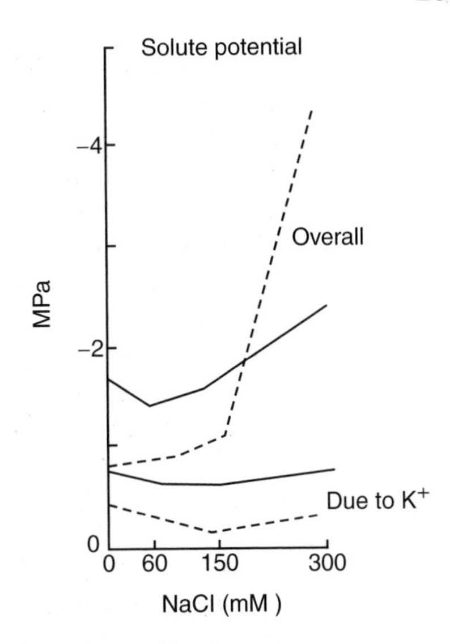

Figure 6.5

Growth, and solute potential of the shoot tissues, of *Juncus maritimus* (solid line) and *J. bufonius* (dashed line) over a range of concentrations of NaCl in the growing medium. *J. maritimus* shows the characteristics of a conservative life style, continuing to grow at a low rate across a wide range of NaCl concentrations, whereas, in *J. bufonius*, there is an association between the sharp inhibition of growth and increase in solute potential caused by ions other than K^+ (from Rozema, 1976)

2. Direct modification of the environment

Some plant species have the ability to alter the environment in their immediate vicinity, particularly the pH of, and oxygen supply to, the rhizosphere, such that they can avoid toxic stress (Plate 6). For example, cation uptake by plant roots generally exceeds anion uptake, stimulating an efflux of H_3O^+ ions into the soil (Darrah, 1993). Where NO_3^- is the main source of N, anion uptake is higher, with the result that the efflux of H_3O^+ is reduced and, under some circumstances, there can be a need for compensatory extrusion of OH^- and HCO_3^- ions, thereby raising rhizosphere pH. In contrast, reliance on NH_4^+ tends to intensify soil acidification. Such differences in nitrogen nutrition, which can result in variation in the pH of the rhizosphere by up to 2 units, explain the responses of calcicoles and calcifuges to acidity under varying nitrogen regimes. Thus, *Rumex acetosa* (neutrophile) and *Scabiosa columbaria* (calcicole) were severely inhibited by low pH in the presence of NH_4^+, but not NO_3^-, whereas *Deschampsia flexuosa* (calcifuge) was inhibited by high pH in the presence of NO_3^-, but insensitive to pH if NH_4^+ was the nitrogen source (Gigon and Rorison, 1972).

Similarly, the ability of cereals to resist Al^{3+} toxicity appears to be related to their nitrogen nutrition; those cultivars which take up NH_4^+ most rapidly, causing the greatest reduction in rhizosphere pH, are the most susceptible to Al (Taylor and Foy, 1985), and resistance is associated with the ability to raise rhizosphere pH (Scott and Fisher, 1989). Such changes in pH, and changes in the availability of metal ions brought about by more specific interactions with plant root exudates, form the basis of the technology of phytoremediation (p. 279).

In a wide range of herbaceous and woody perennials adapted to wetlands, the presence of large continuous air spaces in the root cortex (aerenchyma) facilitates the diffusion of oxygen from the free atmosphere through shoot and root to flooded organs (Fig. 6.6; Armstrong *et al.*, 1994; Jackson and Armstrong, 1999). In some of these species, flooding also stimulates the elongation of leaf blades, petioles and/or stems such that physical contact with the air is maintained (Musgrave and Walters, 1973; Banga *et al.*, 1995). The movement of oxygen *via* stomata, stem lenticels and root aerenchyma is commonly sufficient to oxidize the rhizosphere soil, as clearly shown by the deposits of red ferric hydroxide on the roots and rhizomes of wetland species (Fig. 6.7; Plate 6b). Other morphological features favouring the supply of oxygen to flooded root tissues include superficial mats of adventitious roots which, in trees, can be associated with more vertical 'sinker roots'; and the pneumatophores of mangroves and swamp cypresses (*Taxodium* spp.), whose lenticels above the water level give oxygen direct access to the aerenchyma of submerged portions.

Wetland species differ in the capacity of the aerenchyma system to deliver oxygen to the rhizosphere; in the distance over which such transport can take place; and in the critical oxygen partial pressure required to maintain growth and function (Armstrong and Webb, 1985). Where it is effective, internal ventilation of flooded organs not only ensures the continuation of efficient aerobic respiration (Fig. 6.4), but it also immobilizes and excludes potentially toxic ferrous and manganous ions, and reduces both the loss of NO_3^- by denitrification, and the reduction of sulphate to toxic H_2S in the immediate vicinity of the root. Since roots growing in flooded soil tend to affect the pH as well as the redox status of the rhizosphere (Plate 6), the net effect on the supply of toxic and essential trace elements to the root can be complex. For example, the oxidation of ferrous ions by rice roots growing in a flooded alkaline soil (pH 7.3) not only immobilized the iron but also enhanced the acidification of the rhizosphere; in turn, trace levels of Zn, mobilized by acidification, were adsorbed in exchangeable forms by the precipitated ferric hydroxide (Kirk and Bajita, 1995). Thus, chemical events induced by the root protected the plants against the toxins and wasteful fermentation associated with anaerobic soil conditions, and enhanced their zinc nutrition (Fig. 6.7).

2. Exclusion

In principle, the exclusion of toxins would seem to be the most appropriate mechanism of resistance, but exclusion brings with it problems, both in relation to the ability of the plant to exclude the toxin (recognition; damage at the root surface), and the consequences of exclusion (osmotic effects; deficiency).

1. Recognition

The roots of all plants have highly selective uptake systems which are capable of distinguishing between chemically similar ions, but some ion pairs pose serious problems, particularly K^+/Na^+, Ca^{2+}/Mg^{2+} and phosphate/arsenate. At low

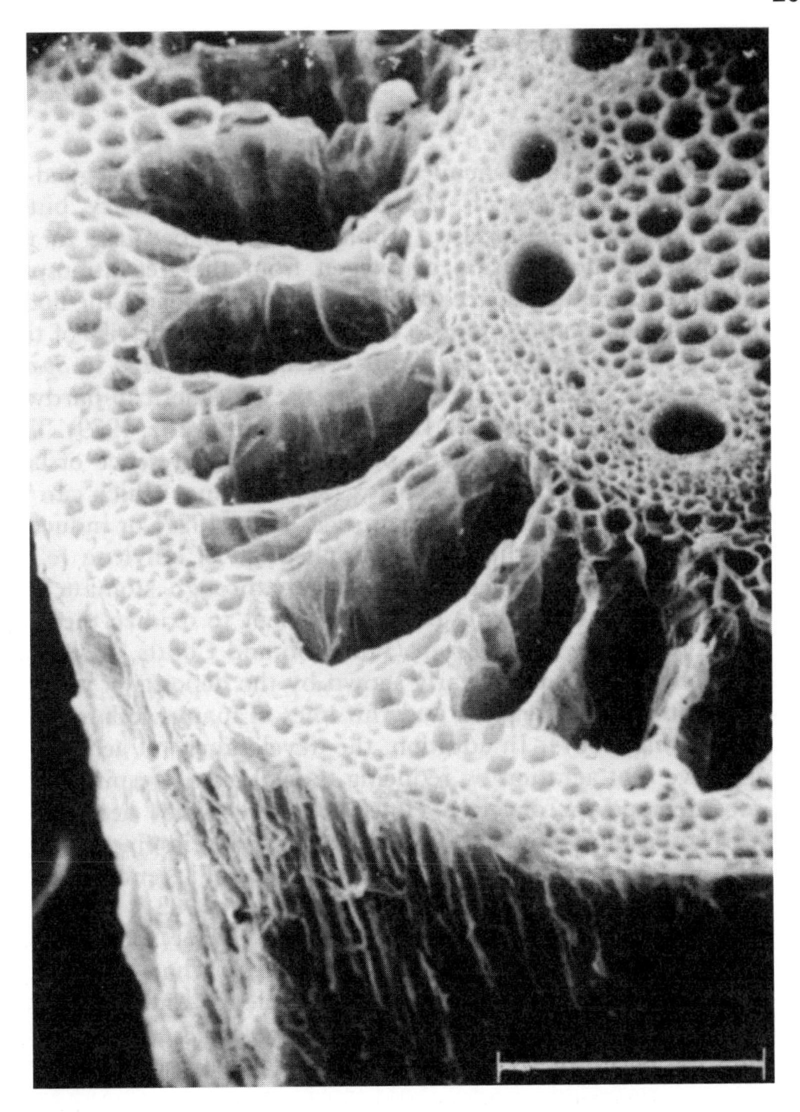

Figure 6.6

Cortical air spaces induced by oxygen deficiency in a nodal root of maize. The parent plant was grown in aerated culture solution before being transferred to unaerated culture solution, which rapidly became oxygen deficient. The original root system died but, within 9 days, had been replaced by new nodal roots with well-developed aerenchyma (cortical tissue containing large air spaces) formed by the controlled collapse of cortical cells. The development of aerenchyma under such conditions can be inhibited by the presence of Ag^+ ions (which suppress the activity of ethylene in plant tissues). Scanning electron micrograph of a transverse section 10 cm from the root tip (bar = 200 μm) kindly supplied by Dr M.C. Drew

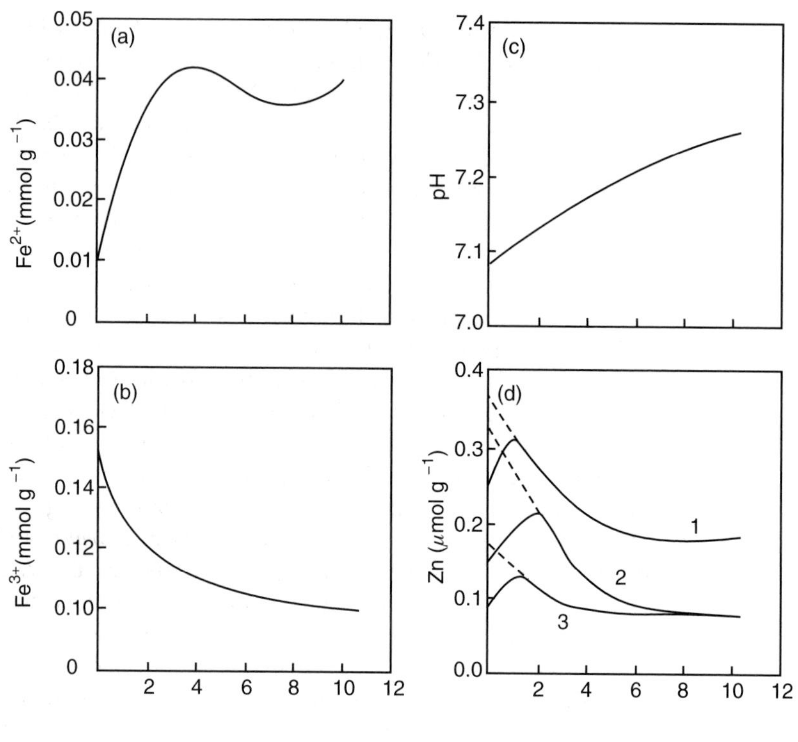

Figure 6.7

Profiles (solid lines) of (a) Fe^{2+}, (b) Fe^{3+}, (c) pH and (d) exchangeable Zn near the rhizoplane of rice roots grown in cylinders of anaerobic soil for 12 days. The types of exchangeable Zn were: 1, Zn in amorphous hydroxides, carbonates and sulphides, and adsorbed on negative charges, extracted with strong acid; 2, Zn complexed with more soluble organic matter, extracted with copper acetate solution; 3, Zn bound in amorphous sesquioxides, extracted with ammonium oxalate. The dashed lines in (d) are the predicted profiles, indicating that Zn was absorbed by the root from each pool of exchangeable Zn (simplified from Kirk and Bajita, 1995)

K^+ concentrations (around 0.5 mM), the selectivity of high-affinity, carrier-mediated, K^+ uptake is high, whereas, at higher concentrations (>1 mM), uptake is *via* lower-affinity ion channels which are generally less selective (Maathuis and Sanders, 1996; Kochian *et al.*, 1993) (p. 98). The lower-affinity system is, however, more effective at selection in many halophytes; for example, plants of the C_4 halophytic grass *Leptochloa fusca*, grown in 100 mM NaCl with 5 mM K^+, had an uptake selectivity of 11.3 in favour of K^+ against Na^+ (Jeschke *et al.*, 1995). This characteristic was linked with rapid cycling of K^+ within the plant, and excretion of absorbed Na^+ by salt glands in the leaves.

Such discrimination at the root surface is by no means universal among halophytes. The accumulation of very high concentrations of cations in the leaves is a feature of the Chenopodiaceae, a family which includes a high proportion of all halophytic dicots as well as glycophytic species (Reimann, 1992); but the discrimination against Na^+ appears to be greater in the *glycophytes* (Fig. 6.8). The halophytes (*Atriplex rosea, A. prostrata* and *Suaeda maritima*) differ in retaining a higher proportion of the absorbed K^+ in the root system, and transporting proportionately more Na^+ to the shoot. This mechanism, which involves recognition of ions, and preferential transport of Na^+ to the root xylem, appears to act to protect root function against toxicity; it is practicable only in species possessing adaptations which enable the leaf canopy to function in the presence of very high loads of Na^+, or to excrete salt (see p. 277).

Since most plants discriminate poorly between Ca^{2+} and Mg^{2+}, toxic levels of Mg accumulate in plants growing in soils with very low Ca : Mg ratios (e.g. Proctor, 1971). Such conditions exist in soils developed from ultramafic or serpentine rocks (see p. 244), and may account for the general paucity of vegetation, although high levels of Ni, Co or Cr, and low levels of macronutrients, also play a part in some habitats (e.g. Nagy and Proctor, 1997). Those plants, such as ecotypes of *Agrostis stolonifera* and *A. canina*, which do grow on serpentine soils, appear to take up the two ions largely in proportion to the external concentrations, in contrast to more susceptible genotypes; their resistance mechanisms must, therefore, be internal.

Advances in molecular genetics and in membrane physiology have revealed that, for certain crop species at least, recognition of the toxin by the root apex plays a major part in resistance to Al^{3+}; exposure to Al^{3+} triggers the opening of anion channels, resulting in efflux of malate, citrate or succinate ions into the rhizosphere, and the associated enhancement of synthesis of organic acid anions. In contrast to Fe, whose uptake is enhanced by chelation, the formation of complexes between Al^{3+} and malate or citrate renders the toxin unavailable for uptake by the root tip (Kochian, 1995; de la Fuente *et al.*, 1997). Here the effectiveness of recognition and exclusion depends on chemical inactivation.

Recognition at the leaf surface does not appear to play a simple role in the exclusion of pollutant gas molecules from the leaf mesophyll; this is not suprising in view of the many cues to which stomata are responding (Winner *et al.*, 1991; see p. 149). Exposure of plants to moderate to high levels of SO_2, NO_x, O_3 and mixtures of these toxins does normally result in stomatal closure (Darrall, 1989; Ashenden *et al.*, 1995; Mansfield, 1999), but this is the end point of a complex signalling response to pollutants which have gained access to leaf cells (e.g. for O_3, Moeder *et al.*, 1999; see also Atkinson *et al.*, 1991; Reiling and Davison, 1995). Interactions among stomatal responses can be important in determining the dose received; thus, plants with undamaged, but partly closed stomata, for example owing to water stress or elevated CO_2 concentration, are more resistant to damage by O_3 (McKee *et al.*, 1997).

2. Damage at the root surface

Even if the root is effective at preventing the entry of a toxic ion into its cells, this does not necessarily mean that damage is prevented. A wide array of metabolically active structures (e.g. proton-translocating ATPases, transporters and ion channels – to name only those involved in ion uptake) are exposed on

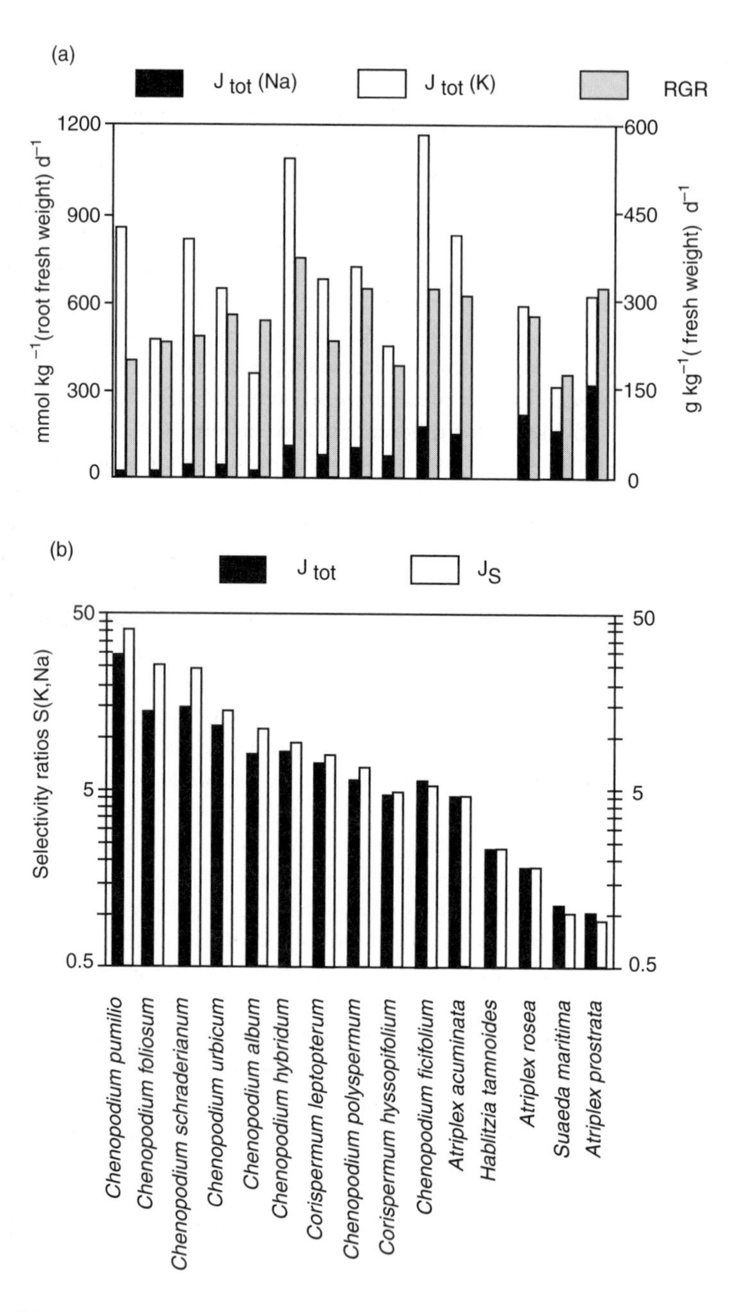

Figure 6.8

(a) the relative growth rate, and uptake of K^+ and Na^+ ions, each on a fresh weight basis, and (b) the ratio of K:Na in uptake to the whole plant (J_{tot}) and to the shoot (J_s), for plants of 15 species of the Chenopodiaceae grown for 2 weeks in sand supplied with 10 mM each of K^+ and Na^+ ions in a complete nutrient solution. Note that the *y*-axis in (b) is logarithmic, and that the data for *H. tamnoides* are incomplete. Each of the halophytes (*A. rosea, S. maritima* and *A. prostrata*) showed poor discrimination against Na^+ both at the root surface and between root and shoot, but only *S. maritima* demonstrated a conservative life style (low growth rate). The other, predominantly glycophytic, species had higher discrimination against Na^+ both at the root surface and between root and shoot (from Reimann and Breckle, 1993)

the outer surfaces of plasmalemmae of root epidermal cells; if the toxin is mobile in the apoplast, such structures on the membranes of all the cells up to the endodermis are potentially-vulnerable. Furthermore, the very act of exclusion will tend to raise the concentration of toxin in the rhizosphere. Examples of root surface damage include the blocking of ion channels by Al^{3+} ions (Pineros and Tester, 1993), and the increased membrane leakiness caused by Cu^{2+} ions (Strange and Macnair, 1991). In contrast, there is evidence of salt-induced *synthesis* of proton-translocating ATPases on the plasmalemmae of the roots of *Atriplex nummularia,* although their normal function at high salt concentrations remains to be demonstrated (Niu *et al.*, 1993).

Evidence for differences between resistant and susceptible plants expressed at the level of surface enzymes is provided by the classic investigation by Wainwright and Woolhouse (1975) of the acid phosphatases on the surfaces of root cells of races of *Agrostis capillaris.* Kinetic analysis indicated that Cu caused non-competitive inhibition of the enzymes of plants susceptible to Cu toxicity, such that the adaptive differences between resistant and susceptible races were related to differences in the inhibition constant (k_i = 1.50 mM Cu^{2+} for the resistant, and 0.54 for the susceptible races, a smaller value signifying that the enzyme has a greater affinity for Cu, and forms a non-functional complex more readily). Such experiments, using root fragments, need to be repeated with purified enzymes and intact organs, to evaluate the potentially protective role of cell walls, and the possible artefacts of an *in vitro* system (Verkleij and Schat, 1990).

3. Deficiencies: ionic imbalance

As a mechanism for resistance to toxicity, exclusion can lead to problems if the potentially toxic ion is also an essential micronutrient; this suggests that exclusion might be more appropriate in dealing with non-essential ions (e.g. Al, Pb, Cr) than Cu, Zn or Fe. There is insufficient evidence to judge whether this is the case, although it should be noted that resistance to Zn in *Agrostis tenuis* (*A. capillaris*) can involve either exclusion (Mathys, 1973) or accumulation (Wu and Antonovics, 1975) of the toxic ion. The related hypothesis, that resistant plants have an intrinsically higher requirement for the relevant micronutrient, is not supported by detailed studies of Cu-resistant *Mimulus guttatus* (Harper *et al.*, 1997, 1999).

The apparently anomalous stimulation of the growth of calcifuge species (Clarkson, 1966; Fig. 6.9) by low concentrations of (apparently non-essential) Al^{3+} ions, can be explained in terms of Al-binding sites on the root surface. When these are unoccupied by Al^{3+}, they tend to bind Fe^{3+} ions which are of similar size, thus causing Fe deficiency. Grime and Hodgson (1969) showed that growth responses to Al tend to be biphasic (Fig. 6.9b, c), with an initial decline at very low concentrations, followed by a minor peak at slightly higher concentrations, apparently caused by the liberation of Fe from binding sites on cell walls. Finally, at higher concentrations, growth is again reduced. The Al concentration for peak growth varied with the pH-resistance of the species or ecotype, from 20–30 μM in calcifuges such as *Nardus stricta* and *Ulex europaeus*, to 5 μM in calcicoles (*Bromus erectus, Scabiosa columbaria*) (Fig. 6.9b, c).

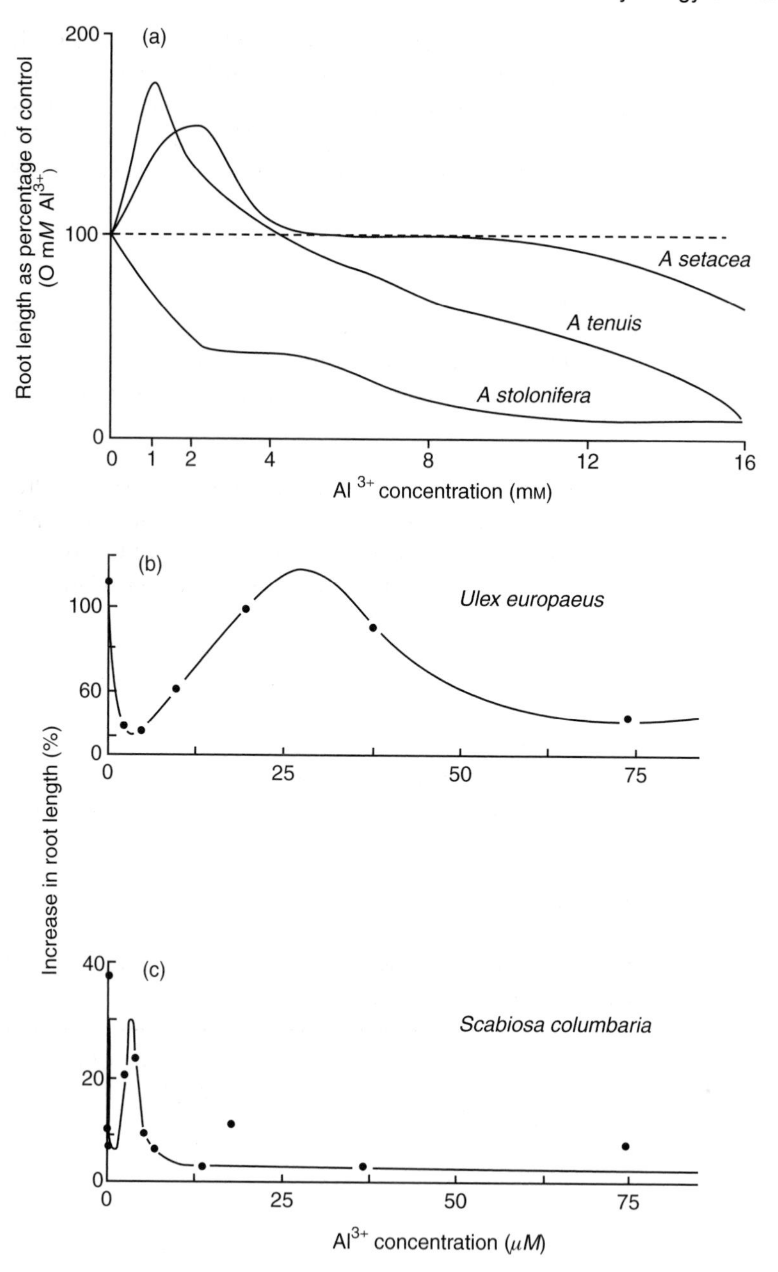

Figure 6.9

The effects of Al^{3+} ions on: (a) the extension of root axes of the pasture grasses *Agrostis setacea* (calcifuge), *A. tenuis (A. capillaris)* (mildly calcifuge), and *A. stolonifera* (mildly calcicole) over 1 week (from Clarkson, 1966); (b) and (c) the extension of root axes of iron-deficient *Ulex europaeus* (gorse) and *Scabiosa columbaria* (small scabious) (redrawn from Grime and Hodgson, 1969). Each species, except *A. stolonifera*, showed stimulation at a low concentration of Al^{3+}, although the critical concentration for stimulation varied considerably (<10μM to 3 mM)

Arsenate poses a unique toxicity problem. It is sufficiently similar chemically to phosphate to replace it in P-containing compounds (and render them ineffective), and to be transported across membranes at the root-soil interface by the same carrier as phosphate. Plants growing on As-contaminated soils achieve resistance to arsenate by reduced activity of the high affinity phosphate transport system, an ability that is under the control of a single major gene (Macnair *et al.*, 1992).

3. Amelioration

If it proves impossible to exclude a toxin, and high internal concentrations are inevitable, then the toxin must be removed from sites of active metabolism in the cytoplasm (amelioration), or tolerated. Possible approaches to amelioration include:

(i) *Localization:* either intra- or extracellularly, and usually in the root system;
(ii) *Chemical inactivation:* such that the toxin is present, but in a combined form of reduced toxicity (e.g. the conversion of SO_2 to SO_4^{2-} in plant leaves);
(iii) *Dilution:* which is of primary significance in relation to salinity (note that many halophytes are succulent; Popp, 1995);
(iv) *Excretion:* either actively *via* shoot glands or the root surface; or passively by abscission of old leaves in which toxins have been accumulated.

1. Localization

Separate and comprehensive chemical analyses of the root and shoot systems of plants are rare. The data compiled by Tyler (1976) (Table 6.4) suggest that, for

Table 6.4

Shoot concentrations, and ratios of root concentration to shoot concentration, for 12 elements in *Anemone nemorosa* (from Tyler, 1976)

		Shoot concentration ($\mu g\ g^{-1}$)	Root: shoot ratio
Widely distributed	Ca	7180	0.8
	Mg	2970	1.3
	K	11,400	1.3
	Rb	35	1.6
Slight root storage	Mn	405	2.3
	Cu	10.5	2.4
	Zn	113	3.6
Major root storage	Al	260	10.7
	Cd	1.24	11.7
	Fe	217	12.4
	Na	242	16.7
	Pb	1.04	62.6

Anemone nemorosa, a widely distributed herb of deciduous woodland, metallic elements can be divided into three groups:

(i) those more or less uniformly distributed throughout the plant, including the macronutrients K, Ca and Mg, which are essential for the normal function of all cells. Rb behaves similarly to K in most plants;
(ii) those showing some accumulation in the root, including three important micronutrients Cu, Zn and Mn;
(iii) non-essential, predominantly toxic, metals, such as Al and Cd, which are *primarily* located in the root. The micronutrient Fe also falls into this class, presumably because of its high availability in soils (compared with Cu, Zn and Mn).

Such preferential accumulation of toxic ions in root systems is a widespread phenomenon (Baker and Walker, 1990) but, without detailed study of the cellular location of the toxins, it is not possible to discriminate between amelioration (i.e. action after uptake, for example, by excretion and immobilization in the apoplast) and exclusion (see above). To add to this complexity, it has been proposed that variation in resistance to Al, Zn and Pb is related to the chemistry of the pectinic acid fraction of the cell wall, and that *extracellular* enzymes might be responsible for renewal of binding sites (e.g. Wainwright and Woolhouse, 1975).

Resolution of such questions may eventually come from an extension of the approach of Brune *et al.* (1995) to plants which are more resistant to toxic metals. Using a commercial cultivar of barley, they showed that exposure to damaging levels of Cd^{2+}, MoO_4^{2-} or Zn^{2+} ions increased the proportion of the metal partitioned to the root, to the leaf epidermis (compared with the mesophyll), and to leaf cell vacuoles (compared with cytoplasm) (Fig. 6.10). There was no evidence of a role for metal-binding proteins or peptides, whose involvement in amelioration seems to be most marked for Cd^{2+} (Macnair, 1993; Kochian, 1995; see below). The distribution of each metal in the leaf, and the related symptoms of toxicity, were interpreted in terms of concentration of the toxic ions in the epidermis *via* transpirational flow, and a degree of sequestration in vacuoles. The pattern for Ni^{2+} was slightly different: there was a lower uptake overall, a reduction in proportional partitioning of the ion to the root, and less concentration of the ion in leaf epidermes and vacuoles (Fig. 6.10).

Even in halophytes, high cytoplasmic concentrations of Na^+ (>100 mM) are considered to be toxic, inhibiting enzymes by disrupting their conformation. In many resistant species, proton-transporting ATPases, acting inwardly at the tonoplast, provide the driving force for the accumulation of Na^+ (by a H^+/Na^+ antiport system) and Cl^- (uniport) in the vacuole at concentrations up to 8 times those in the cytoplasm (Blumwald and Gelli, 1997; Amtmann and Sanders, 1999). To maintain osmotic balance across the tonoplast, the sequestration of salt must be accompanied by a matching accumulation in the cytoplasm of solutes which, unlike Na^+, do not interfere with enzyme reactions at very high concentrations (i.e. 'compatible solutes'; Rhodes and Hanson, 1993) (Fig. 6.11).

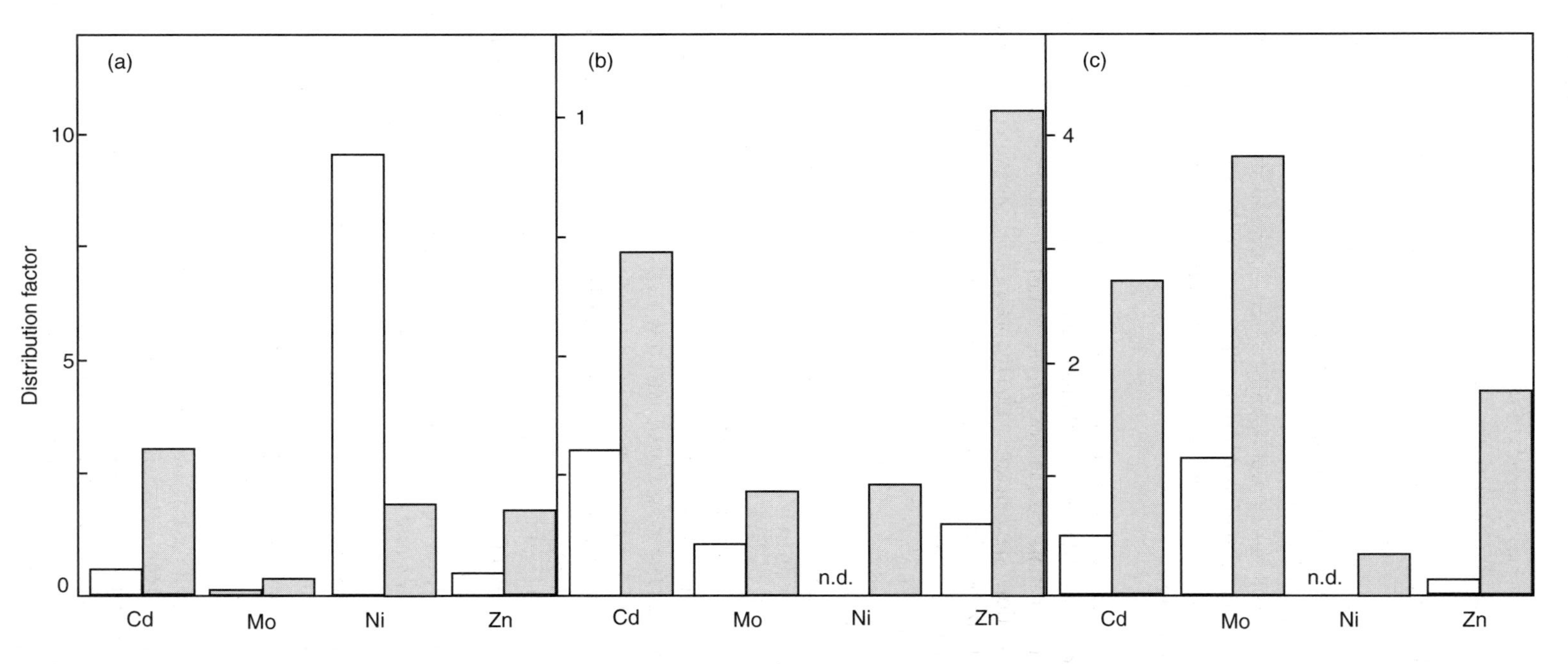

Figure 6.10

The fate of heavy metal ions supplied in culture solution at 1.5 μM (controls), 100 μM (Cd, Ni) or 400 μM (Mo, Zn) to 10-day old barley seedlings, as shown by the distribution factor (i.e. the ratio of metal content in different organs): (a) ratio of root to leaf content; (b) leaf epidermis to mesophyll; and (c) leaf cell vacuole to cytoplasm. Open columns, controls; shaded columns, plants receiving higher supplies of metal ions. The results showed a tendency for Cd, Mo and Zn to be retained in the root, and for the fraction of metal which reached the leaves to be concentrated in the vacuoles of epidermal cells. The pattern was more complex for Ni (from Brune *et al.*, 1995)

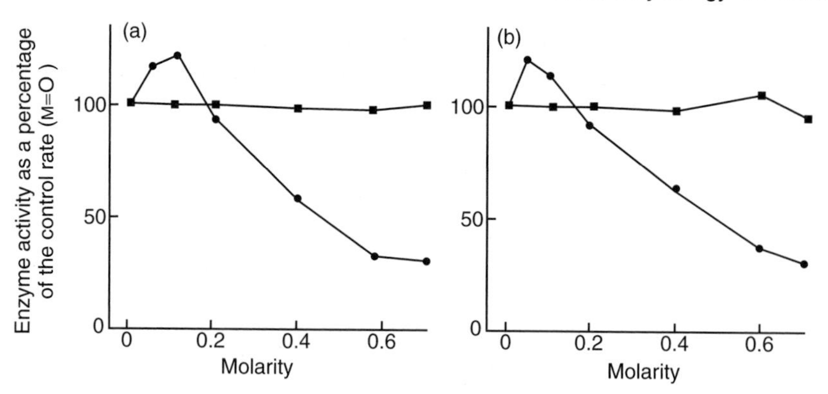

Figure 6.11

The effects of varying the concentration of proline (■) and NaCl (●) on the activity of two enzymes extracted from plants of the halophyte *Triglochin maritima*: (a) glutamate dehydrogenase; (b) nitrate reductase (Reprinted from Stewart, G.R. and Lee, J.A. (1974) The role of proline accumulation in halophytes, *Planta* **120** 279–289. © 1974, with permission from Springer-Verlag, Berlin.)

In an early survey, Stewart and Lee (1974) found that the amino acid proline was used by many halophytes for osmotic balance; in a wide range of species, proline represented, on average, 55% of the amino acid pool in halophytes, compared with 3%, 2% and 4%, respectively in calcicoles, calcifuges and ruderals. Proline accumulation was directly related to the salinity of the medium, at levels up to 700 mM, which were comparable with measured internal levels of NaCl; it had no effect on the activity of nine different enzymes extracted from the halophyte *Triglochin maritima* at concentrations where NaCl was highly inhibitory (e.g. Figure 6.11). This concept of protection by compartmentation of NaCl, and the use of compatible solutes, was further supported by the observation that the ratio of cell proline to vacuolar Na^+ and Cl^- tended to remain constant in halophytes (e.g. Treichel, 1975). In those halophytes which do not accumulate proline (e.g. 1 out of 12 temperate species studied by Stewart and Lee, 1974), its role can be taken by quaternary ammonium compounds (of which, glycinebetaine is the most common, Storey *et al.*, 1977; Rhodes and Hanson, 1993) or by polyhydric alcohols (in *Mesembryanthemum crystallinum*, Bohnert *et al.*, 1995; Popp and Smirnoff, 1995). For example, of the sixteen native South Australian halophytes examined by Poljakoff-Mayber *et al.* (1987), six accumulated proline, eight glycinebetaine and two others unusual amines or alcohols. However, in the same investigation, several drought-resistant species also accumulated compatible solutes, which may also act to protect the conformation of enzymes under dehydration.

Evidence for the compartmentation of NaCl and compatible solutes *within* the cytoplasm of halophytes is equivocal, with accumulation of Cl^- in chloroplasts recorded in *Limonium vulgare* (Larkum, 1968), and *Suaeda maritima* (Harvey *et al.*, 1981), but apparently not in *Mesembryanthemun crystallinum*

(Demmig and Winter, 1986). If confirmed, the implication of such behaviour would be that the photosynthetic machinery is more tolerant of high concentrations of NaCl than are cytoplasmic enzymes.

2. Chemical inactivation within plants

Plant tissues can be exposed to active oxygen and other powerful oxidants (p. 248) originating in the external environment (particularly atmospheric O_3) or generated within their own tissues. Any combination of factors which lead to a low use of intercepted radiant energy by the chloroplasts (radiation saturation of photosynthesis; high irradiance at low temperature; stomatal closure etc.) will tend to cause a build-up of potentially damaging oxidants, but it is now clear that exposure of plants to a very wide range of stresses, as well as natural ageing, can have a similar effect (e.g. drought, low and high temperature, exposure to UV, other air pollutants, post-anoxic stress, metal toxicity and deficiency, and disease stresses; Table 6.5). Resistance to free radicals and oxidants must, therefore, be one of the common, and unifying, aspects of plant resistance to stress, and it provides an explanation for the widely recognized protective effect of exposure to one type of stress against subsequent exposure to another, apparently unrelated, stress (Crawford *et al.*, 1994).

Oxidants are generated in plant tissues throughout life, even when the plant is not under stress, but they are rendered harmless by a range of antioxidant molecules (e.g. the ascorbic acid/tocopherol/glutathione system, flavonoids in general, and certain terpenoids, including carvacrol and thymol; Smirnoff, 1995) and enzymes (particularly superoxide dismutase (SOD), ascorbate peroxidase (APX) and catalase; Allen, 1995). Exposure to stress does tend to increase the production of antioxidants (Smirnoff, 1995), and there is extreme variation among species in the levels of antioxidant agents (Hendry, 1994). The localization of ascorbate in cell walls (Smirnoff, 1996) might indicate a role in intercepting incoming O_3, but Buckland *et al.* (1991) found that the very high concentrations of ascorbic acid found in scurvy grass (*Cochlearia atlantica)* did not provide particular protection under stressful conditions, and that other species appeared to deal more effectively with the risk of oxidation by increased levels of reducing enzymes. Elsewhere, there are reports of stress factors causing *reductions* in enzyme activities (e.g. SOD and glutathione reductase in *Pinus sylvestris* needles exposed to SO_2 and NO_x; Wingsle and Hällgren, 1993). There is, thus, no consensus at present about the adaptive value of differences in antioxidant production and localization.

The adaptation of some species to flooding may involve chemical inactivation of active oxygen. All cell membranes can be damaged by the withdrawal of oxygen, but it is widely observed, in plant and animal systems, that the effects of resupply of oxygen (post-anoxic stress) are more serious. Thus, the rhizomes of (flood-tolerant) *Iris pseudacorus* survive longer under anoxia than those of the closely related (intolerant) *I. germanica*, and only the latter suffer peroxidation of membrane lipids when oxygen supply is re-established, as a consequence of the generation of free radicals (Hunter *et al.*, 1983; Crawford *et al.*, 1994). The biochemical basis for this difference is complex, involving SOD, ascorbic acid and glutathione, and has yet to be thoroughly characterized (Monk *et al.*, 1987; Wollenweber-Ratzer and Crawford, 1994). It is also difficult to evaluate the

Table 6.5

Examples of environmental stresses where active O_2 has been implicated in damage to plants. (Reproduced by permission of the Royal Society of Edinburgh and G.A.F. Hendry (1994) *Proceedings of the Royal Society of Edinburgh, B: Biological Sciences*, **102**, 155–165.)

Environmental stress or event	Plant tissue or species
Natural ageing (senescence)	Seeds
	Leaves
	Petals
	Root nodules
Disease	*Solanum tuberosum* infested with *Phytophthora infestans*
Water stress	
Before germination	*Quercus robur*
During germination	*Zea mays*
Seedlings	*Helianthus annuus*
Mature tissue	*Triticum aestivum*
Temperature	
Freeze-thaw	*Oryza* cell cultures
Low temperature–high light	*Spinacia oleracea*
High temperature	*Helianthus tuberosum*
Atmosphere	
O_3	*Pisum sativum*
O_2-hyperoxia	*Zea mays*; nitrogen-fixing root nodules
O_2-post-anoxia	*Cicer arietinum*
SO_2	*Hordeum vulgare*
SO_2–NO_2	*Pinus sylvestris*
CO_2	*Glycine max*
Nutritional effects/toxicity	
Heavy metals: Cu	*Silene cucubalus*
Cd	*Holcus lanatus*
Al	*Glycine max*
Fe	Various
Ca	*Commelina communis*
NaCl	*Pisum sativum*
Mineral deficiencies	
Ca	*Solanum tuberosum*
Mg	*Phaseolus vulgaris*
K	*Phaseolus vulgaris*
Light	
High irradiance – drought	*Tortula ruraliformis*
High irradiance – nutrient stress	*Phaseolus vulgaris*
Darkness	*Hordeum vulgare*
UV-B	Various

effects of the exposure of surviving tissues to the wide range of chemical species released from sequestration when cells are killed.

Most of the heavy metals which are toxic to plants can be complexed in the plant in some way. Cd can certainly be rendered chemically inactive in cells in the form of complexes with polypeptides called phytochelatins, whose synthesis is induced by exposure to Cd^{2+} ions in the root environment (Rauser, 1990). The complexes can then be excluded from the cytoplasm by transport across the tonoplast (Salt and Rauser, 1995) into the vacuole, where they are probably stabilized by binding to sulphide. The importance of phytochelatins in determining Cd resistance is shown by the *cad1* mutant of *Arabidopsis*, which is both deficient in their production and more sensitive to cadmium than the wildtype (Howden *et al.*, 1995); the CAD1 gene is known to encode the enzyme phytochelatin synthase, in *Arabidopsis*, yeast and bacteria and, possibly, in animals (Ha *et al.*, 1999).

Resistance to Zn appears to be associated with ion transport: van der Zaal *et al.* (1999) showed that when *Arabidopsis* was engineered to overexpress a Zn transporter, analogous to those found in animals, its resistance to Zn was enhanced. Zn can also be sequestered in vacuoles, for example in the zinc hyperaccumulator *Thlaspi caerulescens* (Vasquez *et al.*, 1994); and inactivated by precipitation as zinc phytate (van Steveninck *et al.*, 1990), although this would be a very expensive mechanism, immobilizing six moles of P for every mole of Zn removed. A different chelation mechanism appears to be involved in resistance to copper: in *Arabidopsis*, synthesis of a class of polypeptides known as metallothioneins is induced by exposure to Cu, and the level of expression of the mRNA that encodes one of the metallothioneins has been found to match the degree of copper resistance (Murphy and Taiz, 1995). The mechanisms by which plants bind, and thus minimize the toxic effect of, metals are apparently unique to each metal; polypeptides and amino acids are both used, the former for Cd and Cu as described above, and the latter (specifically histidine) for Ni in hyperaccumulators (Krämer *et al.*, 1996).

3. Metabolic control of the release of toxins

As shown in Fig. 6.4, anaerobic fermentation within plant cells can release a range of potential toxins, primarily ethanol, acetaldehyde and/or lactate. Early production of lactate tends to induce the synthesis of pyruvic decarboxylase and alcohol dehydrogenase (ADH) such that ethanol is the major toxic product (Vartapetian and Jackson, 1997). It is possible that acetaldehyde plays a part in post-anoxic stress. Plants adapted to survive in flooded soils must, therefore, possess adaptations that restrict or divert the production of such toxins; remove them from sites of active metabolism within the tissue; or promote their loss from the tissue. These adaptations can be classed primarily as *amelioration* but they can also involve *escape* and *exclusion.*

Crawford (1982) proposed that there were four possible metabolic responses to tissue anaerobiosis and the resulting toxins:

(i) acceleration of glycolysis/fermentation (the Pasteur effect), leading to the depletion of carbon reserves, the accumulation of potentially toxic metabolites, and rapid death;

(ii) acceleration of glycolysis/fermentation, but associated with the efficient removal of ethanol by diffusion or active efflux into the root medium or in the gas phase *via* the aerenchyma pathway of well-ventilated species (p. 262);
(iii) no acceleration of glycolysis/fermentation, such that carbon reserves are protected, there is no risk of toxicity, but the energy charge will fall sharply under long-term anaerobiosis, with serious implications for cell integrity;
(iv) diversion of fermentation to generate non-toxic products such as malate;

and that wetland plants possessed biochemical adaptations (for example in relation to the induction of ADH and the possession of ADH isoenzymes) which favoured responses (c) and (d).

Subsequent work, largely but not exclusively concentrating on rice and flood-sensitive species, has failed to provide widespread support for such biochemical adaptation; the 'anaerobic stress proteins' synthesized on exposure to anoxia are predominantly the enzymes of fermentation, and show little variation among species (Armstrong *et al.*, 1994). The current consensus (Vartapetian and Jackson, 1997) involves two principal responses, leading to long-term survival of flooding:

(i) the more common *escape* from anoxia (by internal ventilation, and 'adjustments' of plant development such as rapid shoot extension);
(ii) short-term tolerance of fermentation and its toxic products within anaerobic tissues.

Response (ii) is shown characteristically by shoot rather than root tissues, such that the survival of root systems under anoxic conditions is generally very short in resistant and sensitive species alike. Among shoot tissues, high resistance tends to be shown by the underground rhizomes of marsh species such as *Iris pseudacorus* and *Schoenoplectus lacustris*; substantial carbon stores permit them to survive long periods of induced dormancy while retaining the potential for rapid fermentation and growth in spring to re-establish physical contact with the free atmosphere (Brändle, 1991). The resulting ethanol is excreted into the surrounding aqueous medium. Nevertheless, as shown by the flood-sensitivity of *I. germanica*, such adaptations must be accompanied by the ability to deal with the unavoidable post-anoxic stress.

There remains a need for more systematic study of the mechanisms adopted by the wide range of species, most notably wetland trees, which can survive prolonged flooding, commonly under high temperatures. Some survive by virtue of very elaborate systems of ventilation (e.g. *Taxodium* swamp cypresses and mangroves, Scholander *et al.*, 1955; *Nyssa sylvatica*, Hook *et al.*, 1971, Keeley and Franz, 1979). However, the roots of flood-tolerant species from a gallery forest in Brazil showed a range of biochemical adaptations, in some cases associated with no transport of oxygen from the shoot (Joly, 1994). In considering such species and habitats, it may be necessary to revisit some of the concepts of biochemical adaptation.

4. Excretion and abscission

Unlike animals, whose typical response is to excrete toxins, plants generally exercise control over uptake, and activity within their tissues; this approach arises out of the fact that plants absorb their nutrition in its component parts, rather than as a mixed package of food, allowing greater scope for recognition of toxins. Nevertheless, plants are capable of excretion but, because they are stationary, the excreted toxin will tend to accumulate in their immediate environment. In this regard, aquatic plants are at an advantage.

The simplest form of excretion, suitable for plants but for few animal species, involves the shedding of parts which have accumulated high concentrations of the toxin. It is generally true that older leaves have much higher salt or heavy metal contents than young leaves or buds; for example, old leaves of tea bushes, which are adapted to soils of very low pH, may contain up to 3% Al, most of which is in the epidermis (Matsumoto *et al.*, 1976). Plants which accumulate Al in this way are generally found in the more primitive woody families (Chenery and Sporne, 1976); nickel hyperaccumulators, containing similar concentrations of Ni, complexed with citrate or histidine (Lee *et al.*, 1978; Krämer *et al.*, 1996) are also taxonomically primitive (Jaffre *et al.*, 1978). Accumulation of this kind, therefore, appears to be an evolutionarily ancient phenomenon.

Loss of toxin-laden older leaves is also a feature of salt-affected plants, although the effect is probably more marked in sensitive than resistant populations (Cheeseman, 1988). The relationship between toxic accumulation and abscission is, in general, complex, as high levels of salt or heavy metals will hasten senescence even in highly-resistant species. Thus, the Zambian copper flower *Becium homblei*, which is very effective at excluding Cu from the root and shoot, shows early leaf chlorosis and senescence if the Cu level in the shoot rises (Reilly and Reilly, 1973).

Hyperaccumulation of metals (principally of Ni, for example, in the flora of Cuba; Reeves *et al.*, 1999) is now known in around 400 plant species from 45 families, and the metals concerned also include Zn, and possibly Co, Pb, Mn and Cu. Hyperaccumulation of the non-metal selenium has also been well documented, mainly because it is very toxic to grazing livestock. The phenomenon can be dramatic; Ni can make up 26% of the dry matter content of the sap of *Sebertia acuminata*, a tree from New Caledonia, an ancient island massif in the Pacific, north of Australia. However, the role of hyperaccumulation as a method for excretion is doubtful. An alternative explanation is that it provides protection against either pathogens or, more likely, grazing (e.g. in *Thlaspi caerulescens*, a member of the Brassicaceae which grows almost exclusively on zinc-contaminated soils: Pollard and Baker, 1997). Hyperaccumulation may offer exciting potential for the removal of metals from contaminated soils. This technology, known as phytoremediation, is described below (p. 279).

Although a wide range of species does appear to excrete salt into the rhizosphere from the root surface, many halophytes also protect the root tissues by transporting Na^+ preferentially to the leaves in the xylem (see p. 266). The most important route for salt excretion is through salt-secreting structures in the leaf epidermis; these can be relatively simple trichomes (e.g. the 'bladder hairs' of halophytic *Atriplex* spp.; *Halimione portulacoides*) or more complex multicellular glands (e.g. *Tamarix* spp.; *Limonium vulgare*) (Popp, 1995). In *H. portulacoides*, the trichomes are bicellular, and the distal cells contain conspicuous vacuoles, which accumulate Na^+ and Cl^- when the plants are grown in media with a high salt content (e.g. Baumeister and Kloos, 1974). The accumulated salt is discharged mechanically when the trichomes are brushed, and tends to be deposited on the surfaces of the plant, giving rise to characteristic white incrustations. By contrast, *Mesembryanthemum crystallinum* sequesters salt in epidermal bladder cells, on the margins of the epidermis, which are not shed (Adams *et al.*, 1992).

Table 6.6 indicates the astonishing capacity for salt excretion by multicellular salt glands. Although they are permanent anatomical features of the epidermis, the functioning of such glands appears, at least in some species, not to be constitutive but induced by exposure of the plant to NaCl. For example, the activity of a Cl^-stimulated ATPase in the glands of *Limonium vulgare* increased by 300% if incubated *in vitro* with NaCl, but showed no response to Na_2SO_4 (Hill and Hill, 1973). A flexible system of this kind would seem to be of adaptive value, in controlling the metabolic costs of resistance in environments experiencing fluctuation in salt content. The leaf glands of *Tamarix ramosissima* are particularly effective at excreting potentially-toxic ions: up to 65% of leaf Na^+, 82% of Cl^-, 90% of Al^{3+} and 88% of Si could be removed by washing leaf surfaces, but losses of no other element exceeded 4% (Kleinkopf and Wallace, 1974). Nevertheless, even though net photosynthesis was *stimulated* by NaCl at concentrations in the growing medium of up to 200 mM, growth rate

Table 6.6

Excretion of salt by halophytes (from data compiled by Popp, 1995)

Species		NaCl in growing medium (mM)	Rate of excretion, (nmol m^{-2} (leaf surface) s^{-1})			Scale of excretion
			Na^+	Cl^-	NaCl	
Mangroves						
Aegialitis annulata						
young leaves		50	180	200		26% (of the internal salt content of leaves in 24h)
old leaves		50	80	120		2%
young leaves		500	440	420		38%
old leaves		500	280	250		29%
Avicennia marina		field site			119	
Grasses						
Diplachne fusca	19 °C	250	12	9		5 × rate of uptake in 24h
	39 °C	250	258	283		
Spartina alterniflora		690			41	0.5 × rate of uptake
Spartina anglica		500	12			Total turnover of plant content in 6–18d

was lower even at 10 mM, and fell to 32% of the control at 200 mM. Energy for growth was being diverted to support the respiration required to maintain the excretion of salt.

Other examples of excretion include the loss of ethanol in solution or *via* ventilation of flooded tissues (even in flood-sensitive species such as *Helianthus annuus;* Jayasekera *et al.*, 1990); and the extraordinary emission of excess sulphur, in the form of gaseous H_2S, from the shoots of plants exposed to SO_2 pollution (Taylor and Tingley, 1983).

4. Tolerance

The metabolic systems of some prokaryotes show true tolerance of toxins; enzymes extracted from them can operate *in vitro* in the presence of salt or ion concentrations which would destroy those of eukaryotes (e.g. Brown, 1976). Higher plants and animals generally rely on more complex intracellular compartmentation to ensure that their metabolic systems do not experience high concentrations of toxins; such an approach is, however, inappropriate for enzymes exposed on the surfaces of plasmalemmae or cell walls. There are a few recorded cases of induced synthesis of 'more tolerant' cytoplasmic isoenzymes by exposure of plants to high levels of toxins (e.g. of malate dehydrogenase under salinity; Hassan-Porath and Poljakoff-Mayber, 1969); and there are surprising reports of Cu- and Zn-tolerant acid phosphatases on cell walls (Wainwright and Woolhouse, 1975).

Enzyme adaptation can occur even under mild stress. For example, the cultivated oat, *Avena sativa*, is a calcifuge with an Mg^{2+}-activated ATPase system on the surface of the root cells which operates satisfactorily under conditions of low-Ca^{2+}. By contrast, the closely related bread wheat, *Triticum aestivum*, a calcicole, has a Ca^{2+}-activated ATPase which is inhibited by Mg^{2+} ions, but, if grown under low-salt conditions, the ATPase activity is equally stimulated by Ca^{2+} or Mg^{2+} ions (Kylin and Kähr, 1973). Here, changes in enzyme properties have occurred.

Nevertheless, 'true' tolerance is not considered to be a major tool in the adaptation of most species to toxic environments except, perhaps, flooded soils where reliance on short-term tolerance of fermentation products appears to be critically important (p. 275); and ancient mine spoil and soils originating from ultramafic rocks where hyperaccumulation of heavy metals occurs (Baker and Walker, 1990).

5. Phytoremediation: biotechnology to detoxify soils

There has been much excitement at the potential of using plants that are resistant to toxic metals and organic compounds, and can accumulate large quantities in their shoots, to clean up contaminated soils. The market in the United States for the decontamination of polluted land and water by such phytoremediation technology was probably in excess of $100m in 2000, and expanding rapidly. Apart from environmental gains, the cash cost of phytoremediation is much lower than conventional approaches (e.g. Vallack *et al.*, 1998).

A successful phytoremediation system would require plants that:

(i) could survive on the contaminated soil (i.e. they must be resistant);
(ii) could absorb large quantities of the toxin into the roots, even when the toxin was present in soil in a relatively unavailable form;
(iii) could transport the toxin to the shoots, so that it could be harvested and removed; and
(iv) had a high growth rate, so that the absolute amounts of metal transported were large.

In theory all of these are achievable; indeed, some plants are known to have most of these characteristics. For example, *Berkheya coddii*, a member of the Asteraceae from Transvaal, hyperaccumulates Ni, complexing it with histidine, and achieving concentrations of over 35 000 mg kg^{-1} (3.5%). It also grows rapidly and produces a large biomass. Ultramafic soils occur as isolated outcrops in many parts of the world, and typically have unique floras, which include endemic species that can survive on soils that are toxic to all members of the surrounding vegetation. It is likely that the floras of many of the less-characterized areas of ultramafic soils in the world, such as Indonesia and Iran, as well as some that are better known, including Cuba, will contain species with key genes that can be exploited for phytoremediation. The conservation of these floras is therefore of paramount importance.

However, natural ultramafic soils are nearly always nutrient-deficient and support communities of plants that have adapted to those conditions; they tend to be slow growing and of low biomass. These are not ideal characters for phytoremediation. Phytoremediation may therefore be an area where plant biotechnology can make an important contribution to human well-being, without the concerns that seem likely to imperil its application to agricultural systems. Relevant genes from plants that occur naturally on toxic soils, can be transferred to plants that have other desirable properties (fast growth, large size etc.). Rapid progress is being made in identifying genes associated with the characteristics listed above.

One of the most difficult aspects of phytoremediation is that, in many contaminated soils, much of the metal may be in forms that are unavailable for plant uptake. Successful novel plants may, therefore, need to excrete chelating agents such as siderophores from the root surface to increase their ability to capture metals. Such behaviour would normally be the subject of strong negative selection pressures, as it would increase the problem of detoxification for the plant.

Another major area for phytoremediation is the treatment of soils contaminated with organic compounds. It is increasingly recognized that persistent organic pollutants (POPs) pose a major challenge in the field of environmental pollution. Plants can take up and detoxify a wide range of organic compounds, protecting the cytoplasm by storing them in the vacuole. Here, the sulphur-containing tripeptide glutathione appears to play a key role by binding to, and detoxifying, organic toxins; the complexes can then be transported across the tonoplast by a transporter that is apparently very similar to that which confers multidrug resistance on human tumour cells in culture, by exporting drugs from the cell (Coleman *et al.*, 1997).

This is an area where bioremediation by fungi and bacteria has been exploited for some time, and one of the major benefits of involving plants might be that they could be used to maintain and increase microbial populations in contaminated soils, and also to disseminate them widely through soils along root systems. The ectomycorrhizal fungus *Suillus variegatus*, for example, can degrade 2,4,6-trinitrotoluene (TNT) in soil (Meharg *et al.*, 1997). Using natural populations of rhizosphere organisms has one large advantage over any attempt to manipulate micro-organisms to enhance their ability to degrade POPs: it carries with it none of the environmental risks that accompany the release of a genetically modified organism, especially a microbe.

6. The origin of resistance: the genetic basis

Some forms of resistance to toxins are ancient. Saline habitats are certainly as old as the Angiosperms and the high proportion of halophytes in some families, in particular the Chenopodiaceae, including the genera *Chenopodium, Atriplex, Halimione, Salicornia, Suaeda* and *Salsola*, is strongly indicative of an ancient origin. Resistance in these plants relies heavily on methods of removing the salt once it has been transported to the leaves (p. 277). Similarly, Al accumulators, containing more than 1mg Al g^{-1} shoot fresh weight, tend to be primitive on taxonomic grounds (Chenery and Sporne, 1976), suggesting that tissue abscission, a physiologically crude method of excretion, is in evolutionary terms a first attempt.

It is likely that the more sophisticated mechanisms have evolved more recently, in particular, the combinations of resistance mechanisms that characterize most specialist toxicity-resistant species. Resistance to toxic ions can be multilayered, involving an exclusion mechanism at the root surface which reduces the degree of stress, and localization within the tissues of those ions which do penetrate these defences.

Study of the ecological genetics of plant resistance to heavy metals and air pollutants has provided evidence of the rate of development of resistant populations. Characteristically, plants are exposed to heavy metals in restricted areas, at high but relatively constant concentrations; in contrast, although they can receive air pollutants from an industrial source in a similar way, exposure is, typically, geographically widespread at relatively low, but extremely variable concentrations. The experimental approaches to the two types of toxicity must, therefore, take into account the different selection processes which operate on the existing variation in relation to resistance (Shaw, 1990; Taylor *et al.*, 1991). Nevertheless, for both classes of toxin, the process can be very rapid.

In the case of metal resistance, it has been clearly demonstrated that individual plants with a measure of resistance can exist in normal populations not previously exposed to toxicity (Walley *et al.*, 1974). Table 6.7 shows that plants of *Agrostis capillaris* growing on Zn- or Cu-waste from old mines have a high resistance to the appropriate metal, but to that metal only. It is possible, however, by growing commercial seed on contaminated soil to select for a few individuals (1–2%) which have Cu-resistance indices as high as those of mine

Table 6.7

Selection for Cu and Zn resistance in *Agrostis capillaris* (Walley *et al.*, 1974). Plants were grown for four months in soil, copper-mine waste, or zinc-mine waste, and the resistance of survivors measured by rooting tillers in 0.5 g l^{-1} $Ca(NO_3)_2$, containing the appropriate metal. Resistance is measured on a 0–100 scale, with 100 representing insensitivity to the toxin

Population	Treatment	Mean and (maximum) resistance of survivors to	
		Zn	Cu
Pasture	Direct measurement	0.6 (1.7)	2.0 (8.6)
Commercial	Grown on soil	N.D.	5.6 (8.5)
	Grown on Cu-waste	N.D.	48.0 (77.5)
	Grown on Zn-waste	31.8 (41.3)	0.3 (0.9)
Copper mine	Direct measurement	0.8 (0.8)	79.0 (87.5)
	Grown on soil	17.6 (20.2)	85.3 (93.5)
	Grown on Zn-waste	36.4 (47.8)	52.1 (66.9)
Zinc mine	Direct measurement	93.0 (93.0)	3.8 (3.8)

populations. Resistance to zinc appears to be more complex, since the most resistant individual was only about half as resistant as the mine population.

If such effects can be produced in a population in one generation, then rapid development of resistant populations should occur in the wild when new sources of heavy metals are introduced. Thus, resistance to zinc has been observed under galvanized fences and electricity pylons within 25 years (Al-Hiyaly *et al.*, 1990); this confirmed earlier reports that Cu-resistance in lawns of *Agrostis capillaris* around a copper smelter increased progressively over 20 years (Wu and Bradshaw, 1972).

Rapid development of resistant populations can also be observed under the generally more mild stress of air pollution. For example, in a series of field experiments in the North of England, selection for resistance to SO_2-injury of perennial grass swards by exposure to normally polluted air, resulted in population differences within 4–5 years for *Lolium perenne* and *Poa pratensis* (Bell *et al.*, 1991). Breeding experiments have established the genetic nature of such resistance to SO_2 in *L. perenne* (Wilson and Bell, 1990). Where differences in resistance between populations have not been observed, it is assumed that there is insufficient variation for selection to reveal (Bradshaw *et al.*, 1990).

The more detailed genetic basis for resistance to toxins is, however, less understood. Although it was initially thought that the control of resistance was usually polygenic, because of the very variable phenotypes exhibited by resistant plants, there is an increasing number of cases in which single genes have been shown to be responsible for resistance (e.g. to salt, Apse *et al.*, 1999), with the possible involvement of other modifying genes (Macnair, 1993). Resistance is typically low in populations on uncontaminated soils, presumably because of the costs of resistance. When resistance is measured in populations from soils surrounding contaminated sites, steep clines are normally found, with resistant plants disappearing rapidly with distance from the source of toxin (Fig. 6.12). This presumably reflects a balance between gene flow and selection. When Macnair *et al.* (1993) measured Cu resistance in populations of *Mimulus guttatus* at increasing distances from a contaminated stream at the Copperopolis mine in California (Plate 16), they found that resistance itself declined in frequency more slowly than the degree of resistance. In other words, populations at some distance from the contamination contained a high frequency of resistant plants, but these plants were less resistant than those growing nearer the mine. Macnair (1993) interpreted this to mean that selection against the modifier genes, which modulate resistance and determine the degree of resistance, is stronger than against the resistance mechanism itself. This implies that the cost of the core resistance mechanism may be quite low.

In the case of arsenic, however, resistance is found at high frequency even in natural populations of some species growing on uncontaminated soils. Meharg *et al.* (1993) recorded frequencies of 15–70% in populations of *Holcus lanatus* on a wide range of sites. As resistance is achieved by suppression of phosphate uptake (see p. 269), it must be at a very high cost, especially on uncultivated soils which often have low P availability. Paradoxically, As-resistant plants of *H. lanatus*, with suppressed P uptake, have higher shoot P concentrations than sensitive plants. There is evidence that some As-resistant plants may possess a

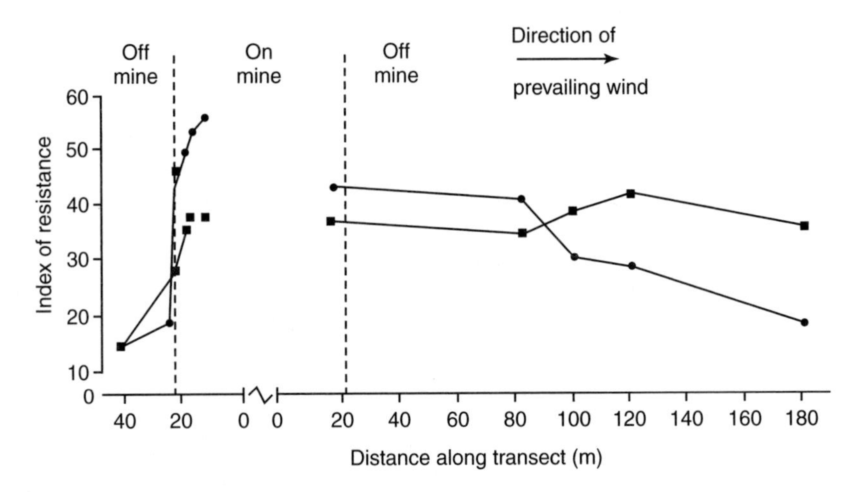

Figure 6.12

Resistance to copper in populations of *Agrostis capillaris* declines away from the copper-contaminated mine at Drws-y-Coed, North Wales. Upwind the decline is very sharp, but downwind more shallow, as wind-borne pollen creates gene flow. However, the degree of resistance is greater in the seed population (■) than in parent plants (●) off the mine, because of selection against resistant plants; the opposite is true on the contaminated mine soils. The *y*-axis represents the 'index of resistance' based on rooting tests of seedlings or tillers in Cu solutions and ranges from 0 to 100 (100 = all individuals develop roots freely in the presence of Cu). Data from McNeilly (1968); figures based on McNair (1993)

gene that promotes P accumulation in the shoots, analogous to the *pho2* mutation in *Arabidopsis* (p. 105); this accumulation leads to inhibition of the P uptake system and, wholly incidentally, to resistance to arsenate. As-resistant *H. lanatus* flower more profusely and earlier than non-resistant plants (Wright *et al.*, 2000), an effect that may be brought about by the high shoot P concentration. Apparently As-resistance is an incidental consequence of a polymorphism for flowering time, controlled by accumulation of shoot P, and expressed also as reduced P and As uptake (Fitter *et al.*, 1998).

7

An Ecological Perspective

1. The individual plant

Environmental physiology is whole-plant physiology. Even though understanding of many of the responses to environmental variables that we have covered in this book requires knowledge of the molecular events that underlie those responses, that knowledge must be integrated into a picture of whole-plant functioning. This is because environmental physiology is in effect the physiology of adaptation, and adaptation is an evolutionary phenomenon, for which the organism itself is the key level of organization. The organism perceives the impacts of its biotic and abiotic environments and, if it is adapted to those impacts, responds by molecular mechanisms, integrating these responses in ways that improve survival. Equally, it is the organism that lies at the root of population and community responses, whose dynamics, in conjunction with resource exchanges between organisms and their environment, determine the functioning of the entire ecosystem. In addition, the organism is the focus of natural selection, the process that gives rise to adaptation, since it is the differential survival and mortality of individual organisms that leads to changes in gene frequency, or evolution. The organism therefore lies at the heart of this ecological and evolutionary nexus (Fig. 7.1).

Environmental physiology, therefore, involves a consideration of optimal patterns of response that are by definition adaptive, as described in Chapter 1. The operation of any physiological process requires resources, whether these be carbon for respiration or for molecular skeletons, nitrogen for enzyme construction, phosphorus for nucleotides or energy transduction or water for the transport of materials. This is why the first part of this book was devoted to describing the ways in which plants respond to limitations in the supply of these resources.

Selection will act on a plant in relation to these resources in two ways, in terms of both the amount the plant acquires and the way in which it uses each resource. For example, a plant growing beneath a leaf canopy is limited in its growth by the low photon flux density (cf. p. 43), and therefore by the small amount of carbon it can fix. Increased fitness is likely (but not certain) to result both from mechanisms which increase quantum efficiency and from patterns of carbon and nitrogen allocation which increase the interception of radiation. Some extreme shade plants from tropical rainforests have curious leaf properties (such as red undersurfaces or blue iridescence) which increase

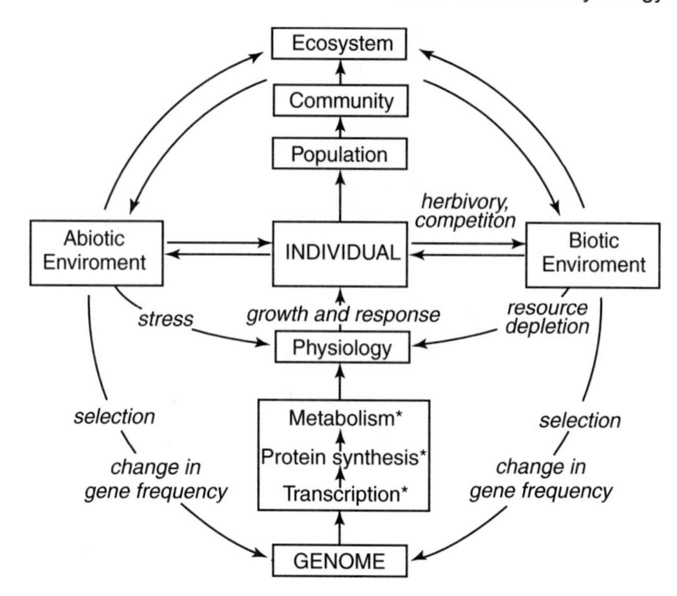

Figure 7.1

The organism as the centre of the ecological–evolutionary nexus. Bold lines indicate evolutionary controls, light lines are the ecological linkages.

photon absorption (Lee and Graham, 1986), and it seems reasonable to propose that these have been selected precisely as a result. A more general response to low irradiation, however, is an increase in allocation of biomass to leaves as opposed to roots, just as plants grown in nutrient-deficient or very dry environments tend to display the opposite allocation pattern (Reynolds and d'Antonio 1996).

For any given set of environmental conditions, one (or possibly several) pattern of allocation of resources to various structures and activities will result in maximum carbon gain; this pattern is defined as optimal. Of course, a plant that achieved this optimal pattern might nevertheless fail to maximize fitness and therefore be selected against, for example because it was too vulnerable to herbivores. Nevertheless, this approach enables us to quantify the physiological and morphological characteristics which contribute to fitness. Iwasa and Roughgarden's (1984) model of resource allocation showed that when the allocation ratio between roots and shoots deviated from the optimum, all future allocation should be to the part that could restore the optimum (leaves in shade, roots in drought) until the optimal trajectory was regained (Fig. 7.2a). Exactly this response is shown by plants of cocksfoot grass *Dactylis glomerata* following defoliation; the optimal trajectory is restored by allocation of resources almost solely to shoot rather than root growth (Fig. 7.2b). The experiment with *D. glomerata* reveals other important features of the process of allocation of resources during plant growth. The plants were grown at either ambient or

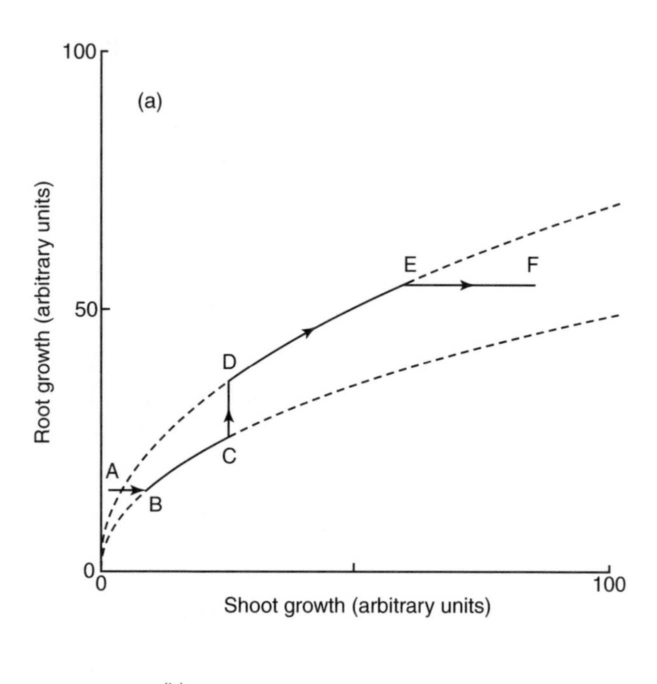

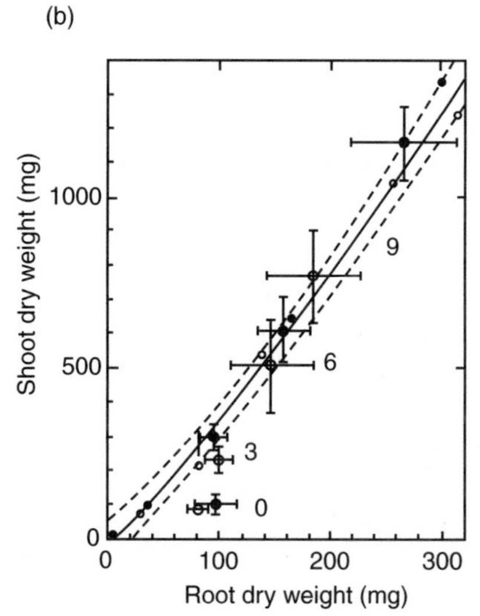

Figure 7.2

(a) Simulation of root and shoot growth under changing environmental conditions. The two dashed lines represent optimal growth paths, the upper for conditions where water limits growth more than *PAR*, the lower for *PAR*-limited growth. In each case the appropriate organ receives greater allocation of biomass. The thick solid line is a growth trajectory for a plant limited by *PAR* from A→C and from E→F, and by water from C→E (from Iwasa and Roughgarden, 1984). (b) Plot of shoot weight against root weight of defoliated plants of the grass *Dactylis glomerata* grown at ambient (solid: 350 $\mu mol\,mol^{-1}$) and elevated (open: 700 $\mu mol\,mol^{-1}$) atmospheric CO_2 concentrations. Smaller symbols with no error bars are means of undefoliated plants with a fitted regression line and confidence limits. Larger symbols with standard error bars are for plants defoliated when 29 d old (2/3 of the leaf area of each plant was removed). The numbers represent the number of days after defoliation. The plants responded immediately by ceasing root growth until the previous growth trajectory (shown by the regression line) was restored (from Farrar, 1996).

elevated atmospheric CO_2 concentrations; although the plants were larger at elevated CO_2, the two treatments did not produce a difference in the relationship between shoot and root growth, which is revealed by plotting ln shoot weight against ln root weight. The slope (k) of such a plot (Fig. 7.3a) is the quotient of the relative growth rates (**R**) of shoot and root, and describes the allometry of growth. If $k = 1$, as in the solid line in Fig. 7.3a, then the biomass ratio of shoot and root will be constant, but if it is not, then the root weight ratio ($RWR = W_R/W_{RS}$) will change progressively as the plant grows (Fig. 7.3b).

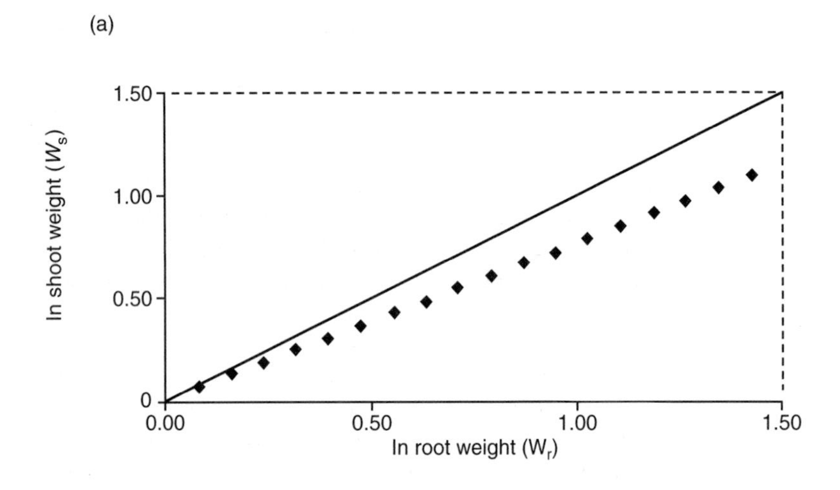

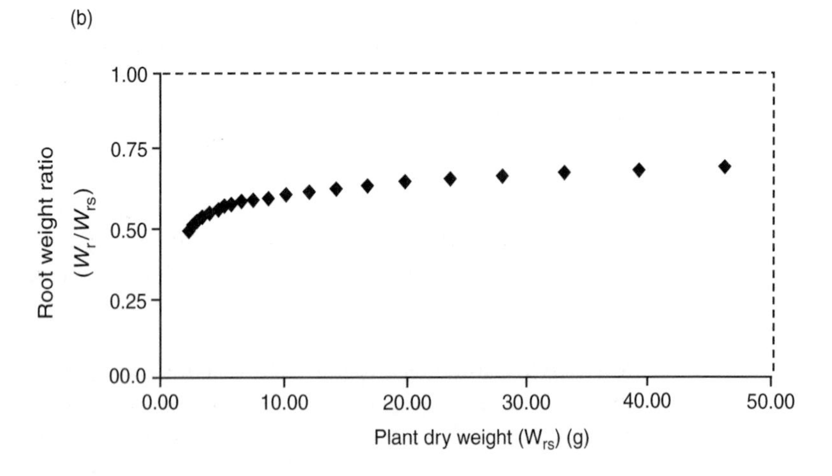

Figure 7.3

(a) An allometric plot showing the relationship between ln shoot (W_s) and ln root (W_r) dry weight, where root and shoot relative growth rates remain constant at 0.20 and 0.15 $g\,g^{-1}\,d^{-1}$, respectively. The solid line represents the relationship where both relative growth rates are 0.2. (b) The ontogenetic drift in root weight ratio (root weight (W_r) divided by whole plant weight (W_{rs})) resulting from the data in (a).

Where $k < 1$, $\mathbf{R}_{shoot} < \mathbf{R}_{root}$, and RWR will increase with time. Hence a faster-growing plant (such as one maintained at elevated atmospheric CO_2) will have a greater value of RWR at any given time, and will appear to be allocating progressively more biomass to the roots, when all that is being observed is an ontogenetic (developmental) trend. This relationship is frequently overlooked in studies of allocation (Farrar, 1999).

Optimal responses to resource acquisition do not simply rely on biomass allocation and the control of growth rate of different parts of the plant. Morphological responses may be as or more significant. Both leaf thickness (typically expressed as specific leaf area; cf. Chapter 2) and root diameter (or its surrogate specific root length; cf. Chapter 3) are key elements of the plasticity of plants. Ryser and Eek (2000) examined the morphology of two grasses, *Dactylis glomerata* and *D. polygama*, when grown under four levels of shading. Total plant biomass was reduced only by severe shading (5.5% of full sunlight), but at intermediate levels of shade (20% and 30%), the plants had lower root biomass although their total root length was unaltered, due to increased specific root length (root length per unit weight). Similarly, they had greater leaf areas at these intermediate shade levels than in full sunlight, because of both a slightly greater leaf weight ratio (leaf weight as a fraction of total plant weight) and greater specific leaf area (leaf area per unit leaf weight).

Although resource allocation is normally perceived in terms of carbon, there is no reason why other plant resources such as nitrogen should not be the critical currency in particular conditions. For example, there is a very close relationship between the maximum photosynthetic rate that a leaf can achieve (A_{max}) and its nitrogen concentration expressed on a mass basis (Fig. 7.4; Field and Mooney 1986, Peterson *et al.*, 1999). This appears to be because the single greatest limitation to the maximum rate of photosynthesis in C_3 plants is the amount of ribulose bisphosphate carboxylase, the CO_2-fixing enzyme. This enzyme is the most abundant protein in the world and typically comprises around half of total leaf protein. The relationship of Fig. 7.4 implies a severe allocation problem for plants; maximizing carbon gain is promoted by high allocation to leaves of both carbon and nitrogen. The former allows the development of large leaf areas that increase the interception of radiation, whereas the latter provides the biochemical machinery for CO_2 assimilation. The acquisition of the nitrogen, however, is a root function, and maximizing that may involve preferential allocation of resources to roots.

Nitrogen is commonly a limiting resource for plants, and the way in which it is distributed within the plant will determine the effectiveness with which it is used. Hirose and Werger (1987) developed a model to predict the optimum distribution of N within a canopy of leaves. This model was based on the known and well-characterized relationships between (i) *PAR* flux density and vertical arrangement of leaf area in a canopy (see p. 32) and (ii) photosynthetic carbon assimilation and *PAR* flux. The latter relationship requires knowledge of the maximum photosynthetic rate (A_{max}), which is a function of leaf nitrogen concentration (Fig. 7.4). From these equations, it is possible to estimate the distribution of N among the leaves within a canopy that maximizes C fixation. Hirose and Werger (1987) compared this optimal distribution with that

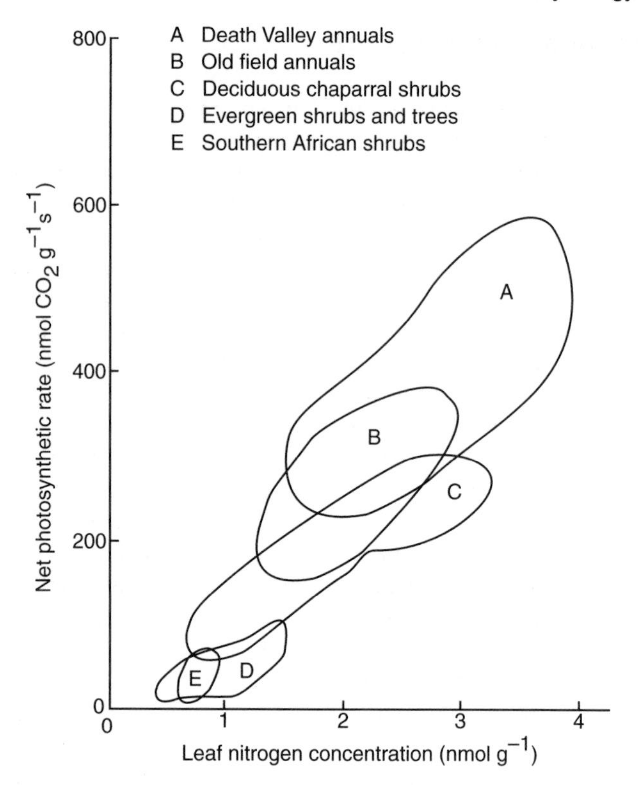

Figure 7.4

The relationship between photosynthetic rate and leaf nitrogen concentration, for a wide range of species, shows that investment of N in leaves is a major constraint. Each line encircles an area containing many data points for each category (from Field and Mooney, 1986).

measured in a stand of a tall herbaceous plant, goldenrod *Solidago altissima*. The measured distribution was suboptimal; it was more uniform than the predicted optimum distribution, but even so it resulted in 20% greater C fixation than would have been achieved by an exactly uniform distribution (N concentration was the same in all leaves), and it fixed only 5% less than the optimum could have achieved. Hirose and Werger (1994) extended this analysis to a mixed stand of wild plants, and showed that, even among the dominant species, nitrogen use efficiency (NUE; dry weight production per unit nitrogen) was highly variable. The grass *Calamagrostis canescens* had a high NUE, but the sedge *Carex acutiformis* had a low NUE, reflecting not just differences in nitrogen concentration, but also in leaf morphology.

There is good evidence, then, that plants do allocate resources in ways that can be seen as optimal and adaptive. However, because of the variety of different growth-forms that plants display, there will never be a single optimum solution for all species, even in a single habitat, and the interactions among plants in competition for resources add a further level of complexity.

2. Interactions among plants

1. Mechanisms of Competition

The optimal pattern of growth and allocation for a plant represents a response to the availability of resources. There are few natural situations in which resource availability is determined solely by physicochemical factors; it is normal for the activity of other plants to affect that availability. By capturing the fundamental resources of all plants (namely photons, nutrient ions, water or, more rarely, CO_2 molecules), a plant will reduce the availability of those resources to its neighbours. That is the basis of competition.

The definition of competition is one of the longest-running semantic debates in biology. It can be viewed either in terms of mechanisms (as above) or in terms of effects. In the latter case, the critical question is whether the growth, fecundity or ultimately the fitness of one organism is reduced by a neighbour. In mechanistic terms, however, competition represents a situation where the supply of a resource is less than the joint requirements of two organisms, and as a result the performance of one or both is impaired.

No two individuals of different species can have identical requirements – this would imply a greater degree of affinity than is found among the genotypes within many species. Nevertheless, because all plants require the same fundamental resources, all have some potential competitors when they are exploiting a resource. Survival, therefore, depends on either partitioning of the resource in some way, thereby avoiding competition, or on the establishment of competitive superiority. If one species possesses competitive superiority over others with respect to all resources, then one might expect that species to eliminate the others in all appropriate habitats. In practice one finds that most plant communities are very diverse; monospecific stands are rather rare in nature, and where they occur (e.g. bracken *Pteridium aquilinum* on hill land in Wales, heather *Calluna vulgaris* on Scottish moors, or creosote bush *Larrea tridentata* in the American south-west), it is often because of human activity. Clearly, then, where species are competing for resources there are processes acting to reduce the intensity of that competition, which implies that resources are being partitioned among the species in some way. Partitioning of a resource is relatively simple to visualize in animals, where food items, for example, can be categorized on the basis of size or type, but for plants resource partitioning must rely on different mechanisms.

Both water and nutrient ions are normally absorbed by roots from soil. In both cases plants generate the supply mechanism which allows them to acquire the resource. Water flows into roots and up stems along a gradient of water potential, created by evaporation from the leaves (cf. Chapter 4). The lowering of water potential at the root surface is, therefore a function of the rate of transpiration, and hence of leaf area. Ignoring other variables such as stomatal conductance and differences in the leaf water potential at which stomata close, it is likely to be the leaf: root surface area ratio which determines relative rates of water uptake in competition. Some ions arrive at the root surface in the consequent mass flow of water in sufficient quantities to satisfy plant demand; this is certainly true of Ca^{2+} in calcareous and of Fe^{3+} in very acid soils (cf. Chapter 3). Competition for these will therefore be governed by the same principles as for water. Competitive superiority is therefore achieved by (i) greater root length density, (ii) greater water inflow, or (iii) physical placement of roots in wetter soil regions (Fig. 7.5b).

For other ions, notably $H_2PO_4^-$, mass flow is quite inadequate to satisfy requirements, and diffusion gradients are set up. For these the situation in Fig. 7.5(c) arises and the controlling factor is the physiological ability to lower the concentration at the root surface. The lower the diffusion coefficient for the ion, the narrower and steeper will be the depletion zones around the roots, as described in Chapter 3. Ions with relatively high diffusion coefficients, such as NO_3^-, are characterized by wide but shallow depletion zones. The higher the diffusion coefficient, the more likely competition is to ensue since depletion zones are wide. Because soil buffering power is low, solution concentrations may be higher for mobile ions, facilitating mass flow or convective transport. In that case, shoot characteristics controlling transpiration will be more important in determining uptake in competition.

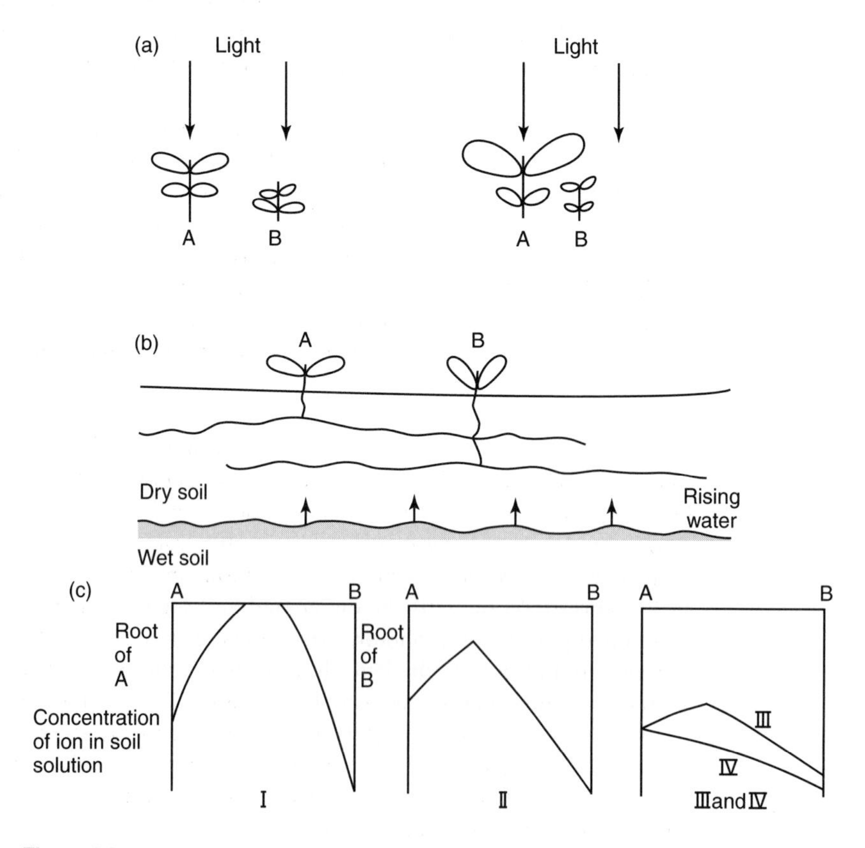

Figure 7.5

Diagrammatic representations of plant factors controlling the development of competition for (a) light, (b) water and (c) nutrient ions. In (c), I–IV represent successive time intervals, with plant B attaining the greater supply due to its ability to lower the root surface concentration further than plant A.

To show that competition for an ion occurs, it is not sufficient simply to demonstrate that the ion limits growth. That could be the result of a very low diffusive flux, which would actually make competition less likely. Two conditions must be satisfied – the ions must not arrive by mass flow at the root surface faster than they are required by the plant, and depletion zones must be wide enough to overlap between adjacent roots. To some extent these conditions counteract one another, since there is a general correlation between ion mobility in soil and soil solution concentration, and so with the contribution of mass flow to uptake (cf. p. 92). They can also be quantified into two dimensionless mathematical expressions explored by Baldwin and Nye (1974) and Baldwin (1975):

(i) the relationship between mass flow (V= the water flux at the root surface, in $m^3 m^{-2} s^{-1} = m s^{-1}$) and plant demand ($\alpha$, root absorbing power, m s^{-1}: see Tinker and Nye (2000) and p. 97) is simply given by the ratio $V:\alpha$. If $V<\alpha$, uptake is greater than supply by mass flow, depletion occurs at the root surface, and competition may ensue. α is itself dependent on the concentration of the ion at the root surface, which in turn depends on D, the diffusion coefficient.

(ii) where depletion occurs, the radius of the depletion zone is approximately given by $2\,(Dt)^{0.5}$, where D is the diffusion coefficient ($m^2\,s^{-1}$) and t is time (s) (Tinker and Nye, 2000), and the extent and occurrence of overlap is controlled by the function DtL_V, where L_V is root density ($m\,m^{-3}$). Baldwin (1975) showed that as DtL_v falls towards zero, the ion becomes increasingly immobile with respect to that particular root system. Uptake of $H_2PO_4^-$ and K^+ will be directly related to this function, and high values of L_v (the only term in the expression determined by the plant) will increase their uptake.

Four parameters of the soil plant system – water flux (V), plant demand (α), root density (L_V), and diffusion coefficients (D) – control uptake in competition. Of these, plants determine all but diffusion coefficient, and the outcome of competition will be controlled by their relative values.

In contrast to the complexity of competition for nutrients and water, that for *PAR* is governed simply by the directional and transitory nature of the photon as an energy source. If a photon strikes a leaf it may be absorbed, transmitted, or reflected, but both of the latter have selective effects, producing radiation of altered spectral quality (see p. 29) and are of little photosynthetic value. Competitive superiority for *PAR* therefore resides in the ability to place leaves in illuminated rather than shaded positions, and so it will be determined by plant characteristics determining foliage height and the rate at which that height is attained (Fig. 7.5a).

2. The occurrence, extent and ecological effects of competition

Competition has traditionally been regarded as a key, perhaps dominant, force in structuring ecological communities, yet among consumer populations it may be of little importance, because predation suppresses populations to a sufficiently low level that resources are not limiting (Hairston *et al.*, 1960).

Detecting competition in a plant community is often difficult. Standard approaches involve either removing species from a community or adding them to it; the effects can then be interpreted as competition, although the mechanism and the resource competed for may be unknown.

Wilson and Tilman (1991) planted seedlings of little bluestem grass *Schizachyrium scoparium* into a prairie grassland in Minnesota, parts of which had been fertilized. In some cases, the existing vegetation was removed; in others it was just the shoots of the surrounding plants that were held back to prevent shading; and in controls, the vegetation was unaltered. Seedlings exposed to full competition in the control plots grew much less well than those in the plots where vegetation was completely removed (Fig. 7.6). Where

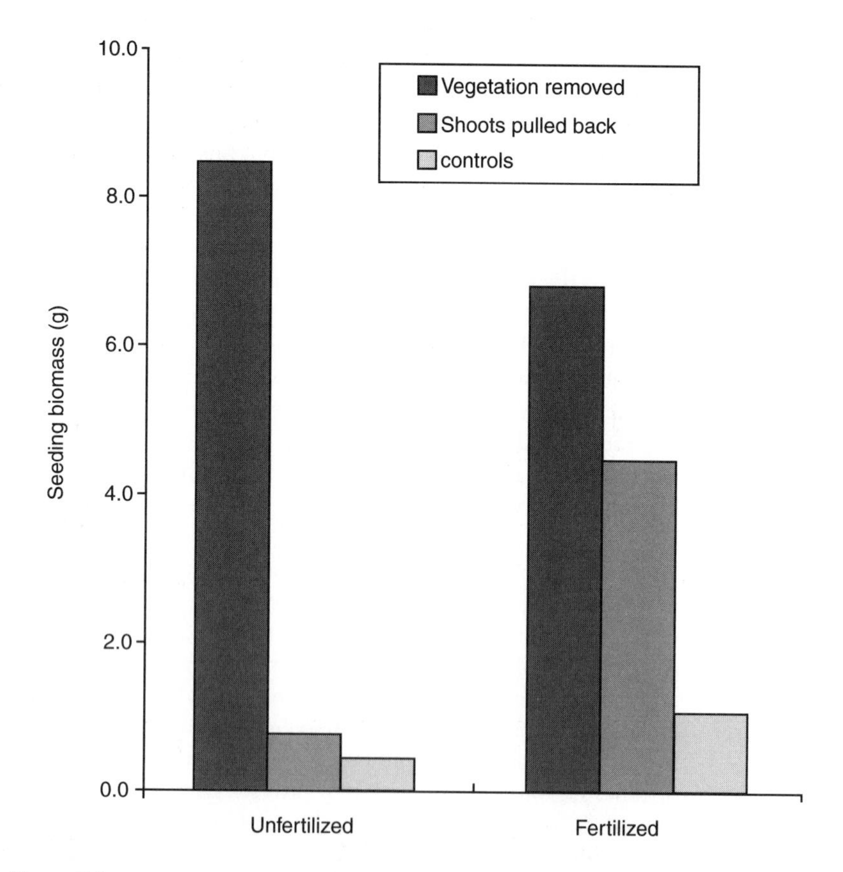

Figure 7.6

The performance of seedlings of the prairie grass *Schizachyrium scoparium* planted into prairie grassland depends on the intensity of competition and on the availability of resources. In the controls, where competition was intense, seedlings grew very poorly. Where all surrounding vegetation was removed, they grew well, in both fertilized and unfertilized fields. Where only the shoots of the surrounding plants were pulled back, competition from established root systems remained severe, and then seedlings grew well only in the fertilized soil (from Wilson and Tilman, 1991).

competition for light alone was prevented, so that below-ground competition between roots continued, seedlings performed very poorly on the natural soil, but on the fertilized soil grew almost as well as if there was no competition at all. Here, nutrients were barely limiting to growth, and so competition between root systems for them was unimportant, whereas on the unfertilized soil, nitrogen was highly deficient, and the plants were almost certainly competing for that.

However, evidence for competition is not always so clear cut. McLellan *et al.* (1997) removed all neighbours from around individual plants of five species in a calcareous grassland in Derbyshire, UK. Removal was done by painting leaves in a ring around the target plant with a systemic, contact herbicide. The radius of the rings was small (from 0 to 50 mm), but this matched the scale of the plants and their actual below-ground neighbourhood size, which had previously been determined experimentally. Four of the five species studied responded by increasing leaf number, more or less linearly in relation to gap size (Fig. 7.7a for examples). However, only two species (a grass *Briza media* and a legume *Lotus corniculatus*) increased shoot biomass, and these were the smallest species. The two largest species (both dicots, *Sanguisorba minor* and *Plantago media*)

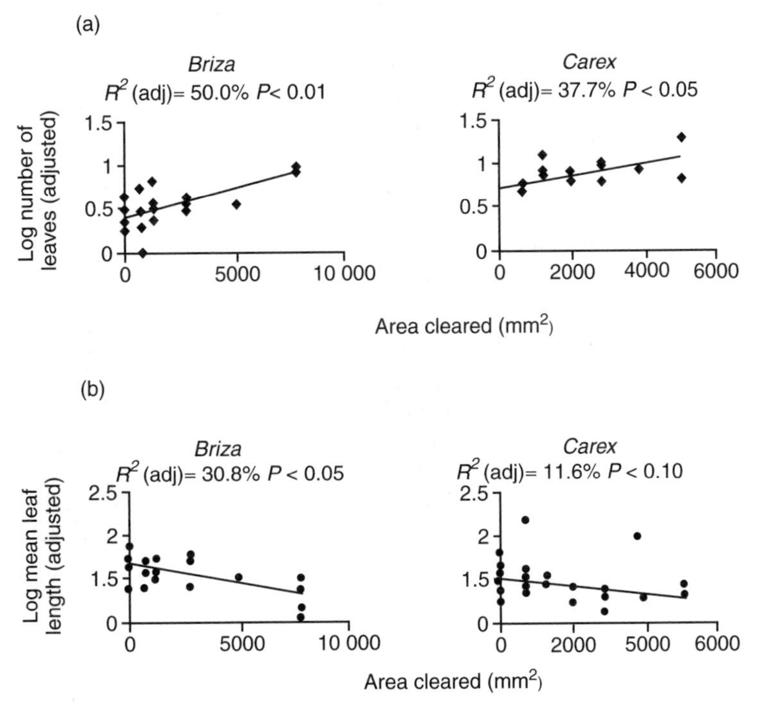

Figure 7.7

(a) The number of leaves produced by individual plants in a calcareous grassland increased in proportion to the size of the gap cleared around each plant, for four of five species tested, of which two are illustrated here. (b) Mean leaf length declined in proportion to gap size; data shown for *Briza media* and *Carex flacca* (from McLellan *et al.*, 1997).

were least affected by competition, perhaps because their growth form is a rosette and this enables them to suppress neighbours effectively. Morphology may be a key determinant of a competitive hierarchy, therefore, and phenotypic plasticity would be expected to be part of a plant's response to competition. One striking effect in the experiment of McLellan *et al.* (1997) was that leaf length (a surrogate for leaf area in this largely linear-leaved species) declined as gap size increased (Fig. 7.7b). This is almost certainly the same response that Ballaré *et al.* (1987) demonstrated as a phytochrome-controlled reaction to shading by neighbours (see p. 138). It means that plants in dense vegetation will have longer leaves and be more likely to overtop the canopy.

In both of these experiments, it is apparent that competition occurred both above and below ground. In the gap experiment, plant performance of most species improved as the gap radius increased. There would ultimately have been a radius beyond which no improvement occurred, because there would be no interactions at such a range, but this was apparently not reached, even though a gap 100 mm across is very rare in grasslands such as that studied by McLellan *et al.* (1997). That plants must simultaneously compete with neighbours for both above- and below-ground resources has important implications for their responses to competition. Optimal growth trajectories require a plant to allocate resources to the function that is most limiting growth; if both light and nutrient supplies are reduced by competition simultaneously, that poses a severe constraint. There is lively debate as to whether there is a trade-off between above- and below-ground competitive abilities. The argument based on resource allocation models would imply that the allocation of resources to, say, roots, because of severe below-ground competition for water, would automatically reduce allocation to shoots and so reduce competitive ability for plants (Tilman, 1987). In contrast, Grime (see Grime *et al.*, 1997) has argued strongly that there are competitive plants that have evolved traits that make them competitive above and below ground, and that these stand in contrast to plants that have evolved to withstand disturbance or harsh environments. These different poles of selection represent primary strategies for plants (see below, p. 306). The two viewpoints are not mutually exclusive; even though some plants are inherently more competitive than others, within that group some are better able to compete for light (because of tall or spreading shoots, for example), and some for water and nutrients (because of larger or better disposed root systems, or interactions with symbionts).

That such fixed patterns of response exist is supported by the fact that species often form stable competitive hierarchies. Mitchley and Grubb (1986) found that the relative abundance of species in chalk grassland in southern England was more or less the same within and between sites and over a 3-year period. When they grew the species in mixtures in pot or field experiments, they were able to recreate the abundance ranking in almost all cases. In other words, some species within this community were inherently the competitive dominants, others the subordinates. Such hierarchies are often found, but they may be strongly affected by resource availability; when Keddy *et al.* (2000) grew 23 species from lake shores under high and low nutrient supply, the general hierarchy was maintained, but there were important shifts of position within the hierarchy of some species (Fig. 7.8). For example, the most competitive species after 2 years under low nutrients was purple loosestrife

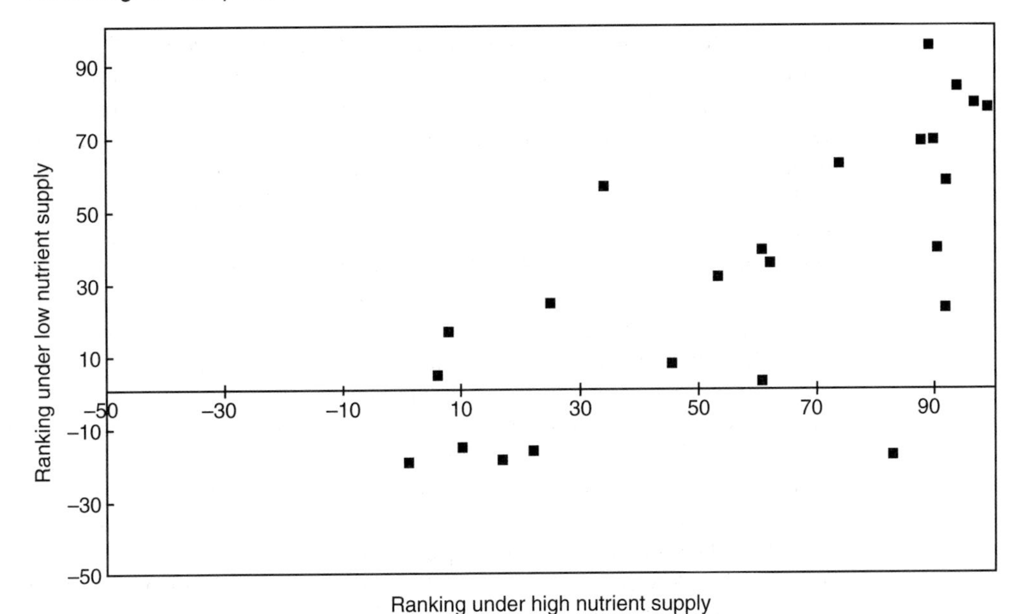

Figure 7.8

Competitive performance, under high and low nutrient supply, of 23 lake shore species after 2 years, relative to a standard test plant, *Penthorum sedoides*. The ranking is broadly maintained, but there are marked deviations for some species. Spearman's correlation coefficient shows that only about the half the variation in competitive performance at one nutrient supply rate is explained by performance at the other (from Keddy *et al.*, 2000).

Lythrum salicaria, a Eurasian plant that is a major invasive weed in north America. At high nutrient supply, *L. salicaria* was eighth in the hierarchy. The maintained hierarchy supports the view that there are inherently more and less competitive species, but the variation between the two nutrient conditions shows that trade-offs are important and that species that are competitive at low nutrients (where root growth will be critical) may be less so at high nutrients (where shoot growth and competition for light may determine the outcome).

3. Interactions between plants and other organisms

Interactions among plants (as among any group of species in the same trophic level) are usually antagonistic; those that occur between plants and other organisms can cover the full spectrum from antagonistic to mutualistic. Mutualistic interactions include pollination relationships with a wide range of animals, and various types of plant–microbe interaction. Antagonistic interactions include parasitism (by bacteria, fungi, insects or other plants) and predation. Exactly what constitutes predation on a plant is not easy to define. Consumption of plant tissue leading to death is clearly predation, but lower intensity grazing may have no detectable impact on plant fitness or even increase it, either by stimulating the conversion of suppressed meristems into reproductive growth and hence increasing seed production (Belsky, 1986), or by removing competitors. Consumption of seeds, for example by the larvae of lepidoptera, might seem a clear case of predation, except that the adult moth is often the mutualistic pollinator of the plant, and pollinates the flower while laying eggs. This remarkable situation is well known in figs, yuccas (Addicott, 1998), but it is known from a number of other species, such as the cactus *Lophocereus schottii* (Holland and Fleming, 1999), globeflower *Trollius europaeus* (Ranunculaceae: Hemborg and Despres, 1999) and several Caryophyllaceae, including bladder campion *Silene vulgaris*. In *S. vulgaris* the flowers are pollinated by noctuid moths (*Hadena* spp.) that simultaneously lay eggs on the developing seed capsule. There is a positive correlation between the percentage of flowers pollinated and the percentage of seed capsules destroyed. The loss of seeds to the larvae is a price the plant pays for pollination, and the relationship can be seen as mutualistic (Petterson, 1991).

Both antagonistic and mutualistic interactions tend to involve the loss of acquired resources to the partner. The difference is simply that in mutualistic interactions, the transaction is reciprocal in some way. In real communities, all these phenomena interact profoundly. For example, interactions among plant species are normally expected to be antagonistic, as a result of resource competition. However, the presence of one species may promote the growth of another, for example by protecting it from grazing, as shown by Rousset and Lepart (2000); seedlings of downy oak *Quercus humilis* growing on calcareous grassland in central France were heavily grazed by sheep. When oak seedlings grew with juniper *Juniperus communis*, which has spiny leaves and is relatively unpalatable to sheep, they survived better, even though their growth rate was reduced by resource competition. To complicate matters further, grazing of oak seedlings by small mammals, presumably rodents, was greater when they were with shrubs such as juniper than when in the open grassland. The balance of positive and negative interactions here is dependent on local factors.

The other major interaction between herbivory and competition is that grazing may suppress competitive dominants and so prevent their excluding subordinate species (Hairston *et al.*, 1960). There has been an excited debate over the role of competition in plant communities, in which Grime (1977) and Tilman (1987) have again represented two poles. In essence the issue is whether competition becomes progressively more important as productivity increases (Grime) or remains a potent force at all levels of productivity, but is primarily for light at high productivity (where by definition soil-based resources must be plentiful) and for water and/or nutrients at low productivity. This question is directly linked to that discussed in the previous section, as to whether there are trade-offs between above- and below-ground competitive ability.

These ideas have been tested on numerous occasions, but the results remain inconclusive. For example, van der Wal *et al.* (2000) examined the performance of a salt-marsh herb *Triglochin maritima* along a natural productivity gradient. The species naturally occurred midway along the gradient, apparently because it was excluded by grazing pressure from the youngest and least productive areas, and by competition for light in the oldest, most productive sites. They found that excluding grazing greatly increased plant biomass (Fig. 7.9), but did so equally at all productivity levels. Competition had a profound effect on ungrazed plant growth, but not on grazed plants, but again did so equally at all productivity levels. The lesson from such studies is that competition and herbivory interact powerfully and that models of competitive hierarchies that do not take into account other

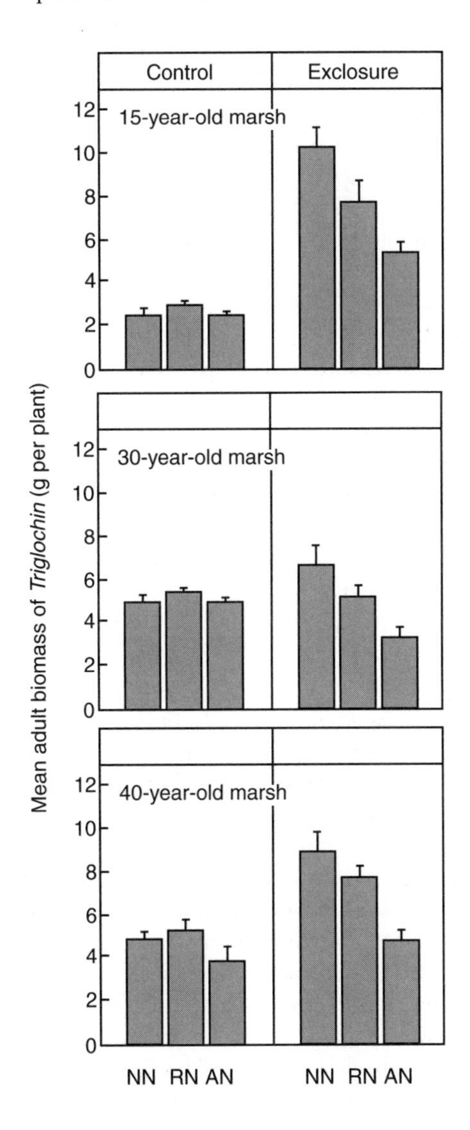

Figure 7.9

The biomass of adult plants of sea arrowgrass *Triglochin maritima* transplanted into saltmarsh turf was greatest when grazing was prevented by an exclosure, but this effect was independent of the intensity of competition (NN, RN, AN) and the age (and hence productivity) of the sward. The competition treatments were: NN, no neighbours (plants removed); RN, root neighbours only (shoots held back); AN, all neighbours (control) (from van der Wal *et al.*, 2000).

Table 7.1

A classification of plant defences to different forms of attack by grazers and parasites. The three types of defence are represented by the rows and the four modes of attack by the columns

Defence type	Mechanism	**Whole tissue grazers** (e.g. mammals, molluscs, insects)	**Xylem feeders** (e.g. fungi (wilts), insects (*Philaenus*), plant hemiparasites (*Rhinanthus*))	**Phloem feeders** (e.g. insects (aphids), plant holoparasites (*Orobanche*))	**Cell contents feeders** (e.g. fungi, some insects (psyllids))
Nutritional inadequacy	Mainly low protein content	Yes, but [N] controls photosynthetic rate, so a trade-off with growth rate	Yes, because xylem has very low organic N concentration	Yes: aphids require endosymbiotic bacteria to survive on imbalanced diet	No, since cells must be metabolically competent
Physical barriers	Hairs, spines, thickened cell walls	Yes; some physical defences can be massive (spines of *Acacia karoo*)	Yes: secondary thickening means that only youngest shoots are vulnerable except to systemic infection	Yes: secondary thickening means that only youngest shoots are vulnerable except to systemic infection	Most attack is by fungi that grow systemically through plant tissue
Chemical toxins	Alkaloids, cyanide, tannins, etc	Yes: both external (e.g. stinging hairs) and internal	Not feasible: toxins would accumulate in leaves. May be defences at plant surface	Not feasible: toxins would accumulate in growing points. May be defences at plant surface	Yes: wide range of defence compounds, mostly induced in response to attack

antagonistic and mutualistic interactions will be inadequate. Such models are starting to appear (see Huisman and Olff, 1998).

Competitively dominant plant species generally achieve dominance by greater size, which requires the acquisition of large amounts of resource. This is a cost of being competitive. Other interactions also impose costs on plants. For example, grazing causes loss of resources, but most plants have some form of defence against grazing and against other interactions that involve loss of resource including parasites and pathogens. Whereas grazers remove tissue from the plant, most parasites remove metabolites. The nature of the defence depends on the nature of the assault (Table 7.1).

The costs of defence vary greatly, but can be large. *Datura wrightii* is a perennial relative of the potato (Solanaceae) that is normally polymorphic for the nature of the hairs or trichomes on its leaves. Hairs are a very common feature of leaf surfaces (Fig. 7.10) and have two main impacts: first they alter the energy balance, both by reducing the loss of heat by convection and water

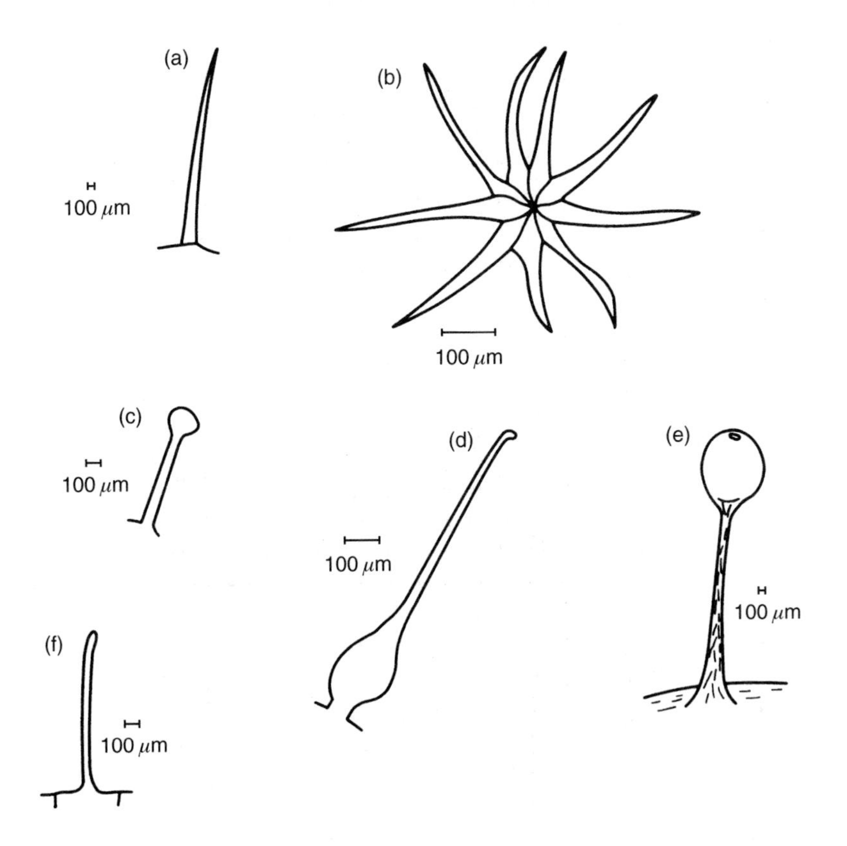

Figure 7.10

Types of hair on plant surfaces. (a) Simple hair from capsule of *Silene dioica*; (b) stellate hair from *Sida* (after Esau); (c) glandular hair from *Epilobium hirsutum*; (d) stinging hair from *Urtica dioica*; (e) glandular hair from *Drosera capensis*, an insectivorous plant; (f) root hair from *Lolium perenne*.

by evaporation and by increasing the reflection of incident radiation (cf. Chapter 4). In addition they may prevent small herbivores from feeding. If the hairs, in addition, are glandular or stinging, they can provide a chemical defence. In *D. wrightii* populations some plants ('velvety') have mostly non-glandular hairs and some ('sticky' plants) have mostly glandular hairs, that produce toxins called acyl sugars; the polymorphism is controlled by a single gene. Elle *et al.* (1999) grew the two types with and without insecticides to remove the risks of herbivory; they found that sticky plants produced 45% fewer viable seeds than velvety plants when they were grown with insecticide. The cost (measured here in fitness terms) of defence is therefore very large. When herbivores were permitted, there was no significant difference in seed production between the two types.

The costs of constitutive resistance mechanisms such as hairs are visually obvious, if often difficult to quantify. Many plants also have inducible defences, especially against systemic parasites such as fungi. The nature of the growth of a fungal pathogen on a plant gives the possibility of advance warning to as yet unaffected parts of the plant; the induced response that infected plants display is known as systemic acquired resistance (SAR; Ryals *et al.*, 1996). SAR can be induced artificially, and Heil *et al.* (2000) did so using wheat plants that were already protected against fungal attack by standard fungicides. The induced defence was therefore of no benefit to the plant, but it resulted in a large reduction in yield (Fig. 7.11). Offsetting these large costs must be large benefits. *Nicotiana* spp. (relatives of tobacco) contain the alkaloid nicotine which is as toxic to insect herbivores as to smokers; the alkaloid is synthesized in response to attack, and that synthesis is triggered by the hormone jasmonic acid. After induction, up to 8% of the plant's nitrogen can be tied up as nicotine, so this is clearly an expensive defence mechanism in that currency. Baldwin (1998) has shown that plants of *N. attenuata* growing in natural habitats that had been treated with the hormone and so had been induced, produced more seeds than uninduced plants if they were attacked by herbivores; however, if they were not attacked, the uninduced plants performed better. For the attacked plant, therefore, the cost is outweighed by the benefit.

In all these cases, the costs of defence have been offset by a benefit from reduced loss of resources to a parasite or a herbivore. In other cases, however, it is that loss that is the cost. The best known mutualistic association between plants and microbes are the nitrogen-fixing nodules formed by rhizobial bacteria on the roots of legumes (Fabaceae) and the mycorrhizal associations that most plants form with fungi (Chapter 3, p. 120). In both cases, the plant provides fixed carbon compounds to the microbial symbiont and the microbe passes a mineral nutrient (N or P) to the plant. The cost here, therefore, is the carbon; the benefit is the mineral nutrient (and, in the case of mycorrhizas, a range of other fitness benefits: Newsham *et al.*, 1995). It is easy to see that the benefits of these mutualistic symbioses to a plant will vary depending on circumstances; a plant growing in a nutrient-rich soil may gain little. In such circumstances, the mutualism may rapidly turn into a parasitic association, with the symbiont gaining carbon from the host without offering any significant benefit (Johnson *et al.*, 1997; Schwartz and Hoeksema, 1998).

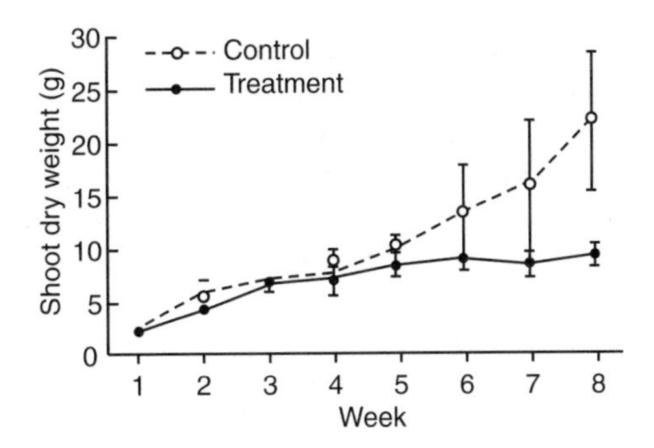

Figure 7.11

Growth of wheat plants in which systemic resistance to fungi had been chemically induced (●) was lower than that of control plants (O), when both were grown with a fungicide to reduce the impact of pathogenic fungi. In these circumstances, the resistance is of no benefit and the reduction in growth represents the cost of resistance (from Heil *et al.*, 2000).

The cost of these mutualistic symbioses, as in the cases of herbivory and pathogens discussed above, can be large. There are a number of estimates of the carbon cost of mycorrhizas, but most are in the region of 10–15% of the carbon fixed by the plant. Nevertheless the arbuscular mycorrhizal symbiosis seems to have existed as a major feature of land plants for over 400 million years. This implies that the benefits are large. A few groups of plants have evolved to become non-mycorrhizal, and some families have evolved their own specialized mycorrhizal types. One type, the ectomycorrhiza, has arisen on several occasions both among plants (Fitter and Moyersoen 1996) and fungi (Hibbett *et al.*, 2000). Importantly, phylogenetic analysis (Hibbett *et al.*, 2000) shows that the fungi involved in ectomycorrhizas have evolved back to a saprotrophic habit on several occasions, showing that this mutualism is evolutionarily unstable. Presumably, evolutionary situations will arise from time to time in which the costs of the association exceed the benefits, to one partner; at that point the mutualism will either break down or become parasitic. In arbuscular mycorrhizas it has been suggested that some fungi are 'cheats', taking carbon from the plant but offering no benefits, and theory suggests that mutualisms may be liable to invasion by such cheats. However, gaining empirical evidence for such a situation is extraordinarily difficult (Fitter, 2001).

4. Strategies

The complexity of the interactions that all plants experience means that any attempt at defining optimal growth strategies must take into account all the factors that control the availability of resources. The optimum growth trajectory discussed above (Fig. 7.2) is defined as that which maximizes carbon gain, but in practice the biological optimum is that which maximizes fitness. Therefore, the reproductive success of the individual must also be taken into account. Reproduction is another activity to be included in any budget of plant resources, together with shoot and root vegetative growth, respiration, defences against animals and pathogens, storage and so on. Resources allocated to reproduction are not available for the acquisition of further resources, as can be seen simply by considering the classical growth equation (Chapter 1) relating relative growth rate (**R**) to leaf area ratio (**F**) and net assimilation rate (**E**):

$$\mathbf{R} = \mathbf{E} \times \mathbf{F} \tag{7.1}$$

or

$$1/W \,.\, \mathrm{d}W/\mathrm{d}t = 1/A_{\mathrm{L}} \,.\, \mathrm{d}W/\mathrm{d}t \times A_{\mathrm{L}}/W \tag{7.2}$$

Growth rate depends on the ratio of leaf area (A_{L}) to plant weight (W) and diversion of resources to reproductive structures reduces growth rate. In this sense, therefore, reproduction and growth are alternatives. In many plants it is also true that their branching structure or architecture restricts the number of growing points they can bear, and here the conversion of a meristem into a flower means that it cannot produce leaves (Fig. 7.12). Of course, many reproductive structures are green, particularly bracts and sepals around flowers, and they may make some contribution to growth (or at least to the carbon requirements of the flower they surround: Bazzaz *et al.*, 1979), but this does not contradict the essential point.

Some plants are annual or ephemeral; they complete their life cycle within a single growing season, or in the extreme within a few weeks, as in some desert plants and in *Arabidopsis thaliana*, a temperate weed and the first plant for which the complete gene sequence was obtained (the Arabidopsis Genome Initiative 2000). They may then devote as much as 50–60% of their resources to seed production. As a result they are capable of rapid changes in population size, and are particularly suited to growth in environments where the adult population is likely to be drastically reduced by hazards outside their range of resistance. In other words, selection acts on their ability to achieve high rates of population increase, described by the parameter r in the classical logistic equation for population growth:

$$\frac{\mathrm{d}N}{\mathrm{d}t} = rN\,\frac{(K - N)}{K} \tag{7.3}$$

Here N is population size, t is time and K is the equilibrium population size. The equation describes a sigmoid growth curve with an asymptote of K; the steepness of the curve is described by r. Such species are often referred to as r-selected. At the other extreme are perennial plants that persist for several, sometimes many years. They may not reproduce at all for the first few years of their life; many trees, for example, have such a juvenile period of from 5–15 years, and even in monocarpic species (those which only reproduce once in their lifetime), the prereproductive phase can be remarkably long (e.g. *Agave deserti*, cf. Chapter 4). Even when they start to produce seeds, trees rarely devote

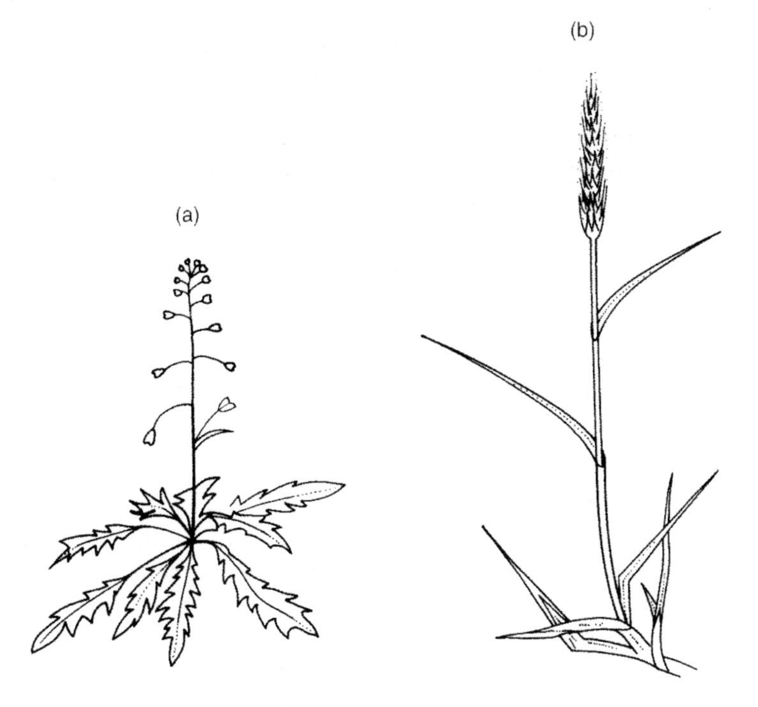

Figure 7.12

Two plant architectures which involve constraints on growth following initiation of reproduction. (a) *Capsella bursa-pastoris* (Brassicaceae) is a typical rosette plant in which leaf production ceases after the production of the almost leafless flowering stem. (b) *Phleum arenarium* (Poaceae) is an annual grass which grows by producing new tillers at the stem base. Each tiller is partially independent and has determinate growth: it produces a few leaves and ends in a flowering spike.

more than a few % of their annual productivity to sexual reproduction. Perennial herbs tend to have higher reproductive allocation rates, largely because they produce less supporting tissue, but maximum values of around 30% are typical. Many perennials reproduce vegetatively as well as sexually, but their total reproductive allocation has a similar ceiling. It may be difficult to distinguish vegetative reproduction (e.g. by a stolon which produces new plants at its nodes; Fig. 1.1b) from normal development, particularly in grasses, which produce tillers, which are effectively new grass plants, though physically linked to the parent.

In all such plants, selection appears to have acted more on their ability to occupy space, resist competition, and maintain activity over a long period. They will, therefore, tend to exhibit only slight changes in population size, which is commonly maintained at or near some equilibrium value denoted by K in equation 7.3 (often termed the 'carrying capacity'). They are therefore referred to as K-selected. This dichotomy between growth-oriented and reproduction-oriented, between *K*-selected and *r*-selected plants, offers a useful evolutionary viewpoint on the constraints which determine plant growth and form. We expect (and find) that plants of disturbed environments are *r*-selected, those of undisturbed and productive sites, such as forests, are *K*-selected.

Such a one-dimensional ordination of plant form is inevitably a caricature of its diversity. In particular it fails totally to distinguish plants of fertile, productive sites where resources provide little limitation to growth, from those of extreme environments, such as deserts, mountains and toxic soils. Both groups apparently fall into the *K*-selected category. They are, however, separated in the alternative, functional classification suggested by Grime (1977, 1979), which has three, rather than two poles. This derives from a two-dimensional ordination of the favourability of the environment and the level of disturbance, in which one extreme (unfavourable, disturbed sites) is uninhabitable, leaving a triangle (Fig. 7.13).

There are three underlying strategies in this scheme, at the three poles of the triangle:

(i) in favourable, disturbed environments the *ruderal* strategy is favoured, corresponding to *r*-selection;
(ii) in favourable, undisturbed sites *competitive* plants occur;
(iii) and in unfavourable, but undisturbed habitats *stress-tolerators* are found.

Various intermediate categories are possible within the scheme, but the relationship of these three strategies to the theme of this book is very close. Ruderals are plants of rapid growth potential, devoting a large proportion of their resources to reproduction, and growing in environments which, because they are recently disturbed, offer little competition for resources from other individuals. In consequence, ruderals can obtain resources without difficulty, and have little recourse to symbioses for nutrient acquisition, such as nitrogen-fixing bacteria or mycorrhizas. Equally, they are short-lived, unlikely to suffer severe pathogen damage, and so devote little to defence. In effect, ruderal species are avoiding the principal stress in their environment, the physical damage caused by disturbance. Intriguingly, if ruderal species are exposed to other types of stress (low nutrients, shade etc.) they tend to respond by further acceleration of the life cycle (a classic example of phenotypic plasticity, cf. Chapter 1), as shown by Stanton *et al.* (2000) for charlock *Sinapis arvensis*.

The competitive strategy, in contrast, though still often associated with rapid growth rate, is found in more persistent plants, which use resources less for seed production and more for support tissue to provide height growth which increases competitive effectiveness for radiation; for root systems to obtain water and ions in competition; for symbioses and defence, and so on. They normally have high rates of turn-over of, for example, leaves (cf. Chapter 3) since, in a competitive situation, the ability to redistribute resources from old leaves to new and better-placed ones may be crucial (Hirose and Werger 1987).

Finally, and perhaps most interestingly, there are stress-tolerators. These are plants of unfavourable environments, where resistance to environmental extremes may be the dominant selective force. In deserts, adaptations that increase water use efficiency are obvious, but may require diversion of

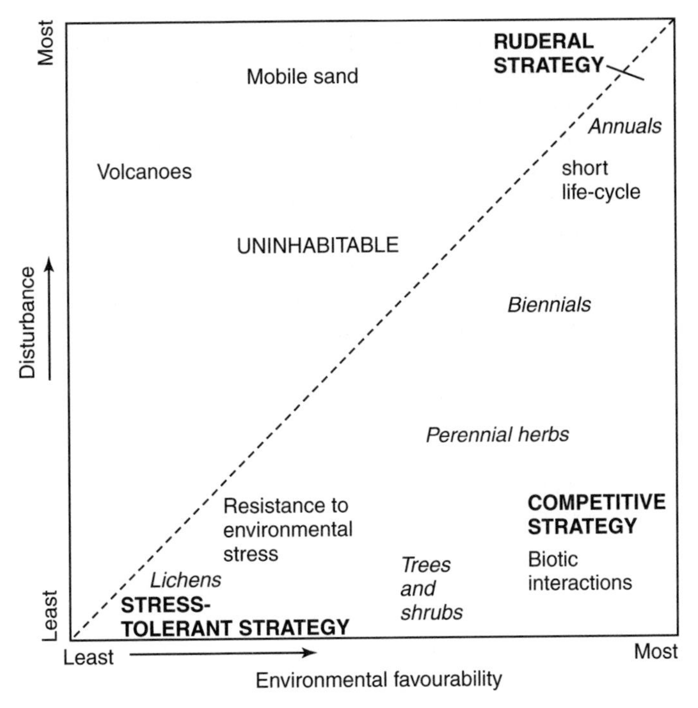

Figure 7.13

Grime's (1977) triangular ordination derives from an ordination using the two axes of disturbance and environmental favourability. Bold type represents the three primary strategies, italics the general position occupied by some major life-forms, and roman type the main selective forces.

resources, or reduce growth potential directly. Very high root weight ratios imply low growth rates, as do morphological adaptations such as reduced leaf blade dimensions (Chapters 4 and 5) or metabolic phenomena such as crassulacean acid metabolism (CAM; Chapters 2 and 4).

The concept of strategies, in the sense that plants with similar ecological behaviour share common sets of characteristics, is not new. The Danish botanist Raunkiaer (1907) produced a scheme of plant life-forms, based on the position of the overwintering organs (seeds, buds below, at or above ground level, in water, etc.), that proved successful and was widely used for comparing floras from different habitats. Another pioneer Danish ecologist, Warming (1909), developed a strictly habitat-based classification, including categories such as lithophytes (plants on rock) and psammophytes (plants on sand), as well as more familiar terms such as halophytes, for plants of saline soils.

The success of Raunkiaer's scheme derived from its ability to relate other aspects of a plant's biology to its habitat preference; in this sense it is a set of strategies. In deserts, for example, therophytes (annuals) are particularly numerous, whereas in grasslands it is hemicryptophytes (buds at ground level) and in the tropics phanerophytes (buds >25 cm above the ground – trees and shrubs) which are most abundant. It is easy to produce arguments based on adaptation to explain these distributions, but they say little about physiology. Schemes such as those of Grime (1979) offer a greater possibility of explaining ecological phenomena on a physiological basis.

5. Dynamics

Defining plant strategies is a powerful method for explaining distributional patterns and vegetation types. It can also be used to understand the dynamics of plant communities, especially because disturbance is a ubiquitous process and one of the major causes of change in communities. Few communities are stable, in the sense that they show unchanging species composition and abundance over long periods of time. Even what are often known as climax communities (for example, forests) are typically very dynamic, but the life spans of the individual trees are long relative to the time-scale over which environmental change occurs. Consequently, each generation of trees develops in a distinctive environment. In other cases, disturbance will be more severe and more wide-ranging, perhaps resulting in a major shift in the environmental conditions. The processes by which communities recover from such disturbance are known as ecological succession.

The environment of early successional communities (i.e. those that develop on a site after disturbance) differs in many ways from that of communities which have not been disturbed for a long period. Where the disturbance has been simply the removal of the vegetation, which then typically regenerates rapidly (secondary succession), the main environmental effect is on the microclimate, since the soil is more or less undisturbed. If the mature community is of trees, these create a subcanopy environment of low radiation flux density, low windspeed, little temperature fluctuation, and normally little risk of water deficit. In these conditions, seedlings of many trees can establish much more readily than is sometimes possible in exposed, recently disturbed environments. The effects can sometimes be unexpected: Finegan (1984) found that oak *Quercus robur* seeds failed to colonize a bare chalk quarry floor because they were too visible and so eaten by birds. Once hawthorn *Crataegus monogyna* became established, however, the acorns could be hidden in the hawthorn leaf litter, and oaks appeared. The constraint was not the expected one, which might have been water or nutrient shortage due to the lack of soil development on the bare chalk, but a biotic interaction.

If the environmental disturbance is more severe and a new, undeveloped substrate, such as moraine gravel, lava or sand, is colonized, then there are enormous differences in soil factors between early and late stages of succession. In the classic study at Glacier Bay, Crocker and Major (1955) found a 10-fold increase in soil nitrogen over the first 100 years of succession (Fig. 7.14), largely because of the activity of nitrogen-fixing alders. This was accompanied by a reduction in pH of over 3 units and a large increase in soil organic matter and therefore soil water-holding capacity. Obviously the early colonists – plants such as fireweed *Epilobium angustifolium* and mountain avens *Dryas drummondii* –

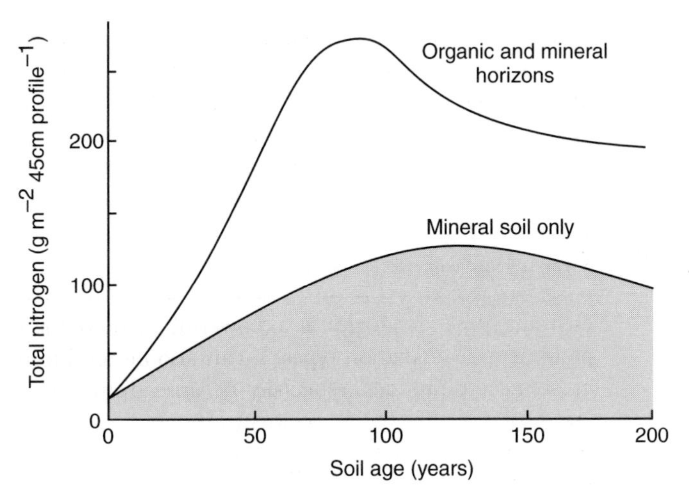

Figure 7.14

Soil nitrogen concentration increased markedly, in the period from 50–100 years after exposure of glacial moraine at Glacier Bay, Alaska, corresponding to the period of dominance of N-fixing alders. The later decline occurred when spruce replaced alder (from Crocker and Major, 1955).

experienced a totally different environment from that of the spruces and firs that eventually came to occupy the site.

At one level these changes can be seen in terms of Grime's strategies (see p. 306, above). In secondary successions, the initial colonists are typically fast-growing plants, capable of taking advantage of the relatively fertile soil and lack of competition; in other words, they are *ruderals*. In primary successions, however, where soil water and nutrient levels are initially low and other environmental factors, such as temperature, are often extreme, it is slow-growing plants resistant to such stresses that predominate. The classic primary colonists are lichens, which provide excellent examples of *stress-tolerance*. Whichever starting point a succession may derive from, the mature communities that result are ones in which the level of utilization of resources, and hence competition for those resources, is high. The plants that dominate these therefore typify the *competitive* strategy.

At the physiological level of analysis similar patterns emerge. Early successional species are by definition shorter-lived than later species, and this implies, as we have seen, faster growth rates. Rapid growth is typically associated with a well-defined suite of morphological and physiological characteristics (van der Werf *et al.*, 1993). In morphology, leaf thickness (measured as specific leaf area, SLA) is more important than allocation phenomena; in physiology, we find high maximum photosynthetic and respiratory rates. Bazzaz and Pickett (1980) proposed a similar set of attributes that characterize pioneer species in tropical forests, namely rapid growth achieved by high photosynthetic rates and low density of wood (allowing a tall stem to be built rapidly). Other linked traits included seeds capable of remaining dormant for long periods and being stimulated into germination by light, as well as a low dependence of the young seedlings on symbionts, especially mycorrhizal fungi. This hypothesis was tested by Veenendaal *et al.* (1996) on a large set of west African trees. In this group of 15 species, the pioneer species had consistently higher growth rate, net assimilation rate and specific leaf area than shade-tolerant species characteristic of later stages in the succession (Fig. 7.15).

Early successional species must be productive because of the intense competition that can occur in such habitats. Later species may have to devote more resources to support, defence and other non-growth functions. In addition, there is an architectural problem to be considered. Horn's (1971) theory of monolayer and multilayer tree canopies has already been described (Chapter 2, p. 38). As he points out, the greatest productivity is attained by exposing the largest leaf area to radiation, and this is achieved by the multilayer canopy, which lets a high proportion of the incident radiation penetrate to the lower layers. In doing so, however, it also permits shade-resistant saplings of other species to grow beneath the canopy, with the result that it is susceptible to invasion and replacement. The monolayer trees, with a low total leaf area produced by a poorly penetrable canopy, are therefore much less productive but relatively immune to invasion, until the death of an individual.

Increasingly, community change is being driven by human activity rather than by natural processes. In ecosystems where human impact is limited, natural disturbance by fires and storms may be the major force determining community dynamics. On a longer time-scale, climate can change; for example, the long-term warming that occurred after the most recent glacial retreat in the Northern Hemisphere 12 000–15 000 years ago has been studied in great detail. Plants have evolved in relation to these types of change but humanity is increasing both the scale and intensity of change. Habitat conversion is occurring at an astonishing rate, with huge areas of forest being converted into pasture. In most economically developed countries, natural habitats now exist as a patchwork of islands surrounded by 'deserts' of intensively farmed land in which few species can survive and this pattern is spreading throughout the world. Two specific drivers of change, however, have impacts that can be understood only by applying the science described in this book; they are the accumulation of carbon dioxide in the atmosphere and the deposition of fixed nitrogen from the atmosphere.

The likely impacts of a rising atmospheric CO_2 concentration on plants are described in Chapter 2, and those of the consequent increases in temperature in Chapter 5. A major issue facing land managers and conservationists, as well as those who try to predict the longer-term interactions of the climate system with the world's vegetation, is the extent to which we can predict the composition of communities on the basis of what we know of the responses of individual plants to these environmental changes. One possible route is to predict the behaviour of functional types of plants (for example, the different strategies described above). These might enable linkages to be made between plant ecophysiology on the one hand, which by definition must be studied at the level of the individual plant, and community, ecosystem and even biome-level processes (Diaz and Cabido, 1997).

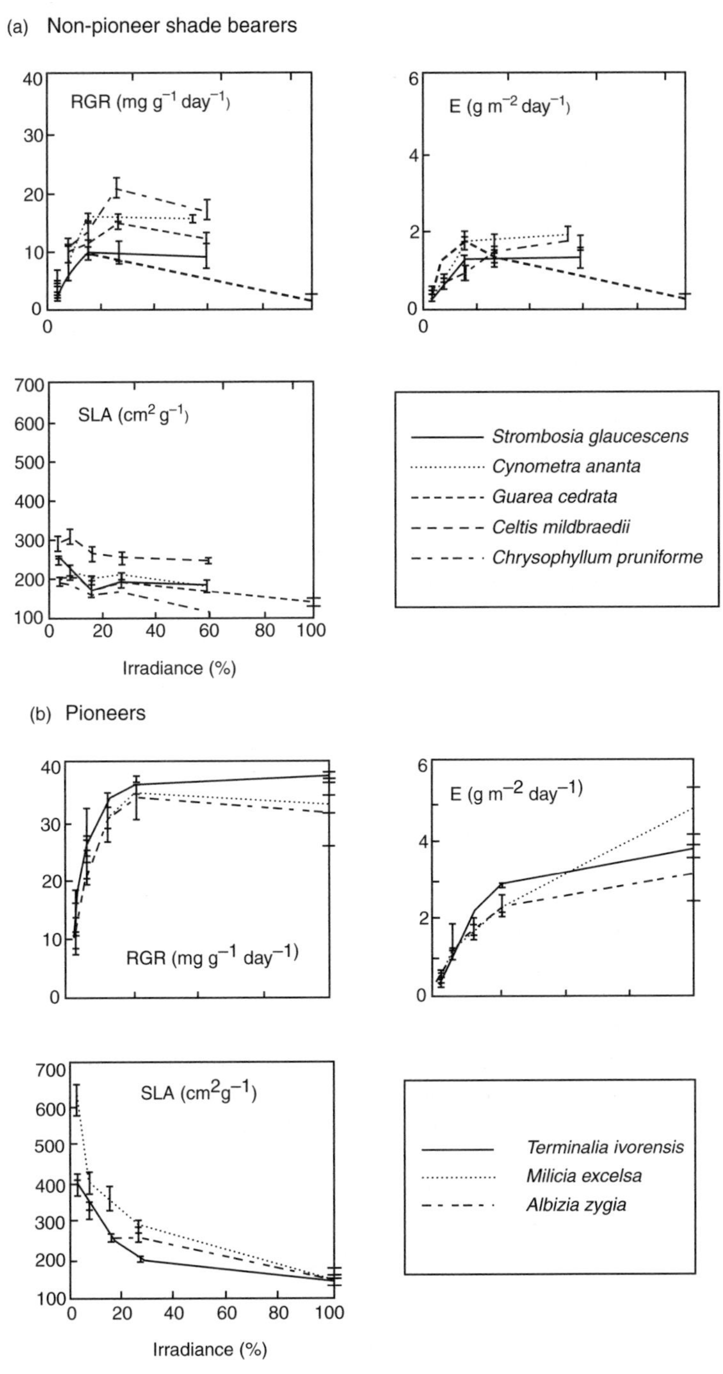

Figure 7.15

The relative growth rate (RGR), net assimilation rate (E) and specific leaf area (SLA) of (a) five non-pioneer, shade-resistant tree species and (b) three pioneer tree species from Ghana grown in varying depths of shade. The shade-resistant species had lower growth and assimilation rates, but were less sensitive to shading (from Grace (2001) using data of Veenendaal *et al.*, 1996).

Physiological responses, therefore, underlie the most fundamental ecological processes, and these processes can be studied at a series of levels. In this book we have described the physiological level and related it to the ecological. The differentiation we have described is related, often remarkably precisely, to environmental differences, and we have presented it as adaptive, because we view the differentiation as the result of evolutionary events.

References

Abrams, M.D., Kubiske, M.E. and Mostoller, S.A. (1994). Relating wet and dry year ecophysiology to leaf structure in contrasting temperate tree species. *Ecology* **75**, 123–133.

Ackerly, D.D. (1999). Comparative plant ecology and the role of phylogenetic information. In: Press, M.C., Scholes, J.D. and Barker, M.G. (eds.). *Physiological Plant Ecology*, pp. 391–413. Blackwell Science, Oxford.

Ackerly, D.D. and Donoghue, M.J. (1998). Leaf size, sapling allometry and Corner's rules: a phylogenetic study of correlated evolution in maples (*Acer*). *Am. Nat.* **152**, 767–791.

Adams, P., Thomas, J.C., Vernon, D.M., Bohnert, H.J. and Jensen, R.G. (1992). Distinct cellular and organismic responses to salt stress. *Plant Cell Physiol.* **33**, 1215–1223.

Addicott, J.F. (1998). Regulation of mutualism between yuccas and yucca moths: population level processes. *Oikos* **81**, 119–129.

Addison, P.A. and Bliss, L.C. (1984). Adaptations of *Luzula confusa* to the polar semi-desert environment. *Arctic* **37**, 121–132.

Aerts, R. (1996). Nutrient resorption from senescing leaves of perennials: are there general patterns? *J. Ecol.* **84**, 597–608.

Aerts, R. and Chapin, F.S. (1999). The mineral nutrition of wild plants revisited: a re-evaluation of processes and patterns. *Adv. Ecol. Res.* **30**, 2–67.

Alexander, C. and Hadley, G. (1984). The effect of mycorrhizal infection of *Goodyera repens* and its control by fungicide. *New Phytol.* **97**, 391–400.

Alexander, I.J. (1991). Systematics and ecology of ectomycorrhizal legumes. In: Stirton, C.H. and Zarucchi, J.L. (eds.). *Advances in Legume Biology*, pp. 607–624. *Monographs in Systematic Botany from the Missouri Botanic Garden* **29**.

Alexander, N.L., Flint, H.I. and Hammer, P.A. (1984). Variation in cold-hardiness of *Fraxinus americana* stem tissue according to geographic origin. *Ecology* **65**, 1087–1092.

Al-Hiyaly, S.A., McNeilly, T. and Bradshaw, A.D. (1990). The effect of zinc contamination from electricity pylons. Contrasting patterns of evolution in five grass species. *New Phytol.* **114**, 183–190.

Allen, R.A. (1995). Dissection of oxidative stress tolerance using transgenic plants. *Plant Physiol.* **107**, 1049–1054.

Allen, S.E. (1989) *Chemical Analysis of Ecological Materials*, 2nd Edn. Blackwell Scientific Publications, Oxford.

Amen, R.D. (1966). The extent and role of seed dormancy in alpine plants. *Q. Rev. Biol.* **41**, 271–281.

Amtmann, A. and Sanders, D. (1999). Mechanisms of Na^+ uptake by plant cells. *Adv. Bot. Res.* **29**, 75–112.

Anderson, M.C. (1964). Studies of the woodland light climate. I. The photographic computation of light conditions. *J. Ecol.* **52**, 27–42.

Anderson, M.C. (1970). Radiation climate, crop architecture and photosynthesis. In: Setlik, I. (ed.). *Prediction and Measurement of Photosynthetic Productivity*, pp. 71–78. Centre for Agricultural Publishing and Documentation, Wageningen.

Anderson, M.C. and Miller, E.E. (1974). Forest cover as a solar camera: penumbra effect in plant canopies. *J. Appl. Ecol.* **1**, 691–698.

Andrew, C.S. and Vanden Berg, P.J. (1973). The influence of aluminium on phosphate sorption by whole plants and excised roots of some pasture legumes. *Aust. J. Agric. Res.* **24**, 341–351.

Andrews, C.J. (1996). How do plants survive ice? *Ann. Bot.* **78**, 529–536.

Apse, M.P., Aharon, G.S., Snedden, W.A. and Blumwald, E. (1999). Salt tolerance conferred by overexpression of a vacuolar Na^+/H^+ antiport in *Arabidopsis*. *Science* **285**, 1256–1258.

Arabidopsis Genome Initiative (2000). Analysis of the genome sequence of the flowering plant *Arabidopsis thaliana*. *Nature* **408**, 796–815.

Armstrong, W. and Webb, T. (1985). A critical oxygen pressure for root extension in rice. *J. Exp. Bot.* **36**, 1573–1582.

Armstrong, W., Brändle, R. and Jackson, M.B. (1994). Mechanisms of flood tolerance in plants. *Acta Bot. Neerl.* **43**, 307–358.

Ashenden, T.W., Bell, S.A. and Rafarel, C.R. (1995). Responses of white clover to gaseous pollutants and acid mist: implications for setting critical levels and loads. *New Phytol.* **130**, 89–96.

Ashenden, T.W., Hunt, R., Bell, S.A., Williams, T.G., Mann, A., Booth, R.E. and Poorter, L. (1996). Responses to SO_2 pollution in 41 British herbaceous species. *Funct. Ecol.* **10**, 483–490.

Asher, C.J. and Loneragan, J.F. (1967). Response of plants to phosphate concentration in solution culture. 1. Growth and phosphate content. *Soil Sci.* **103**, 225–233.

Asman, W.A.H., Sutton, M.A. and Schjørring, J.K. (1998). Ammonia: emission, atmospheric transport and deposition. *New Phytol.* **139**, 27–48.

Assmann, S.M. (1999). The cellular basis of guard cell sensing of rising CO_2. *Plant, Cell Env.* **22**, 629–637.

Atherley, A.G. and Jenkins, G.I. (1997). Mechanisms underlying plant acclimation to low temperatures. *AgBiotech* **9**(**4**), 77N–80N.

Atkin, O.K. (1996). Reassessing the nitrogen relations of Arctic plants. *Plant, Cell Env.* **19**, 695–704.

Atkin, O.K. and Day, D.A. (1990). A comparison of the respiratory processes and growth rates of selected Australian alpine and related lowland plant species. *Aust. J. Plant Physiol.* **17**, 517–526.

Atkin, O.K., Botman, B. and Lambers, H. (1996). The causes of intrinsically slow growth in alpine plants: an analysis based on the underlying carbon economies of alpine and lowland *Poa* species. *Funct. Ecol.* **10**, 698–707.

Atkin, O.K., Schortemeyer, M., McFarlane, N. and Evans, J.R. (1999). The response of fast and slow-growing *Acacia* species to elevated atmospheric CO_2: an analysis of the underlying components of relative growth rate. *Oecologia* **120**, 544–554.

Atkinson, C.J., Wookey, P.A. and Mansfield, T.A. (1991). Atmospheric pollution and the sensitivity of stomata in barley leaves to abscisic acid and carbon dioxide. *New Phytol.* **117**, 535–541.

Aung, L.H. (1974). Root-shoot relationships. In: Carson, E.W. (ed.). *The Plant Root and Its Environment*, pp. 29–61. University Press, Virginia.

Bagshaw, R., Vaidyanathan, L.V. and Nye, P.H. (1972). The supply of nutrient ions to plant roots in soil. V. *Plant and Soil* **37**, 617–626.

Baker, A.J.M. and Walker, P.L. (1990). Ecophysiology of metal uptake by tolerant plants. In: Shaw, A.J. (ed.). *Heavy Metal Tolerance of Plants*, pp. 155–177. CRC Press, Boca Raton, Florida.

Baker, H.G. (1972). Seed weight in relation to environmental conditions in California. *Ecology* **53**, 997–1010.

Baldocchi, D. and Colineau, S. (1994). The physical nature of solar radiation in heterogeneous canopies: spatial and temporal attributes. In: Caldwell, M.M. and Pearcy, R.W. (eds.). *Exploitation of Environmental Heterogeneity by Plants*, pp. 21–72. Academic Press, New York.

Baldwin, I.T. (1998). Jasmonate-induced responses are costly but benefit plants under attack in native populations. *Proc. Natl. Acad. Sci. USA* **95**, 8113–8118.

Baldwin, J.P. (1975). A quantitative analysis of the factors affecting plant nutrient uptake from some soils. *J. Soil Sci.* **26**, 195–206.

Baldwin, J.P. and Nye, P.H. (1974). Uptake of solutes by multiple root systems from soil. IV. A model to calculate the uptake by a developing root system or root hair system of solutes with concentration variable diffusion coefficients. *Plant and Soil* **40**, 703–706.

Ball, M.C. (1994). The role of photo-inhibition during tree seedling establishment at low temperatures. In: Baker, N.R. and Bowyer, J.R. (eds.). *Photoinhibition of Photosynthesis – from Molecular Mechanisms to the Field*, pp. 367–378. Bios Scientific Publishers, Oxford.

Ballaré, C.L., Sanchez, R.A., Scopel, A.L., Casal, J.J. and Ghersa, C.M. (1987). Early detection of neighbour plants by phytochrome perception of spectral changes in reflected sunlight. *Plant, Cell Env.* **10**, 551–557.

Banga, M., Blom, C.W.P.M. and Voesenek, L.A.C.J. (1995). Flood-induced leaf elongation in *Rumex* species: effects of water table and water movements. *New Phytol.* **131**, 191–198.

Bannister, P. (1971). The water relations of heath plants from open and shaded habitats. *J. Ecol.* **59**, 51–64.

Baon, J.B., Smith, S.E. and Alston, A.M. (1994). Growth response and phosphorus uptake of rye with long and short root hairs – interactions with mycorrhizal infection. *Plant and Soil* **167**, 247–254.

Barber, S.A. (1974). Influence of the plant root on ion movement in the soil. In: Carson, E.W. (ed.). *The Plant Root and Its Environment*, pp. 525–563. University Press, Virginia.

Barley, K. (1970). The configuration of the root system in relation to nutrient uptake. *Adv. Agron.* **22**, 159–201.

Barnes, P.W. (1985). Adaptation to water stress in the big bluestem – sand bluestem complex. *Ecology* **66**, 1908–1920.

Barrs, H.D. (1968). Determination of water deficits in plant tissues. In: Kozlowski, T.T. (ed.). *Water Deficits and Plant Growth*, vol. 1, pp. 235–368. Academic Press, New York and London.

Bartels, D., Schneider, K., Terstappen, G., Piatkowski, D. and Salamini, F. (1990). Molecular cloning of abscisic acid modulated genes which are induced during desiccation of the resurrection plant *Craterostigma plantagineum*. *Planta* **181**, 27–34.

Bartlett, R.J. and James, B.R. (1993). Redox chemistry of soils. *Adv. Agron.* **50**, 151–208.

Bates, T.R. and Lynch, J.P. (1996). Stimulation of root hair elongation in *Arabidopsis thaliana* by low phosphorus availability. *Plant, Cell Env.* **19**, 529–538.

Baumeister, W. and Kloos, G. (1974). Über die Salzsekretion bei *Halimione portulacoides* (L.) Aellen. *Flora* **163**, 310–326.

Baxter, R., Bell, S.A., Sparks, T.H., Ashenden, T.W. and Farrar, J.F. (1995). Effects of elevated CO_2 concentrations on three montane grass species. III. Source leaf metabolism and whole plant carbon partitioning. *J. Exp. Bot.* **46**, 917–929.

Baylis, G.T.S. (1975). The magnolioid mycorrhiza and mycotrophy in root systems derived from it. In: Sanders, F.E., Mosse, B. and Tinker, P.B. (eds.). *Endomycorrhizas*, pp. 373–390. Academic Press, London and New York.

Bazzaz, F.A. and Pickett, S.T.A. (1980). Physiological ecology of tropical succession: a comparative review. *Annu. Rev. Ecol. Systemat.* **11**, 287–310.

Bazzaz, F.A., Carlson, R.W. and Harper, J.L. (1979). Contribution to reproductive effort by photosynthesis of flowers and fruits. *Nature* **279**, 554–555.

Beerling, D.J. (1994). Predicting leaf gas exchange and $\delta^{13}C$ responses to the past 30 000 years of global environmental change. *New Phytol.* **128**, 425–433.

Beerling, D.J. and Woodward, F.I. (1997). Changes in land plant function over the Phanerozoic: reconstructions based on the fossil record. *Bot. J. Linn. Soc.* **124**, 137–153.

Begg, J.E. (1980). Morphological adaptations of leaves to water stress. In: Turner, N.C. and Kramer, P.J. (eds.). *Adaptation of Plants to Water and High Temperature Stress*, pp. 33–42. Wiley-Interscience, New York.

Bell, D.T., Plummer, J.A. and Taylor, S.K. (1993). Seed germination ecology in southwestern Australia. *Bot. Rev.* **59**, 24–73.

Bell, J.N.B., Ashmore, M.R. and Wilson, G.B. (1991). Ecological genetics and chemical modifications of the atmosphere. In: Taylor, G.E., Pitelka, L.F. and Clegg, M.T. (eds.). *Ecological Genetics and Air Pollution*, pp. 33–59. Springer-Verlag, New York.

Belsky, A.J. (1986). Does herbivory benefit plants – a review of the evidence. *Am. Nat.* **127**, 870–892.

Bennert, W.H. and Mooney, H.A. (1979). The water relations of some desert plants in Death Valley, California. *Flora* **168**, 405–427.

Berner, R.A. (1993). Palaeozoic atmospheric CO_2: importance of solar radiation and plant evolution. *Science* **261**, 68–70.

Berry, J.A. and Raison, J.K. (1981). Responses of macrophytes to temperature. *Encyclopaedia of Plant Physiology* **12A**, 278–338.

Bewley, J.D. and Krochko, J.E. (1982). Desiccation-tolerance. *Encyclopaedia of Plant Physiology* **12B**, 325–378.

Beyers, J.L., Reichers, G.H. and Temple, P.J. (1992). Effects of long-term ozone exposure and drought on the photosynthetic capacity of Ponderosa pine (*Pinus ponderosa* Laws.). *New Phytol.* **122**, 81–90.

Bhat, K.K.S. and Nye, P.H. (1973). Diffusion of phosphate to plant roots in soil. 1. Quantitative autoradiography of the depletion zone. *Plant and Soil* **38**, 161–175.
Billings, W.D. (1974). Arctic and alpine vegetation: plant adaptations to cold summer climates. In: Ives, J.D. and Barry, R.G. (eds.). *Arctic and Alpine Environments*, pp. 403–443. Methuen, London.
Billings, W.D. and Mooney, H.A. (1968). The ecology of arctic and alpine plants. *Biol. Rev.* **43**, 481–529.
Billings, W.D., Godfrey, P.J., Chabot, B.F. and Bourque, D.P. (1971). Metabolic acclimation to temperature in arctic and alpine ecotypes of *Oxyria digyna*. *Arc. Alp. Res.* **3**, 277–289.
Björkman, O. (1968). Further studies on differentiation of photosynthetic properties in sun and shade ecotypes of *Solidago virgaurea*. *Physiol. Plant.* **21**, 84–99.
Björkman, O. (1980). The response of photosynthesis to temperature. In: Grace, J., Ford, E.D. and Jarvis, P.G. (eds.). *Plants and their Atmospheric Environment*, pp. 273–301. Blackwell Scientific Publications, Oxford.
Björkman, O. (1981). Responses to different quantum flux densities. *Encyclopaedia of Plant Physiology* **12A**, 57–108.
Björkman, O. and Holmgren, P. (1963). Adaptability of the photosynthetic apparatus to light intensity in ecotypes from exposed and shaded habitats. *Physiol. Plant.* **16**, 889–914.
Björkman, O., Pearcy, R. and Nobs, M. (1971). Hybrids between *Atriplex* species with and without β-carboxylation photosynthesis. *Carnegie Inst. Wash. Yearbook* **69**, 640–648.
Björkman, O., Boardman, N.K., Anderson, M.C., Goodchild, D.J. and Pyliotis, N.A. (1972). Effect of light intensity during growth of *Atriplex patula* on the capacity of photosynthetic reactions, chloroplast components and structure. *Carnegie Inst. Wash. Yearbook* **71**, 115–135.
Black, C.C. (1971). Ecological implications of dividing plants into groups with distinct photosynthetic production capacities. *Adv. Ecol. Res.* **7**, 87–114.
Black, M. (1969). Light controlled germination of seeds. In: *Dormancy and Survival. Symp. Soc. Exp. Biol.* **23**, 193–218.
Black, M. and Wareing, P.F. (1955). Growth studies in woody species. VII. Photoperiodic control of germination in *Betula pubescens* Ehrh. *Physiol. Plant.* **8**, 300–316.
Blackman, G.E. and Rutter, A.J. (1946). Physiological and ecological studies in the analysis of plant environment. I. The light factor and the distribution of the bluebell (*Scilla non-scripta*) in woodland communities. *Ann. Bot.* **10**, 361–390.
Blackman, G.E. and Wilson, G.L. (1951). Physiological and ecological studies in the analysis of plant environment. II. The constancy for different species of a logarithmic relationship between net assimilation rate and light intensity, and its ecological significance. *Ann. Bot.* **15**, 63–94.
Blackman, V.H. (1919). The compound interest law and plant growth. *Ann. Bot.* **33**, 353–360.
Blatt, M.R. and Grabov, A. (1997). Signal redundancy, gates and integration in the control of ion channels for stomatal movement. *J. Exp. Bot.* **48**, 529–537.
Blumwald, E. and Gelli, A. (1997). Secondary inorganic ion transport at the tonoplast. *Adv. Bot. Res.* **25**, 401–417.
Boddey, R.M., Urquiaga, S., Reis, V. and Döbereiner, J. (1991). Biological nitrogen fixation associated with sugar cane. *Plant and Soil* **137**, 111–117.
Bodner, M. and Beck, E. (1987). Effect of supercooling and freezing on photosynthesis in freezing-tolerant leaves of afroalpine 'giant rosette' plants. *Oecologia* **72**, 366–371.
Bohnert, H.J., Nelson, D.E. and Jensen, R.G. (1995). Adaptations to environmental stress. *Plant Cell* **7**, 1099–1111.
Bole, J.B. (1973). Influence of root hairs in supplying soil phosphorus to wheat. *Can. J. Soil Sci.* **53**, 169–175.
Bond, G. (1976). The results of the IBP survey of root nodule formation in non-leguminous angiosperms. In: Nutman, P.S. (ed.). *Symbiotic Nitrogen Fixation in Plants*, vol. **7**, pp. 443–474. Cambridge University Press, Cambridge.
Boorman, L. (1967). Biological flora of the British Isles: *Limonium vulgare* Mill. and *L. humile* Mill. *J. Ecol.* **55**, 221–232.
Borghetti, M., Grace, J. and Raschi, A. (eds.) (1993). *Water Transport in Plants under Climatic Stress.* Cambridge University Press, Cambridge.
Borland, A.M. (1996). A model for the partitioning of photosynthetically fixed carbon during the C_3-CAM transition in *Sedum telephium*. *New Phytol.* **134**, 433–444.

Borland, A.M., Tecsi, L.I, Leegood, R.C. and Walker, R.P. (1998). Inducibility of crassulacean acid metabolism (CAM) in *Clusia* species; physiological/biochemical characterisation and intercellular localization of carboxylation and decarboxylation processes in three species which exhibit different degrees of CAM. *Planta* **205**, 342–351.

Borland, A.M., Griffiths, H., Maxwell, C., Broadmeadow, M.S.J., Griffiths, N.M. and Barnes, J.D. (1992). On the ecophysiology of the Clusiaceae in Trinidad: expression of CAM in *Clusia minor* during the transition from wet to dry season and characterisation of three endemic species. *New Phytol.* **122**, 349–357.

Borland, A.M., Griffiths, H., Broadmeadow, M.S.J., Fordham, M.C. and Maxwell, C. (1994). Carbon-isotope composition of biochemical fractions and the regulation of carbon balance in leaves of the C_3-crassulacean acid metabolism intermediate *Clusia minor* L. growing in Trinidad. *Plant Physiol.* **106**, 493–501.

Bornman, J.F. and Sundby-Emanuelsson, C. (1995). Response of plants to UV-B radiation: some biochemical and physiological effects. In: Smirnoff, N. (ed.). *Environment and Plant Metabolism*, pp. 245–262. Bios Scientific Publishers, Oxford.

Bowler, C., Van Montague, M. and Inze, D. (1992). Superoxide dismutase and stress tolerance. *Annu. Rev. Plant Physiol. Plant Mol. Biol.* **43**, 83–116.

Bowman, D.M.J.S. (1998). The impact of Aboriginal landscape burning on the Australian biota. *New Phytol.* **140**, 385–410.

Bowman, W.D., Theodose, T.A., Schardt, J.C. and Conant, R.T. (1993). Constraints of nutrient availability on primary production in two alpine tundra communities. *Ecology* **74**, 2085–2097.

Boysen-Jensen, P. and Müller, D. (1929). Die maximale Ausbeute und der tägliche Verlauf der Kohlensäureassimilation. *Jahrb. Wiss. Bot.* **70**, 493–502.

Bradshaw, A.D., McNeilly, T. and Putwain, P.D. (1990). The essential qualities. In: Shaw, A.J. *Heavy Metal Tolerance in Plants*, pp.323–334. CRC Press, Boca Raton, Florida.

Bradstock, R.A. and Auld, T.D. (1995). Soil temperatures during experimental bushfires in relation to intensity: consequences for legume germination and fire management in south-eastern Australia. *J. Appl. Ecol.* **32**, 76–84.

Brady, N.C. (1974). *The Nature and Properties of Soils* 8th Edn. Macmillan, New York.

Brändle, R. (1991). Flooding resistance in rhizomatous amphibious plants. In: Jackson, M.B., Davies, D.D. and Lambers, H. *Plant Life under Oxygen Stress*, pp. 35–46. SPB Academic Publishing BV, The Hague.

Briones, O., Montana, C. and Excurra, E. (1996). Competition between three Chihuahuan desert species: evidence from plant size–distance relations and root distribution. *J. Veg. Sci.* **7**, 453–460.

Brookes, P.C., Powlson, D.S. and Jenkinson, D.S. (1985). The microbial biomass in soil. In: Fitter, A.H., Atkinson, D., Read, D.J. and Usher, M.B. *Ecological Interactions in Soil*, pp. 123–125. Blackwell Scientific Publications, Oxford.

Brown, A.D. (1976). Microbial water stress. *Bact. Rev.* **40**, 803–846.

Brown, J.C. (1972). Competition between phosphate and the plant for Fe from Fe^{++} ferrozine. *Agron. J.* **64**, 240–243.

Brown, M.E. (1975). Rhizosphere micro-organisms – opportunists, bandits or benefactors. In: Walker, N. (ed.). *Soil Microbiology*, pp. 21–38. Butterworth, London.

Brown, N.A.C., Jamieson, H. and Botha, P.A. (1994). Stimulation of seed germination in South African species of Restionaceae by plant-derived smoke. *Plant Growth Reg.* **15**, 93–100.

Brown, P.H., Welch, R.M. and Cary, E.E. (1987). Nickel: a micronutrient for higher plants. *Plant Physiol.* **85**, 801–803.

Brownell, P.F. and Crosland, C.J. (1972). Requirement for sodium as a micro-nutrient for species having the C_4 dicarboxylic photosynthetic pathway. *Plant Physiol.* **49**, 794–797.

Browse, J. and Somerville, C. (1991). Glycerolipid synthesis: biochemistry and regulation. *Annu. Rev. Plant Physiol. Plant Mol. Biol.* **42**, 467–506.

Brune, A., Urbach, W. and Dietz, K.-J. (1995). Differential toxicity of heavy metals is partly related to a loss of preferential extraplasmic compartmentation: a comparison of Cd-, Mo-, Ni- and Zn-stress. *New Phytol.* **129**, 403–409.

Buckland, S.M., Price, A.H. and Hendry, G.A.F. (1991). The role of ascorbate in drought-treated *Cochlearia atlantica* Pobed. and *Armeria maritima* (Mill.) Willd. *New Phytol.* **119**, 155–160.

Bull, K.R. (1991). The critical loads/levels approach to gaseous pollutant emission control. *Environ. Pollut.* **69**, 105–123.

Cakmak, I. and Marschner, H. (1993). Effect of zinc nutritional status on activities of superoxide radical and hydrogen peroxide scavenging enzymes in bean leaves. *Plant and Soil* **155/6**, 127–130.

Caldwell, M.M., Dawson, T.E. and Richards, J.H. (1998). Hydraulic lift: consequences of water efflux from the roots of plants. *Oecologia* **113**, 151–161.

Callaghan, T.V. and Lewis, M.C. (1971). The growth of *Phleum alpinum* L. in contrasting habitats at a sub-antarctic station. *New Phytol.* **70**, 1143–1154.

Campbell, B.D. and Grime, J.P. (1989). A comparative study of plant responsiveness to the duration of episodes of mineral nutrient enrichment. *New Phytol.* **112**, 261–267.

Campbell, B.D., Grime, J.P. and Mackey, J.M.L. (1991). A trade-off between scale and precision in resource foraging. *Oecologia* **87**, 532–538.

Campbell, G.S., Jungbauer, J.D., Bristow, K.L. and Hungerford, R.D. (1995). Soil temperatures and water content beneath a surface fire. *Soil Sci.* **159**, 363–374.

Canny, M. (1995). A new theory for the ascent of sap – cohesion supported by tissue pressure. *Ann. Bot.* **75**, 343–357.

Canny, M. (1998). Applications of the compensatory pressure theory of water transport. *Am. J. Bot.* **85**, 897–909.

Caradus, J.R. (1981). Effect of root hair length on white clover *Trifolium repens* cv Tamar grown over a range of phosphorus levels. *N.Z. J. Agric. Res.* **24**, 353–358.

Carlson, P.S. (1980). *The Biology of Crop Productivity*. Academic Press, New York.

Caswell, H., Reed, F., Stephenson, F.N. and Werner, P.A. (1973). Photosynthetic pathways and selective herbivory: a hypothesis. *Am. Nat.* **107**, 465–480.

Chapin, D.M., Bliss, L.C. and Bledsoe, L.J. (1991). Environmental regulation of nitrogen fixation in a high arctic lowland ecosystem. *Can. J. Bot.* **69**, 2744–2755.

Chapin, F.S. (1980). The mineral nutrition of wild plants. *Annu. Rev. Ecol. Systemat.* **11**, 233–250.

Chapin, F.S. (1983). Direct and indirect effects of temperature on arctic plants. *Polar Biol.* **2**, 47–52.

Chapin, F.S. and Shaver, G.R. (1985). Individualistic growth response of tundra plant species to environmental manipulations in the field. *Ecology* **66**, 564–576.

Chapin, F.S. and Shaver, G.R. (1989). Lack of latitudinal variations in graminoid storage reserves. *Ecology* **70**, 269–272.

Chapin, F.S., Moilanen, L. and Kielland, K. (1993). Preferential use of organic nitrogen for growth by a non-mycorrhizal arctic sedge. *Nature* **361**, 150–152.

Chapin, F.S., Bret-Harte, M.S., Hobbie, S.E. and Zhong, H. (1996). Plant functional types as predictors of transient responses of arctic vegetation to global change. *J. Veg. Sci.* **7**, 347–358.

Chase, M., Soltis, D.E., Olmstead, R.G. *et al.* (1993). Phylogenetics of seed plants: an analysis of nucleotide sequences from the plastid gene rbcL. *Ann. Missouri Bot. Gdn.* **80**, 528–580.

Chazdon, R.L. and Pearcy, R.W. (1986a). Photosynthetic responses to light variation in rainforest species. I. Induction under constant and fluctuating light conditions. *Oecologia* **69**, 517–523.

Chazdon, R.L. and Pearcy, R.W. (1986b). Photosynthetic responses to light variation in rainforest species. II. Carbon gain and photosynthetic efficiency during lightflecks. *Oecologia* **69**, 524–531.

Cheeseman, J.C. (1988). Mechanisms of salinity tolerance in plants. *Plant Physiol.* **87**, 547–550.

Chenery, E.A. and Sporne, K.P. (1976). A note on the evolutionary status of aluminium accumulation among dicotyledons. *New Phytol.* **76**, 551–554.

Chippindale, H.G. (1932). The operation of interspecific competition in causing delayed growth of grasses. *Ann. Appl. Biol.* **19**, 221–242.

Christie, E.K. and Moorby, J. (1975). Physiological responses of arid grasses. I. The influence of phosphorus supply on growth and phosphorus absorption. *Aust. J. Agric. Res.* **26**, 423–436.

Christie, J.M., Reymond, P., Powell, G.K., Bernasconi, P., Raibekas, A.A., Liscum, E. and Briggs, W.R. (1998). *Arabidopsis* NPH1: a flavoprotein with the properties of a photoreceptor for phototropism. *Science* **282**, 1698–1701.

Clack, T., Matthews, S. and Sharrock, R.A. (1994). The phytochrome apoprotein family in *Arabidopsis* is encoded by five genes: the sequences and expression of PHYD and PHYE. *Plant Mol. Biol.* **25**, 413–427.

Clarkson, D.T. (1966). Aluminium tolerance within the genus *Agrostis*. *J. Ecol.* **54**, 167–178.

Clarkson, D.T., Hall, K.C. and Roberts, J.K.M. (1980). Phospholipid composition and fatty acid desaturation in the roots of rye during acclimatization to low temperature: positional analysis of fatty acids. *Planta* **149**, 464–471.

Clarkson, D.T., Robards, A.W. and Sanderson, J. (1971). The tertiary endodermis in barley roots: fine structure in relation to radial transport of ions and water. *Planta* **96**, 296–305.

Clarkson, D.T., Sanderson, J. and Russell, R.S. (1968). Ion uptake and root age. *Nature* **220**, 805–806.

Clement, C.R., Hopper, M.J. and Jones, L.H.P. (1978). The uptake of nitrate by *Lolium perenne* from flowing nutrient solution. I. Effect of NO_3^- concentration. *J. Exp. Bot.* **109**, 453–464.

Cockburn, W. (1985). Variation in photosynthetic acid metabolism in vascular plants: CAM and related phenomena. *New Phytol.* **101**, 3–24.

Coen, E. (1999). *The Art of Genes: How Organisms Make Themselves.* Oxford University Press, Oxford.

Coleman, J.O.D., Blake-Kalff, M.M.A. and Davies, T.G.E. (1997). Detoxification of xenobiotics by plants: chemical modification and vacuolar compartmentation. *Trends Plant Sci.* **2**, 144–151.

Collier, D.E. (1996). No difference in leaf respiration rates among temperate, subarctic, and arctic species grown under controlled conditions. *Can. J. Bot.* **74**, 317–320.

Colls, J.J., Geissler, P.A. and Baker, C.K. (1992). Use of a field release system to distinguish the effects of dose and concentration of sulphur dioxide on winter barley. *Agric. Ecosyst. Env.* **38**, 3–10.

Comstock, J.P. (1999). Why Canny's theory doesn't hold water. *Am. J. Bot.* **86**, 1077–1081.

Connor, D.J. and Jones, T.R. (1985). Response of sunflower to strategies of irrigation. 2. Morphological and physiological responses to water stress. *Field Crops Res.* **12**, 91–103.

Cook, C.D.K. (1972). Phenotypic plasticity with particular reference to three amphibious plant species. In: Heywood, V.H. (ed.). *Taxonomy and Ecology*, pp. 97–111. Systematics Association Special Volume **5**. Academic Press, London and New York.

Coombe, D.E. (1966). The seasonal light climate and plant growth in a Cambridgeshire wood. In: *Light as an Ecological Factor. Symp. Br. Ecol. Soc.* **6**, 148–166.

Cooper, J.P. and Breeze, A.L. (1971). Plant breeding: forage grasses and legumes. In: Wareing, P.F. and Cooper, J.P. (eds.). *Potential Crop Production*, pp. 295–318. Heinemann, London.

Corbett, A.L., Krannitz, P.G. and Aarssen, L.W. (1992). The influence of petals on reproductive success in the arctic poppy (*Papaver radicatum*). *Can. J. Bot.* **70**, 200–204.

Cosgrove, D.J. (1993). Wall extensibility: its nature, measurement and relationship to plant cell growth. *New Phytol.* **124**, 1–23.

Cosgrove, D.J. (2000). Loosening of plant cell walls by expansins. *Nature* **407**, 321–326.

Cotrufo, M.F. and Ineson, P. (1996). Elevated CO_2 reduces field decomposition rates of *Betula pendula* (Roth.) leaf litter. *Oecologia* **106**, 525–530.

Coupland, R.T. and Johnson, R.E. (1965). Rooting characteristics of native grassland species in Saskatchewan. *J. Ecol.* **53**, 475–507.

Cowan, I.R. (1982). Regulation of water use in relation to carbon gain in higher plants. *Encyclopaedia of Plant Physiology* **12B**, 589–613.

Cowling, R.M. and Lamont, B.B. (1985). Seed release in *Banksia*: the role of wet-dry cycles. *Aust. J. Ecol.* **10**, 169–171.

Coxson, D.S. and Kershaw, K.A. (1983). The ecology of *Rhizocarpon superficiale*. 1. The rock surface boundary layer microclimate. *Can. J. Bot.* **61**, 3009–3018.

Cramer, G.R. and Nowak, R.S. (1992). Supplemental manganese improves the relative growth, net assimilation and photosynthetic rates of salt-stressed barley. *Physiol. Plant.* **84**, 600–605.

Crawford, N.M. and Glass, A.D.M. (1998). Molecular and physiological aspects of nitrate uptake in plants. *Trends Plant Sci.* **3**, 389–395.

Crawford, R.M.M. (1982). Physiological responses to flooding. *Encyclopaedia of Plant Physiology* **12B**, 453–477.

Crawford, R.M.M. and Palin, M.A. (1981). Root respiration and temperature limits to the north–south distribution of four perennial maritime plants. *Flora* **171**, 338–354.

Crawford, R.M.M., Hendry, G.A.F. and Goodman, B.A. (eds.). (1994). *Oxygen and Environmental Stress in Plants. Proc. R. Soc. Edin.* **B102**.

Crawford, R.M.M., Lindsay, D.A., Walton, J.C. and Wollenweber-Ratzer, B. (1994). Towards the characterization of radicals formed in rhizomes of *Iris germanica*. *Phytochemistry*. **37**, 979–985.

Cresswell, E. and Grime, J.P. (1981). Induction of a light requirement during seed development and its ecological consequences. *Nature* **291**, 583–585.

Criddle, R.S., Hopkin, M.S., McArthur, E.D. and Hansen, L.D.(1994). Plant distribution and the temperature coefficient of metabolism. *Plant Cell Env.* **17**, 233–243.

Crocker, R.L. and Major, J. (1955). Soil development in relation to vegetation and surface age, Glacier Bay, Alaska. *J. Ecol.* **43**, 427–448.

Cross, J.R. (1975). Biological flora of the British Isles: *Rhododendron ponticum*. *J. Ecol.* **63**, 345–359.

Crossley, G.K. and Bradshaw, A.D. (1968). Differences in response to mineral nutrients of populations of ryegrass, *Lolium perenne* L. and orchard grass *Dactylis glomerata* L. *Crop Sci.* **8**, 383–387.

Crush, J.R. (1974). Plant growth responses to vesicular arbuscular mycorrhiza. VII. Growth and nodulation of some herbage legumes. *New Phytol.* **73**, 743–749.

Cumming, J.R. and Weinstein, L.H. (1990). Utilization of $AlPO_4$ as a phosphorus source by ectomycorrhizal *Pinus rigida* Mill. seedlings. *New Phytol.* **116**, 99–106.

Cushman, J.C. and Bohnert, H.J. (1999). Crassulacean acid metabolism: molecular genetics. *Annu. Rev. Plant Physiol. Plant Mol. Biol.* **50**, 305–332.

Dahl, E. (1951). On the relation between summer temperature and the distribution of alpine vascular plants in the lowlands of Fennoscandia. *Oikos* **3**, 22–52.

Daniel, P.P., Woodward, F.I., Bryant, J.A. and Etherington, J.R. (1985). Nocturnal accumulation of acid in leaves of wall pennywort (*Umbilicus rupestris*) following exposure to water stress. *Ann. Bot.* **55**, 217–223.

Daniels, M.J., Chaumont, F., Mirkov, J.E. and Chrispeels, J. (1996). Characterisation of a new vacuolar membrane aquaporin sensitive to mercury at a unique site. *Plant Cell* **8**, 587–599.

Daram, P., Brunner, S., Amrhein, N. and Bucher, M. (1998). Functional analysis and cell specific expression of a phosphate transporter from tomato. *Planta* **206**, 225–233.

Darrah, P.R. (1993). The rhizosphere and plant nutrition: a quantitative approach. In: Barrow, N.J. (ed.). *Plant Nutrition – from Genetic Engineering to Field Practice*, pp. 3–22. Kluwer Academic Publications, Dordrecht.

Darrall, N.M. (1989). The effect of air pollutants on physiological processes in plants. *Plant, Cell and Env.* **12**, 1–30.

Davidson, R.L. (1969). Effect of root-leaf temperature differentials on root–shoot ratios in some pasture grasses and clover. *Ann. Bot.* **33**, 561–569.

Davies, M.S. (1974). Physiological differences among populations of *Anthoxanthum odoratum* L. collected from the Park Grass Experiment, Rothamsted. III. Response to phosphorus. *J. Appl. Ecol.* **11**, 699–707.

Davies, W.J. and Gowing, D.J.G. (1999). Plant responses to small perturbations in soil water status. In: Press, M.C., Scholes, J.D. and Barker, M.G. (eds.) *Physiological Plant Ecology*, pp. 67–89. Blackwell Science, Oxford.

Davies, W.J., Tardieu, F. and Trejo, C.L. (1994). How do chemical signals work in plants that grow in drying soil? *Plant Physiol.* **104**, 309–314.

Daxer, H. (1934). Über die Assimilations-ökologie der Waldbodenflora. *Jahrb. Wiss. Bot.* **80**, 363–420.

de la Fuente, J.M., Ramirez-Rodriguez, V., Cabrera-Ponce, J.L. and Herrera-Estrella, L. (1997). Aluminum-tolerance in transgenic plants by alteration of citrate synthesis. *Science* **276**, 1566–1568.

Delhaize, E. and Randall, P.J. (1995). Characterization of a phosphate-accumulator mutant of *Arabidopsis thaliana*. *Plant Physiol.* **107**, 207–213.

Demming, B. and Winter, K. (1986). Sodium, potassium, chloride and proline concentrations of chloroplasts isolated from a halophyte, *Mesembryanthemum crystallinum* L. *Planta* **168**, 421–426.

Demmig-Adams, B. and Adams, W.W. (1996). Xanthophyll cycle and light stress in nature: uniform response to excess direct sunlight among higher plant species. *Planta* **198**, 460–470.

Diatloff, E., Asher, C.J. and Smith, F.W. (1999). The effects of rare earth elements on the growth and nutrition of plants. *Rare Earths* **98**, 354–360.

Diaz, S. and Cabido, M. (1997). Plant functional types and ecosystem function in relation to global change. *J. Veg. Sci.* **8**, 463–474.

Dinkelaker, B., Hengeler, C. and Marschner, H. (1995). Distribution and function of proteid roots and other root clusters. *Acta Bot.* **108**, 183–200.

Dittmer, H.J. (1940). A quantitative study of the subterranean members of soybean. *Soil Conserv.* **6**, 33–34.

Döbereiner, J. and Day, J.M. (1976). Associative symbioses and free-living systems. In: Newton, W.E. and Nyman, C.J. (eds.). *Proceedings of the 1st International Symposium on Nitrogen Fixation*, pp. 518–538. Washington State University Press, Pullman.

Drake, B.G., Muehe, M., Peresta, G., Gonzales-Meler, M.A. and Matamala, R. (1997). Acclimation of photosynthesis, respiration and ecosystem carbon flux of a Chesapeake Bay wetland after eight years exposure to elevated CO_2. *Plant and Soil* **187**, 111–118.

Drew, M.C. (1975). Comparison of the effects of a localized supply of phosphate, nitrate, ammonium and potassium on the growth of the seminal root system, and the shoot, in barley. *New Phytol.* **75**, 479–490.

Drew, M.C. (1992). Soil aeration and plant root metabolism. *Soil Sci.* **154**, 259–268.

Drew, M.C. and Lynch, J.M. (1980). Soil anaerobiosis, micro-organisms and root function. *Annu. Rev. Phytopathol.* **18**, 37–66.

Drew, M.C. and Nye, P.H. (1969). The supply of nutrient ions by diffusion to plant roots in soil. II. *Plant and Soil* **31**, 407–424.

Drew, M.C. and Saker, L.R. (1975). Nutrient supply and the growth of the seminal root system in barley. II. *J. Exp. Bot.* **26**, 79–90.

Drew, M.C., Saker, L.R. and Ashley, T.W. (1973). Nutrient supply and the growth of the seminal root system in barley. I. *J. Exp. Bot.* **24**, 1189–1202.

Dudley, S.A. and Schmitt, J. (1995). Genetic differentiation in morphological responses to simulated foliage shade between populations of *Impatiens capensis* from open and woodland sites. *Funct. Ecol.* **9**, 665–666.

Duncan, W.G. and Ohlrogge, A.J. (1958). Principles of nutrient uptake from fertiliser bands. II. *Argon. J.* **50**, 605–608.

Eagles, C.F., Williams, J. and Louis, D.V. (1993). Recovery after freezing in *Avena sativa, Lolium perenne* and *Lolium multiflorum*. *New Phytol.* **123**, 477–483.

Eastmond, P.J. and Ross, J.D. (1997). Evidence that the induction of crassulacean acid metabolism by water stress in *Mesembryanthemum crystallinum* (L) involves root signalling. *Plant, Cell and Env.* **20**, 1559–1565.

Edwards, W.R.N., Jarvis, P.G., Grace, J. and Moncrieff, J.B. (1994). Reversing cavitation in tracheids of *Pinus sylvestris* L. under negative water potentials. *Plant, Cell Env.* **17**, 389–397.

Ehleringer, J.R. (1978). Implications of quantum yield differences on the distribution of C_3 and C_4 grasses. *Oecologia* **31**, 255–267.

Ehleringer, J.R. (1980). Leaf morphology and reflectance in relation to water and temperature stress. In: Turner, N.C. and Kramer, P.J. (eds.). *Adaptations of Plants to Water and High Temperature Stress*, pp. 295–308. Wiley-Interscience, New York.

Ehleringer, J.R. (1988). Comparative ecophysiology of *Encelia farinosa* and *Encelia frutescens*. 1. Energy balance considerations. *Oecologia* **76**, 553–561.

Ehleringer, J.R. (1993a). Gas-exchange implications of isotopic variation in arid-land plants. In: Smith, J.A.C. and Griffiths, H. (eds.). *Water Deficits: Plant Responses from Cell to Community*, pp. 265–284. Bios Scientific Publishers, Oxford.

Ehleringer, J.R. (1993b). Variation in leaf carbon isotope discrimination in *Encelia farinosa*: implications for growth, competition, and drought survival. *Oecologia* **95**, 340–346.

Ehleringer, J.R. and Field, C.B. (eds.) (1993). *Scaling Physiological Processes: Leaf to Globe*, Academic Press, San Diego.

Ehleringer, J.R. and Forseth, I. (1980). Solar tracking by plants. *Science* **210**, 1094–1098.

Ehleringer, J.R. and Monson, R.K. (1993). Evolutionary and ecological aspects of photosynthetic pathway variation. *Annu. Rev. Ecol. Systemat.* **24**, 411–439.

Ehleringer, J.R. and Werk, K.S. (1986). Modifications of solar radiation absorption patterns and implications for carbon gain at the leaf level. In: Givnish, T.J. (ed.). *On the Economy of Plant Form and Function*. Cambridge University Press, Cambridge.

Ehleringer, J.R., Hall, A.E. and Farquhar, G.D. (1993). Water use in relation to productivity. In: Ehleringer, J.R., Hall, A.E. and Farquhar, G.D. (eds.). *Stable Isotopes and Plant Carbon-Water Relations*, pp. 3–8. Academic Press, San Diego.

Ehleringer, J.R., Schwinning, S. and Gebauer, R. (1999). Water use in arid land ecosystems. In: Press, M.C., Scholes, J.D. and Barker, M.G. (eds.). *Physiological Plant Ecology*, pp. 347–365. Blackwell Science, Oxford.

Eickmeier, W.G., Casper, C. and Osmond, C.B. (1993). Chlorophyll fluorescence in the resurrection plant *Selaginella lepidophylla* (Hook. & Grev.) Spring during high light and desiccation stress, and evidence for zeaxanthin-associated photoprotection. *Planta* **189**, 30–38.

Einsmann, J.C., Jones, R.H., Pu, M. and Mitchell, R.J. (1999). Nutrient foraging traits in 10 co-occurring plant species of contrasting life forms. *J. Ecol.* **87**, 609–619.

Elle, E., Van Dam, N.M. and Hare, J.D. (1999). Cost of glandular trichomes, a 'resistance' character in *Datura wrightii* Regel (Solanaceae). *Evolution* **53**, 22–35.

Emerson, R. and Arnold, W. (1932). A separation of the reactions in photosynthesis by means of intermittent light. *J. Gen. Physiol.* **19**, 391–420.

Enright, N.J. and Lamont, B.B. (1989). Seed banks, fire season, safe sites and seedling recruitment in five co-occurring *Banksia* species. *J. Ecol.* **77**, 1111–1122.

Epstein, E. (1973). Mechanisms of ion transport through plant cell membranes. *Int. Rev. Cytol.* **34**, 123–168.

Epstein, E. (1999). Silicon. *Annu. Rev. Plant Physiol. Plant Mol. Biol.* **50**, 641–664.

Erdei, L. and Trivedi, S. (1991). Caesium/potassium selectivity in wheat and lettuce of different K^+ status. *J. Plant Physiol.* **138**, 696–699.

Ericsson, T. (1995). Growth and root: shoot ratio of seedlings in relation to nutrient availability. *Plant and Soil* **168/169**, 205–214.

Evans, G.C. and Hughes, A.P. (1961). Plant growth and the aerial environment. I. Effect of artificial shading on *Impatiens parviflora. New Phytol.* **60**, 150–180.

Evans, J.R. (1989). Photosynthesis and nitrogen relationships in leaves of C3 plants. *Oecologia* **78**, 9–19.

Farley, R.A. and Fitter, A.H. (1999a). Temporal and spatial variation in soil resources in a deciduous woodland. *J. Ecol.* **87**, 688–696.

Farley, R.A. and Fitter, A.H. (1999b). The responses of seven co-occurring woodland herbaceous perennials to localized nutrient-rich patches. *J. Ecol.* **87**, 849–859.

Farquhar, G.D., Ehleringer, J.R. and Hubick, K.T. (1989). Carbon isotope discrimination and photosynthesis. *Annu. Rev. Plant Physiol. Plant Mol. Biol.* **40**, 503–537.

Farrant, J.M., Cooper, K., Kruger, L.A. and Sherwin, H.W. (1999). The effect of drying rate on the survival of three desiccation-tolerant angiosperm species. *Ann. Bot.* **84**, 371–379.

Farrar, J.F. (1996). Regulation of root weight ratio is mediated by sucrose. *Plant and Soil* **185**, 13–19.

Farrar, J.F. (1999). Acquisition, partitioning and loss of carbon. In: Press, M.C., Scholes, J.D. and Barker, M.G. (eds.). *Physiological Plant Ecology*, pp. 25–43. Blackwell Science, Oxford.

Ferrari, G. and Renosto, F. (1972). Comparative studies on the active transport by excised roots of inbred and hybrid maize. *J. Agric. Sci., Cambridge* **79**, 105–108.

Field, C. and Mooney, H.A. (1986). The photosynthesis-nitrogen relationship in wild plants. In: Givnish, T.J. (ed.). *On the Economy of Plant Form and Function*, pp. 25–55. Cambridge University Press, Cambridge.

Finegan, B. (1984). Forest succession. *Nature* **312**, 109–114.

Firn, R.D. (1994). Phototropism. In: Kendrick, R.E. and Kronenberg, G.H.M. (eds.). *Photomorphogenesis in Higher Plants*, pp. 659–681. Kluwer, Dordrecht.

Fitter, A.H. (1976). Effects of nutrient supply and competition from other species on root growth of *Lolium perenne* in soil. *Plant and Soil* **45**, 177–189.

Fitter, A.H. (1977). Influence of mycorrhizal infection on competition for phosphorus and potassium by two grasses. *New Phytol.* **79**, 119–125.

Fitter, A.H. (1985). Functional significance of root morphology and root system architecture. In: Fitter, A.H., Atkinson, D., Read, D.J. and Usher, M.B. (eds.). *Ecological Interactions in Soil*, pp. 87–106. Blackwell Scientific Publications, Oxford.

Fitter, A.H. (1991). Costs and benefits of mycorrhizas: implications for functioning under natural conditions. *Experientia* **47**, 350–355.

Fitter, A.H. (1994). Architecture and biomass allocation as components of the plastic response of root systems to soil heterogeneity. In: Caldwell, M.M. and Pearcy, R.W. (eds.). *Exploitation of Environmental Heterogeneity by Plants*, pp. 305–323. Academic Press, New York.

Fitter, A.H. (1996). Characteristics and functions of root systems. In: Waisel, Y., Eshel, A. and Kafkafi, U. (eds.). *Plant Roots: the Hidden Half*, pp. 1–20. Marcel Dekker, New York.

Fitter, A.H. (1999). Roots as dynamic systems: the deveopmental ecology of roots and root systems. In: Press, M., Scholes, J.D. and Barker, M.G. (eds.). *Plant Physiological Ecology*, British Ecological Society Symbposium No. 39, pp. 115–131. Blackwell Scientific Publications, Oxford.

Fitter, A.H. (2001). Specificity, links and networks in the control of diversity in plant and microbial communities. In: Press, M.C., Huntly, N. and Levin, S. (eds.). *Ecology: Achievement and Challenge*, pp. 95–114. Blackwell Scientific Publications, Oxford.

Fitter, A.H. and Ashmore, C.J. (1974). Response of *Veronica* species to a simulated woodland light climate. *New Phytol.* **73**, 997–1001.

Fitter, A.H. and Bradshaw, A.D. (1974). Root penetration of *Lolium perenne* on colliery shale in response to reclamation treatments. *J. Appl. Ecol.* **11**, 609–616.

Fitter, A.H. and Moyersoen, B. (1996). Evolutionary trends in root-microbic symbioses. *Phil. Trans. R. Soc. Lond.* **B 351**, 1367–1375.

Fitter, A.H. and Strickland, T.R. (1991). Architectural analyses of plant root systems. II. Influence of nutrient supply on architecture in contrasting plant species. *New Phytol.* **118**, 383–389.

Fitter, A.H., Graves, J.D., Wolfenden, J., Self, G.K., Brown, T.K., Bogie, D. and Mansfield, T.A. (1997). Root production and turnover and carbon budgets of two contrasting grasslands under ambient and elevated atmospheric carbon dioxide concentrations. *New Phytol.* **137**, 247–255.

Fitter, A.H., Wright, W.J., Wiliamson, L., Belshaw, M., Fairclough, J. and Meharg, A.A. (1998). The phosphorus nutrition of wild plants and the paradox of arsenate tolerance: does leaf phosphate concentration control flowering? In: Lynch, J.P. and Deikman, J. (eds.). *Phosphorus in Plant Biology: Regulatory Roles in Molecular, Cellular, Organismic and Ecosystem Processes*, pp. 39–51. American Society of Plant Physiologists, Rockville.

Flowers, T.J. (1972). Salt tolerance in *Suaeda maritima* (L.) Dum. The effect of sodium chloride on growth, respiration, and soluble enzymes in a comparative study with *Pisum sativum* L. *J. Exp. Bot.* **23**, 310–321.

Fogel, R. (1985). Roots as primary producers in below-ground ecosystems. In: Fitter, A.H., Atkinson, D., Read, D.J. and Usher, M.B. (eds.). *Ecological Interactions in Soil.* Blackwell Scientific Publications, Oxford.

Forde, B.G. and Clarkson, D.T. (1998). Nitrate and ammonium nutrition of plants: physiological and molecular perspectives. *Adv. Bot. Res.* **30**, 1–90.

Fowler, D. and Cape, J.N. (1982). Air pollutants in agriculture and horticulture. In: Unsworth, M.H. and Ormrod, D.P. (eds.). *Effects of Gaseous Air Pollution in Agriculture and Horticulture*, pp. 1–26. Butterworth, London.

Fowler, D., Flechard, C., Skiba, U., Coyle, M. and Cape, J.N. (1998). The atmospheric budget of oxidised nitrogen and its role in ozone formation and deposition. *New Phytol.* **139**, 11–23.

Fox, M.D. and Fox, B.J. (1986). The effect of fire frequency on the structure and floristic composition of a woodland understorey. *Aust. J. Ecol.* **11**, 77–85.

Foy, C.D. (1988). Plant adaptation to acid, aluminium-toxic soils. *Comm. Soil Sci. Plant Anal.* **19**, 959–987.

Franco, A.C. and Nobel, P.S. (1988). Interactions between seedlings of *Agave deserti* and the nurse plant *Hilaria rigida*. *Ecology* **69**, 1731–1740.

Fried, M. and Broeshart, H. (1967). *The Soil–Plant System.* Academic Press, New York and London.

Furuya, M. and Schäfer, E. (1996). Photoperception and signalling of induction reactions by different phytochromes. *Trends Plant Sci.* **1**, 301–307.

Gahoonia, T.S. and Nielsen, N.E. (1998). Direct evidence on participation of root hairs in phosphorus (^{32}P) uptake from soil. *Plant and Soil* **198**, 147–152.

Gange, A. (2000). Arbuscular mycorrhizal fungi, collembola and plant growth. *Trends Ecol. Evol.* **15**, 369–372.

Garcia-Huidobro, J., Monteith, J.L. and Squire, G.R. (1982). Time, temperature and germination of pearl millet (*Pennisetum typhoides* S. & H.). 1. Constant temperature. *J. Exp. Bot.* **33**, 288–296.

Gardner, W.K., Parberry, D.G. and Barber, D.A. (1982). The acquisition of phosphorus by *Lupinus albus* L. 1. Some characteristics of the soil–root interface. *Plant and Soil* **68**, 19–32.

Gardner, W.R. (1960). Dynamic aspects of water availability to plants. *Soil Sci.* **89**, 63–73.

Garland, J.L. (1996). Patterns of potential C source utilisation by rhizosphere communities. *Soil Biol. Biochem.* **28**, 223–230.

Garner, W.W. and Allard, H.A. (1920). Effect of length of day on plant growth. *J. Agric. Res.* **18**, 553–606.

Garnier, E., Koch, G.W., Roy, J. and Mooney, H.A. (1989). Responses of wild plants to nitrate availability. Relationships between growth rate and nitrate uptake parameters, a case study with two *Bromus* species, and a survey. *Oecologia* **79**, 542–550.

Gates, D.M. (1968). Transpiration and leaf temperature. *Annu. Rev. Plant. Physiol.* **19**, 211–238.

Gauhl, E. (1969). Leaf factors affecting the rate of light-saturated photosynthesis in ecotypes of *Solanum dulcamara. Carnegie Inst. Wash. Yrbk* **68** 633–636.

Gauhl, E. (1976). Photosynthetic response to varying light intensity in ecotypes of *Solanum dulcamara* L. from shaded and exposed habitats. *Oecologia* **22**, 275–286.

Gauhl, E. (1979). Sun and shade ecotypes in *Solanum dulcamara* L.: Photosynthetic light dependence characteristics in relation to mild water stress. *Oecologia* **39**, 61–70.

Gauslaa, Y. (1984). Heat resistance and energy budget in different Scandinavian plants. *Holarctic Ecol.* **7**, 1–78.

Geller, G.N. and Smith, W.K. (1982). Influence of leaf size, orientation and arrangement on temperature and transpiration in three high-elevation, larger-leaved herbs. *Oecologia* **53**, 227–234.

George, M.F., Hong, S.G. and Burke, M.J. (1977). Cold hardiness and deep supercooling of hardwoods: its occurrence in provenance collections of red oak, yellow birch, black walnut and black cherry. *Ecology* **58**, 674–680.

Gerretsen, F.C. (1948). The influence of micro-organisms on the phosphate intake by the plant. *Plant and Soil* **1**, 51–81.

Ghersa, C.M., Benech Arnold, R.L. and Martinez-Ghersa, M.A. (1992). The role of fluctuating temperatures in germination and establishment of *Sorghum halapense.* Regulation of germination at increasing depths. *Funct. Ecol.* **6**, 460–468.

Gigon, A. and Rorison, I.H. (1972). The response of some ecologically distinct plant species to nitrate and ammonium nitrogen. *J. Ecol.* **60**, 93–102.

Giller, K.E. and Day, J.M. (1985). Nitrogen fixation in the rhizosphere: significance in natural and agricultural systems. In: Fitter, A.H., Atkinson, D., Read, D.J. and Usher, M.B. (eds.). *Ecological Interactions in Soil*, pp. 127–147. Blackwell Scientific Publications, Oxford.

Glass, A.D.M. (1973). Influence of phenolic acids on ion uptake. I. Inhibition of phosphate uptake. *Plant Physiol.* **51**, 1037–1041.

Glass, A.D.M. (1974). Influence of phenolic acids on ion uptake. III. Inhibition of potassium absorption. *J. Exp. Bot.* **25**, 1104–1113.

Glass, A.D.M., Siddiqi, M.Y., Ruth, T.J. and Rufty, T.W. (1990). Studies on the uptake of nitrate in barley. II. Energetics. *Plant Physiol.* **93**, 1585–1589.

Godwin, H., Clowes, D.R. and Huntley, B. (1974). Studies in the ecology of Wicken Fen. V. Development of fen carr. *J. Ecol.* **62**, 197–214.

Good, R. (1964). *The Geography of Flowering Plants.* Longman, London.

Grabherr, G., Gottfried, A., Gruber, A. and Pauli, H. (1995). Patterns and current changes in alpine plant diversity. In: Chapin, F.S. and Körner, C. (eds.). *Arctic and Alpine Biodiversity*, pp. 167–181. Springer-Verlag, Berlin.

Grace, J. (1977). *Plant Response to Wind.* Academic Press, London.

Grace, J. (1983). *Plant–Atmosphere Relations.* Chapman and Hall, London.

Grace, J. (2001). Environmental controls of gas exchange in tropical rain forests. In: Press, M., Huntly, N. and Levin, S. (eds.). *Ecology: Achievement and Challenge.* Blackwell Scientific Publications, Oxford.

Grace, J. and Wilson, J. (1976). The boundary layer over a *Populus* leaf. *J. Exp. Bot.* **27**, 231–241.

Grace, J., Allen, S. and Wilson, C. (1989). Climate and meristem temperatures of plant communities near the tree-line. *Oecologia* **79**, 198–204.

Graham, E.A. and Nobel, P.S. (1999). Root water uptake, leaf water storage and gas exchange of a desert succulent: implications for root system redundancy. *Ann. Bot.* **84**, 213–223.

Gravett, A. and Long, S.P. (1990). Intraspecific variation in susceptibility to photo-inhibition during chilling of *Cyperus longus* L. populations from Europe. In: Baltscheffsky, M. (ed.). *Current Research in Photosynthesis*, vol. 2, pp. 475–478. Kluwer, Dordrecht.

Greaves, M.P. and Darbyshire, J.F. (1972). The ultrastructure of the mucilaginous layer on plant roots. *Soil Biol. Biochem.* **4**, 443–449.

Green, D.G. and Warder, F.G. (1973). Accumulation of damaging concentrations of phosphorus by leaves of Selkirk wheat. *Plant and Soil* **38**, 567–572.

Greer, D.H., Laing, W.A. and Campbell, B.D. (1995). Photosynthetic responses of thirteen pasture species to elevated CO_2 and temperature. *Aust. J. Plant Physiol.* **22**, 713–722.

Grime, J.P. (1966). Shade avoidance and shade tolerance in flowering plants. In: *Light as an Ecological Factor'. Symp. Br. Ecol. Soc.* **6**, 187–207.

Grime, J.P. (1977). Evidence for the existence of three primary strategies in plants and its relevance to ecological and evolutionary theory. *Am. Nat.* **111**, 1169–1194.

Grime, J.P. (1979). *Plant Strategies and Vegetation Processes*, Wiley, London.

Grime, J.P. and Hodgson, J.G. (1969). An investigation of the ecological significance of lime-chlorosis by means of large-scale comparative experiments. In: *Ecological Aspects of the Mineral Nutrition of Plants. Br. Ecol. Soc. Symp.* **9**, 67–100.

Grime, J.P. and Hunt, R. (1975). Relative growth rate: its range and adaptive significance in a local flora. *J. Ecol.* **63**, 393–422.

Grime, J.P. and Jeffery, D.W. (1965). Seedling establishment in vertical gradients of sunlight. *J. Ecol.* **53**, 621–642.

Grime, J.P., Thompson, K., Hunt, R. and 31 others (1997). Integrated screening validates primary races of specialisation in plants. *Oikos* **79**, 259–281.

Grubb, P.J. and Suter, M.B. (1971). The mechanism of acidification of soil by *Calluna* and *Ulex* and the significance for conservation. In: *The Scientific Management of Animal and Plant Communities for Conservation. Br. Ecol. Soc. Symp.* **11**, 115–135.

Grulke, N.E. and Bliss, L.C. (1988). Comparative life history characteristics of two high arctic grasses, Northwest Territories. *Ecology* **69**, 484–496.

Gulmon, S.L. and Mooney, H.A. (1977). Spatial and temporal relationships between two desert shrubs *Atriplex hymenelytra* and *Tidestromia oblongifolia*. *J. Ecol.* **65**, 831–838.

Gupta, P.L. and Rorison, I.H. (1975). Seasonal differences in the availability of nutrients down a podzolic profile. *J. Ecol.* **63**, 521–534.

Gusta, L.V., Tyler, N.J. and Chen, T.H.-H. (1983). Deep undercooling in woody taxa growing north of the −40 °C isotherm. *Plant Physiol.* **72**, 122–128.

Gutterman, Y. (1980/81). Annual rhythm and position effects in the germinability of *Mesembryanthemum nodiflorum*. *Isr. J. Bot.* **29**, 93–97.

Gutterman, Y. (1994). Long-term seed position influences on seed germinability of the desert annual, *Mesembryanthemum nodiflorum* L. *Isr. J. Plant Sci.* **42**, 197–205.

Gutterman, Y. (1998). Ecological strategies of desert annual plants. In: Ambasht, R.S. (ed.). *Modern Trends in Ecology and Environment*, pp. 203–231. Backhuys Publishers, Leiden.

Guy, C.L. (1990). Cold acclimation and freezing stress tolerance: role of protein metabolites. *Annu. Rev. Plant Physiol. Plant Mol. Biol.* **41**, 187–223.

Ha, S.B., Smith, A.P., Howden, R., Dietrich, W.M., Bugg, S., O'Connell, M.J., Goldsbrough, P.B. and Cobbett, C.S. (1999). Phytochelatin synthase genes from *Arabidopsis* and the yeast *Schizosaccharomyces pombe*. *Plant Cell* **11**, 1153–1163.

Håbjørg, A. (1978a). Photoperiod ecotypes in Scandinavian trees and shrubs. *Meld. Norges Landbrukshøgskole* **57** (33), 2–20.

Håbjørg, A. (1978b). Climatic control of floral differentiation and development in selected latitudinal and altitudinal ecotypes of *Poa pratensis*. *Meld. Norges Landbrukshøgskole* **57**(17), 1–21.

Hairston, N.G., Smith, F.E. and Slobodkin, L.B. (1960). Community structure, population control and competition. *Am. Nat.* **94**, 421–425.

Harley, J.L. and Lewis, D.H. (1969). The physiology of ectotrophic mycorrhizas. *Adv. Microbiol. Physiol.* **3**, 53–58.

Harper, F.A., Smith, S.E. and Macnair, M.R. (1997). Can an increased copper requirement in copper tolerant *Mimulus guttatus* explain the cost of tolerance? 1. Vegetative growth. *New Phytol.* **136**, 455–467.

Harper, F.A., Smith, S.E. and Macnair, M.R. (1999). Can an increased copper requirement in copper tolerant *Mimulus guttatus* explain the cost of tolerance? 2. Reproductive phase. *New Phytol.* **140**, 637–654.

Harper, J.L. (1969). The role of predation in vegetational diversity. In: *Diversity and Stability in Ecological Systems, Brookhaven Symp. Biol.* **22**, 48–62.

Harper, J.L. (1986). Modules, branches and the capture of resources. In: Jackson, J.B.C., Buss, L.E. and Cook, R.E. (eds.). *Population Biology and the Evolution of Clonal Organisms*, pp. 1–34. Yale University Press, New Haven.

Harper, J.L. and Benton, R.A. (1966). The behaviour of seeds in soil. II. The germination of seeds on the surface of a water supplying substrate. *J. Ecol.* **54**, 151–166.

Harper, J.L., Williams, J.T. and Sagar, G.R. (1965). The behaviour of seeds in soil. I. The heterogeneity of soil surfaces and its role in determining the establishment of plants from seed. *J. Ecol.* **53**, 273–286.

Harris, G.A. and Wilson, A.M. (1970). Competition for moisture among seedlings of annual and perennial grasses as influenced by root elongation at low temperature. *Ecology* **51**, 530–534.

Harrison-Murray, R.S. and Clarkson, D.T. (1973). Relationships between structural development and absorption of ions by the root system of *Cucurbita pepo*. *Planta* **114**, 1–16.

Harvey, D.M., Hall, J.L., Flowers, T.J. and Kent, B. (1981). Quantitative ion localisation within *Suaeda maritima* leaf mesophyll cells. *Planta* **151**, 550–560.

Hassan-Porath, E. and Poljakoff-Mayber, A. (1969). The effect of salinity on the malic dehydrogenase of pea roots. *Plant Physiol.* **44**, 103–104.

Havill, D.C., Lee, J.A. and Stewart, G.R. (1974). Nitrate utilisation by species from acid and calareous soils. *New Phytol.* **73**, 1221–1232.

Havström, M., Callaghan, T.V. and Janasson, S. (1993). Differential growth responses of *Cassiope tetragona*, an arctic dwarf shrub, to environmental perturbations among three contrasting high- and sub-arctic sites. *Oikos* **66**, 389–402.

Hay, R.K.M. (1981). Timely planting of maize: a case history from the Lilongwe Plain. *Trop. Agric.* **58**, 147–155.

Hay, R.K.M. (1990). The influence of photoperiod on the dry-matter production of grasses and cereals. *New Phytol.* **116**, 233–254.

Hay, R.K.M. and Kemp, D.R. (1992). The prediction of leaf canopy expansion in the leek from a simple model dependent upon primordial development. *Ann. Appl. Biol.* **120**, 537–545.

Hay, R.K.M. and Kirby, E.J.M. (1991). Convergence and synchrony: a review of the coordination of development in wheat. *Aust. J. Agric. Res.* **42**, 661–700.

Hay, R.K.M. and Walker, A.J. (1989). *An Introduction to the Physiology of Crop Yield*. Longman, London.

Hayman, D.S. and Mosse, B. (1972). Plant growth responses to vesicular-arbuscular mycorrhiza. III. Increased uptake of labile P from soil. *New Phytol.* **71**, 41–47.

Heck, W.W. (1984). Defining gaseous pollution problems in North America. In: Koziol, M.J. and Whatley, F.R. (eds.). *Gaseous Air Pollutants and Plant Metabolism*, pp. 35–48. Butterworth, London.

Hegarty, T.W. and Ross, H.A. (1980/81). Investigations of control mechanisms of germination under water stress. *Isr. J. Bot.* **29**, 83–92.

Heide, O.M. (1992). Flowering strategies of the high arctic and high alpine snow bed grass *Phippsia algida*. *Physiol. Plant.* **85**, 606–610.

Heide, O.M., Hay, R.K.M. and Baugeröd, H. (1985). Specific daylength effects on leaf growth and dry matter production in high-latitude grasses. *Ann. Bot.* **55**, 579–586.

Heil, M., Hilpert, A., Kaiser, W. and Linsenmair, K.E. (2000). Reduced growth and seed set following chemical induction of pathogen defence: does systemic acquired resistance (SAR) incur allocation costs? *J. Ecol.* **88**, 645–654.

Hellkvist, J., Richards, G.P. and Jarvis, P.G. (1974). Vertical gradients of water potential and tissue water relations in Sitka Spruce trees measured with the pressure chamber. *J. Appl. Ecol.* **11**, 637–667.

Hemborg, A.M. and Despres, L. (1999). Oviposition by mutualistic seed-parasitic pollinators and its effects on annual fitness of single- and multi-flowered host plants. *Oecologia* **120**, 427–436.

Henderson, S., von Caemmerer, S., Farquhar, G.D., Wade, L. and Hammer, G. (1998). Correlation between carbon isotope discrimination and transpiration efficiency in lines of the C_4 species *Sorghum bicolor* in the glasshouse and the field. *Aust. J. Plant Physiol.* **25**, 111–123.

Hendry, G.A.F. (1994). Oxygen and environmental stress in plants: an evolutionary context. *Proc. R. Soc. Edin.* **102B**, 155–165.

Henry, G.H.R., Freedman, B. and Svoboda, J. (1986). Effects of fertilization on three tundra plant communities of a polar desert oasis. *Can. J. Bot.* **64**, 2502–2507.

Hewitt, E.J. (1967). *Sand and Water Culture Methods Used in the Study of Plant Nutrition*. C.A.B., London.

Hibbett, D.S., Gilbert, L.-B. and Donoghue, M.J. (2000). Evolutionary instability of ectomycorrhizal symbioses in basidiomycetes. *Nature* **407**, 506–508.

Hiesey, W.M., Nobs, M.A. and Bjørkman, O. (1971). Experimental studies on the nature of species. V. Biosystematics, genetics, and physiological ecology of the *Erythranthe* section of *Mimulus*. *Carnegie Inst. Wash. Publ.* **628**.

Hill, B.S. and Hill, A.E. (1973). Enzymatic approaches to chloride transport in the *Limonium* salt gland. In: Anderson, W.P. (ed.). *Ion Transport in Plants*, pp. 379–384. Academic Press, London and New York.

Hirose, T. and Werger, M.J.A. (1987). Maximizing daily canopy photosynthesis with respect to the leaf nitrogen allocation pattern in the canopy. *Oecologia* **72**, 520–526.

Hirose, T. and Werger, M.J.A. (1994). Photosynthetic capacity and nitrogen partitioning among species in the canopy of a herbaceous plant community. *Oecologia* **100**, 203–212.

Hodge, A., Paterson, E., Thornton, B., Millard, P. and Killham, K. (1997). Effects of photon flux density on carbon partitioning and rhizosphere carbon flow of *Lolium perenne*. *J. Exp. Bot.* **48**, 1797–1805.

Hodge, A., Stewart, J., Robinson, D., Griffiths, B.S. and Fitter, H. (1999a). Plant, soil fauna and microbial responses to N-rich organic patches of contrasting temporal availability. *Soil Biol. Biochem.* **31**, 1517–1530.

Hodge, A., Robinson, D., Griffiths, B.S. and Fitter, A.H. (1999b). Why plants bother: root proliferation results in increased nitrogen capture from an organic patch when two grasses compete. *Plant, Cell Env.* **22**, 811–820.

Hodgkinson, K.C. (1992). Water relations and growth of shrubs before and after fire in a semi-arid woodland. *Oecologia* **90**, 467–473.

Holbrook, N.M. and Putz, F.E. (1996). From epiphyte to tree: differences in leaf structure and leaf water relations associated with the transition in growth form in eight species of hemiepiphytes. *Plant, Cell Env.* **19**, 631–642.

Holbrook, N.M., Burns, M.J. and Field, C.B. (1995). Negative xylem pressure in plants: a test of the balancing pressure technique. *Science* **270**, 1193–1194.

Holford, I.C.R., Wedderburn, R.M. and Mattingly, G.E.G. (1974). A Langmuir two-surface equation as a model for phosphate absorption by soils. *J. Soil Sci.* **25**, 242–255.

Holland, J.N. and Fleming, T.H. (1999). Mutualistic interactions between *Upiga virescens* (Pyralidae), a pollinating seed-consumer, and *Lophocereus schottii* (Cactaceae). *Ecology* **80**, 2074–2084.

Holmes, M.G. and Smith, H. (1977). The function of phytochrome in the natural environment. II. The influence of vegetation canopies on the spectral energy distribution of natural daylight. *Photochem. Photobiol.* **25**, 539–546.

Holmgren, P. (1968). Leaf factors affecting light-saturated photosynthesis in ecotypes of *Solidago virgaurea* from exposed and shaded habitats. *Physiol. Plant.* **21**, 676–698.

Hook, D.D., Brown, C.L. and Kormanik, P.P. (1971). Inductive flood tolerance in swamp tupelo (*Nyssa sylvatica* var *biflora* (Walt.) Sarg.). *J. Exp. Bot.* **22**, 78–89.

Hope-Simpson, J.F. (1938). A chalk flora of the Lower Greensand, and its use in determining the calcicole habit. *J. Ecol.* **26**, 218–235.

Horn, H. (1971). *The Adaptive Geometry of Trees.* University Press, Princeton.

Houghton, J.T., Jenkins, G.J. and Ephraums, J.J. (1990). *Climatic Change: the IPCC Assessment.* Cambridge University Press, Cambridge.

Howard, R.J. and Mendelssohn, I.A. (1999). Salinity as a constraint on growth of oligohaline marsh macrophytes. II. Salt pulses and recovery potential. *Am. J. Bot.* **86**, 795–806.

Howarth, C.J. and Ougham, H.J. (1993). Gene expression under temperature stress. *New Phytol.* **125**, 1–26.

Howden, R., Goldsborough, P.B., Anderson, C.R. and Cobbett, C.S. (1995). Cadmium-sensitive *cad1* mutants of *Arabidopsis thaliana* are phytochelatin deficient. *Plant Physiol.* **107**, 1059–1066.

Hsiao, T.C. (1973). Plant responses to water stress. *Annu. Rev. Plant Physiol.* **24**, 519–570.

Hsiao, T.C., Acevedo, E., Fereres, E. and Henderson, D.W. (1976). Water stress, growth and osmotic adjustment. *Phil. Trans. R. Soc. London* **B273**, 479–500.

Huber, S.C., Bachmann, M. and Huber, J.L. (1996). Post-translational regulation of nitrate reductase activity: a role for Ca^{2+} and 14-3-3 proteins. *Trends Plant Sci.* **1**, 432–438.

Hughes, A.P. (1959). Effects of the environment on leaf development in *Impatiens parviflora* D.C. *J. Linn. Soc.(Bot)* **56**, 161–165.

Hughes, M.A. and Dunn, M.A. (1996). The molecular biology of plant acclimation to low temperature. *J. Exp. Bot.* **47**, 291–305.

Huisman, J. and Olff, H. (1998). Competition and facilitation in multispecies plant–herbivore systems of productive environments. *Ecol. Lett.* **1**, 25–29.

Hulten, E. (1962). *The Circumpolar Plants. II. Dicotyledons. K. svenska. Vetensk. Akad. Handl. ser.* **5**, 13(1).

Huner, N.P.A., Öquist, G. and Sarhan, F. (1998). Energy balance and acclimation to light and cold. *Trends Plant Sci.* **3**, 224–230.

Hunt, R. (1982). *Plant Growth Curves.* Edward Arnold, London.

Hunt, R. and Parsons, T. (1974). A computer program for deriving growth functions in plant growth-analysis. *J. Appl. Ecol.* **11**, 297–307.

Hunt R., Hand, D.W., Hannah, M.A. and Neal, A.M. (1991). Response to CO_2 enrichment in 27 herbaceous species. *Funct. Ecol.* **5**, 410–421.

Hunter, M.I.S., Hetherington, A.M. and Crawford, R.M.M. (1983). Lipid peroxidation – a factor in anoxia intolerance in *Iris* species. *Phytochemistry* **22**, 1145–1147.

Huntley, B. (1991). How plants respond to climatic change: migration rates, individualism and the consequences for plant communities. *Ann. Bot.* **67 (Supplement 1)**, 15–22.

Hutchinson, T.C. (1967). Comparative studies of the ability of species to withstand prolonged periods of darkness. *J. Ecol.* **55**, 291–299.

Ingram, J. and Bartels, D. (1996). The molecular basis of dehydration tolerance in plants. *Annu. Rev. Plant Physiol. Plant Mol. Biol.* **47**, 377–403.

Idso, S.B. and Kimball, B.A. (1993). Tree growth in carbon dioxide enriched air and its implications for global carbon cycling and maximum levels of atmospheric CO_2. *Global Biogeochem. Cycles* **7**, 537–555.

Iino, M. (1990). Phototropism: mechanisms and ecological implications. *Plant, Cell Env.* **13**, 633–650.

Ineson, P., Cotrufo, M.F., Bol, R., Harkness, D.D. and Blum, H. (1996). Quantification of soil carbon inputs under elevated CO_2: C_3 plants in a C_4 soil. *Plant and Soil* **187**, 345–350.

Ishikawa, M. and Sakai, A. (1981). Freezing avoidance mechanisms by supercooling in some *Rhododendron* flower buds with reference to water relations. *Plant Cell Physiol.* **22**, 953–967.

Israel, D.W. (1987). Investigation of the role of phosphorus in symbiotic dinitrogen fixation. *Plant Physiol.* **84**, 835–840.

Ives, J.D. and Barry, R.G. (Eds.) (1974). *Arctic and Alpine Environments.* Methuen, London.

Iwasa, Y. and Roughgarden, J. (1984). Shoot/root balance of plants: optimal growth of a system with many vegetative organs. *Theor. Popul. Biol.* **25**, 78–105.

Jackson, L.W.R. (1967). Effect of shade on leaf structure of deciduous tree species. *Ecology* **48**, 498–499.

Jackson, M.B. (1994). Root-to-shoot communication in flooded plants: involvement of abscisic acid, ethylene, and 1-aminocyclopropane-1-carboxylic acid. *Agron. J.* **86**, 775–782.

Jackson, M.B. and Armstrong, W. (1999). Formation of aerenchyma and the processes of plant ventilation in relation to soil flooding and submergence. *Plant Biol.* **1**, 274–287.

Jackson, R.B. and Caldwell, M.M. (1993). Geostatistical patterns of soil heterogeneity around individual perennial plants. *J. Ecol.* **81**, 683–692.

Jaffre, T., Brooks, R.R., Lee, J. and Reeves, D. (1978). *Serbetia acuminata*: a nickel-accumulating plant from New Caledonia. *Science* **193**, 579–580.

Jakobsen, I., Abbott, L.K. and Robson, A.D. (1992a). External hyphae of vesicular–arbuscular mycorrhizal hyphae associated with *Trifolium subterraneum* L. 1. Spread of hyphae and phosphorus inflow into roots. *New Phytol.* **120**, 371–380.

Jakobsen, I., Abbott, L.K. and Robson, A.D. (1992b). External hyphae of vesicular–arbuscular mycorrhizal hyphae associated with *Trifolium subterraneum* L. 2. Hyphal transport of ^{32}P over defined distances. *New Phytol.* **120**, 509–516.

James, J.C., Grace, J. and Hoad, S.P. (1994). Growth and photosynthesis of *Pinus sylvestris* at its altitudinal limit in Scotland. *J. Ecol.* **82**, 297–306.

Jarvis, A.J., Mansfield, T.A. and Davies, W.J. (1999). Stomatal behaviour, photosynthesis and transpiration under rising CO_2. *Plant, Cell Env.* **22**, 639–648.

Jarvis, P.G. (1964). The adaptability to light intensity of seedlings of *Quercus petraea* (Matt.) Liebl. *J. Ecol.* **52**, 545–571.

Jayasekera, G.A.U., Reid, D.M. and Yeung, E.C. (1990). Fates of ethanol produced during flooding of sunflower roots. *Can. J. Bot.* **68**, 2408–2414.

Jeschke, W.D., Klagges, S., Hilpert, A., Bhatti, A.S. and Sarwar, G. (1995). Partitioning and flows of ions and nutrients in salt-treated plants of *Leptochloa fusca* L. Kunth. 1. Cations and chloride. *New Phytol.* **130**, 23–35.

Johanson, U., Gehrke, C., Björn, L.O. and Callaghan, T.V. (1995). The effects of enhanced UV-B radiation on the growth of dwarf shrubs in a subarctic heathland. *Funct. Ecol.* **9**, 713–719.

Johnson, D.A. and Tieszen, L.L. (1976). Aboveground biomass allocation, leaf growth and photosynthesis patterns in tundra plant forms in arctic Alaska. *Oecologia* **24**, 159–173.

Johnson, I.R., Melkonian, J.J., Thornley, J.H.M. and Riha, S.J. (1991). A model of water flow through plants incorporating shoot/root 'message' control of stomatal conductance. *Plant, Cell Env.* **14**, 531–544.

Johnson, N.C., Graham, J.H. and Smith, F.A. (1997). Functioning of mycorrhizal associations along the mutualism- parasitism continuum. *New Phytol.* **135**, 757–586.

Joly, C.A. (1994). Flooding tolerance: a reinterpretation of Crawford's metabolic theory. *Proc. R. Soc. Edin.* **102B**, 343–354.

Jones, D.L. and Darrah, P.R. (1993a). Influx and efflux of amino-acids from *Zea mays* roots and their implications for N-nutrition and the rhizosphere. *Plant and Soil* **156**, 87–90.

Jones, D.L. and Darrah, D.L. (1993b). Re-sorption of organic compounds by roots of *Zea mays* L. and its consequences in the rhizosphere. 2. Experimental and model evidence for simultaneous exudation and re-sorption of soluble C compounds. *Plant and Soil* **153**, 47–59.

Jones, H.G. (1993). Drought tolerance and water-use efficiency. In: Smith, J.A.C. and Griffiths, H. (eds.). *Water Deficits: Plant Responses from Cell to Community*, pp. 193–203. Bios Scientific Publishers, Oxford.

Jones, H.G. and Corlett, J.E. (1992). Current topics in drought physiology. *J. Agric Sci., Cambridge* **119**, 291–296.

Jones, H.G. and Sutherland, R.A. (1991). Stomatal control of xylem embolism. *Plant, Cell Env.* **14**, 607–612.

Jones, M.D., Durall, D.M. and Tinker, P.B. (1991). Fluxes of carbon and phosphorus between symbionts in willow ectomycorrhizas and their changes with time. *New Phytol.* **119**, 99–106.

Jones, M.H., Fahnestock, J.T., Walker, D.A., Walker, M.D. and Welker, J.M. (1998). Carbon dioxide fluxes in moist and dry arctic tundra during the snow-free season: responses to increases in summer temperature and winter snow accumulation. *Arc. Alp. Res.* **30**, 373–380.

Jordan, C.F. and Kline, J.R. (1977). Transpiration of trees in tropical rainforests. *J. Appl. Ecol.* **14**, 853–860.

Juniper, B.E. and Roberts, R.M. (1966). Polysaccharide synthesis and the fine structure of root cap cells. *J. R. Micro. Soc.* **85**, 63–72.

Junttila, O. and Robberecht, R. (1993). The influence of season and phenology on freezing tolerance in *Silene acaulis* L., a subarctic and arctic cushion plant of circumpolar distribution. *Ann. Bot.* **71**, 423–426.

Jurik, T.W. and Akey, W.C. (1994). Solar-tracking leaf movements in velvetleaf (*Abutilon theoprastis*). *Vegetatio* **112**, 93–99.

Jurik, T.W., Zhang, H. and Pleasants, J.M. (1990). Ecophysiological consequences of non-random leaf orientation in the prairie compass plant *Silphium lacinatum*. *Oecologia* **82**, 180–186.

Katznelson, H. (1946). The 'rhizosphere effect' of mangels on certain groups of soil micro-organisms. *Soil Sci.* **62**, 343–354.

Keddy, P., Gaudet, C. and Fraser, L.H. (2000). Effects of low and high nutrients on the competitive hierarchy of 26 shoreline plants. *J. Ecol.* **88**, 413–423.

Keeley, J.E. (1991). Seed germination and life history syndromes in the Californian chaparral. *Bot. Rev.* **57**, 81–115.

Keeley, J.E. (1996). Aquatic CAM photosynthesis. In: Winter, K. and Smith, J.A.C. (eds.). *Crassulacean Acid Metabolism: Biochemistry, Eco-physiology and Evolution*, pp. 281–295. *Ecological Studies* **114**. Springer-Verlag, Berlin.

Keeley, J.E. and Fotheringham, C.J. (1997). Trace gas emissions and smoke-induced seed germination. *Science* **276**, 1248–1250.

Keeley, J.E. and Franz, E.H. (1979). Alcoholic fermentation in swamp and upland populations of *Nyssa sylvatica*: temporal changes in adaptive strategy. *Am. Nat.* **113**, 587–592.

Kelly, K.M., Van Staden, J. and Bell, W.E. (1992). Seed coat structure and dormancy. *Plant Growth Reg.* **11**, 201–209.

Kevan, P.G. (1975). Sun-tracking solar furnaces in high-arctic flowers: significance for pollination and insects. *Science* **189**, 723–726.

Kimball, B.A., Pinter, P.J., Garcia, R.L., LaMorte, R.L., Wall, G.W., Hunsaker, D.J., Wechsung, G., Wechsung, F. and Kartschall, T. (1995). Productivity and water use of wheat under free-air CO_2 enrichment. *Global Change Biol.* **1**, 429–442.

King, T.J. (1975). Inhibition of seed germination under leaf canopies in *Arenaria serpyllifolia, Veronica arvensis*, and *Cerastium holosteoides*. *New Phytol.* **75**, 87–90.

Kirk, G.J.D. and Bajita, J.B. (1995). Root-induced iron oxidation, pH changes and zinc solubilization in the rhizosphere of lowland rice. *New Phytol.* **131**, 129–137.

Kleinkopf, G.E. and Wallace, A. (1974). Physiological basis for salt tolerance in *Tamarix ramosissima. Plant Sci. Lett.* **3**, 157–163.

Knight, J.D., Livingston, N.J. and Van Kessel, C. (1994). Carbon isotope discrimination and water-use efficiency of six crops grown under wet and dryland conditions. *Plant, Cell Env.* **17**, 173–179.

Knutson, R.M. (1974). Heat production and temperature regulation in eastern skunk cabbage. *Science* **186**, 746–747.

Kochenderfer, J.N. (1973). Root distribution under some forest types native to West Virginia. *Ecology* **54**, 445–449.

Kochian, L.V. (1995). Cellular mechanisms of aluminium toxicity and resistance in plants. *Annu. Rev. Plant Physiol. Plant Mol. Biol.* **46**, 237–260.

Kochian, L.V., Garvin, D.F., Shaff, J.E., Chilcott, T.C. and Lucas, W.J. (1993). Towards an understanding of the molecular basis of plant K^+ transport: characterization of cloned K^+ transport DNAs. In: Barrow, N.J. (ed.). *Plant Nutrition – From Genetic Engineering to Field Practice*, pp. 121–124. Kluwer Academic Publications, Dordrecht.

Kolb, K.J. and Davis, S.D. (1994). Drought tolerance and xylem embolism in co-occurring species of coastal sage and chaparral. *Ecology* **75**, 648–659.

Koller, D. and Hadas, A. (1982). Water relations in the germination of seeds. *Encyclopaedia of Plant Physiology* **12B**, 401–431.

Körner, C. (1982). CO_2 exchange in the alpine sedge *Carex curvula* as influenced by canopy structure, light and temperature. *Oecologia* **53**, 98–104.

Körner, C. (1989). The nutritional status of plants from high altitudes. *Oecologia* **81**, 379–391.

Körner, C. (1999a). Alpine plants: stressed or adapted? In: Press, M.C., Scholes, J.D. and Barker, M.G. (eds.). *Physiological Plant Ecology*, pp. 297–311. Blackwell Science, Oxford.

Körner, C. (1999b). *Alpine Plant Life.* Springer-Verlag, Berlin.

Körner, C. and Larcher, W. (1988). Plant life in cold climates. In: Long, S.P. and Woodward, F.I. (eds.). *Plants and Temperature*, pp. 25–57. Company of Biologists, Cambridge, UK.

Körner, C. and Woodward, F.I. (1987). The dynamics of leaf extension in plants with diverse altitudinal ranges. 2. Field studies in *Poa* species between 600 and 3200 m altitude. *Oecologia* **72**, 279–283.

Körner, C., Neumayer, M., Menendez-Riedl, S.P. and Smeets-Scheel, A. (1989). Functional morphology of mountain plants. *Flora* **182**, 353–383.

Krämer, U., Cotter-Howels, J.D., Charnock, J.M., Baker, A.J.M. and Smith, J.A. (1996). Free histidine as a metal chelator in plants that accumulate nickel. *Nature* **379**, 635–638.

Krapp, A., Fraisier, V., Scheible, W.R., Quesada, A., Gojon, A., Stitt, M., Caboche, M. and Daniel-Vedele, F. (1998). Expression studies of *Nrt2 : 1Np*, a putative high-affinity nitrate transporter: evidence for its role in nitrate uptake. *Plant J.* **14**, 723–731.

Krause, G.H. and Weis, G.H. (1991). Chlorophyll fluorescence and photosynthesis. *Annu. Rev. Plant Physiol. Plant Mol. Biol.* ***42***, 313–349.

Kroh, G.C. and Beaver, D.L. (1978). Insect response to mixture and monoculture patches of Michigan old-field annual herbs. *Oecologia* **31**, 269–275.

Kullman, L. (1988). Holocene history of the forest-alpine tundra ecotone in the Scandes Mountains (central Sweden). *New Phytol.* **108**, 101–110.

Kummerow, J. (1980). Adaptation of roots in water-stressed native vegetation. In: Turner, N.C. and Kramer, P.J. (eds.). *Adaptation of Plants to Water and High Temperature Stress*, pp. 57–73. Wiley-Interscience, New York.

Kylin, A. and Kähr, M. (1973). The effect of magnesium and calcium on adenosine triphosphatases from wheat and oat roots at different pH. *Physiol. Plant.* **28**, 452–457.

Lambers, H. (1997). Respiration and the alternative oxidase. In: Foyer, C.H. and Quick, P. *A Molecular Approach to Primary Metabolism in Plants*, pp. 295–309. Taylor and Francis, London.

Lamont, B.B. (1993). Why are hairy root clusters so abundant in the most nutrient-impoverished soils of Australia? *Plant and Soil* **155/156**, 269–272.

Lane, S.D., Martin, E.S. and Garrod, J.F. (1978). Lead toxicity effects on indole-3-ylacetic acid induced cell elongation. *Planta* **144**, 79–84.

Lange, O.L., Kappen, L. and Schulze, E.-D. (1976). *Water and Plant Life. Ecological Studies* **19**. Springer-Verlag, Berlin.

Lange, O.L. and Zuber, M. (1977). *Frerea indica*, a stem succulent CAM plant with deciduous C_3 leaves. *Oecologia* **31**, 67–72.

Larcher, W. (1975). *Physiological Plant Ecology*. Springer-Verlag, Berlin.

Larigauderie, A. and Körner, C. (1995). Acclimation of leaf dark respiration to temperature in alpine and lowland plant species. *Ann. Bot.* **76**, 245–252.

Larkum, A.W.D. (1968). Ionic relations of chloroplasts *in vivo*. *Nature* **218**, 447–449.

Larson, R.A., Garrison, W.J. and Carlson, R.W. (1990). Differential responses of alpine and non-alpine *Aquilegia* species to increased ultraviolet-B radiation. *Plant, Cell Env.* **13**, 983–987.

Lajtha, K. (1994). Nutrient uptake in eastern deciduous tree seedlings. *Plant and Soil* **160**, 193–199.

Lawlor, D.W. (1995). The effects of water deficit on photosynthesis. In: Smirnoff, N. (ed.). *Environment and Plant Metabolism*, pp. 129–160. Bios Scientific Publishers, Oxford.

Leake, J.R. (1994). The biology of mycoheterotrophic ('saprophytic') plants. *New Phytol.* **127**, 171–216.

Lee, D.W. and Graham, R. (1986). Leaf optical properties of rainforest sun and extreme shade plants. *Am. J. Bot.* **73**, 1100–1108.

Lee, J., Reeves, R.D., Brooks, R.R. and Jaffre, T. (1978). The relation between nickel and citric acid in some nickel-accumulating plants. *Phytochemistry*. **17**, 1033–1035.

Lee, J.A. (1999a). Arctic plants: adaptation and environmental change. In: Press, M.C., Scholes, J.D. and Barker, M.G. (eds.). *Physiological Plant Ecology*, pp. 313–329. Blackwell Science, Oxford.

Lee, J.A. (1999b). The calcicole-calcifuge problem revisited. *Adv. Bot. Res.* **29**, 1–30.

Lee, J.A. and Caporn, S.J.M. (1998). Ecological effects of atmospheric reactive nitrogen deposition on semi-natural terrestrial ecosystems. *New Phytol.* **139**, 127–134.

Lee, R.B. and Ratcliffe, R.G. (1993). Subcellular distribution of inorganic phosphate and levels of nucleoside triphosphate in mature maize roots at low external phosphate concentrations: measurements with ^{31}P NMR. *J. Exp. Bot.* **44**, 587–598.

Leinonen, I., Repo, T. and Hänninen, H. (1997). Changing environmental effects on frost hardiness of Scots Pine during dehardening. *Ann. Bot.* **79**, 133–138.

Leishman, M.R. and Westoby, M. (1994). The role of seed size in seedling establishment in dry soil conditions – experimental evidence from semi-arid species. *J. Ecol.* **82**, 249–258.

Levitt, J. (1980). *Responses of Plants to Environmental Stress.* 2nd Edn, Vol. 2. Academic Press, New York.

Lewis, M.C. (1972). The physiological significance of variation in leaf structure. *Sci. Prog., Oxf.* **60**, 25–51.

Lin, C., Yang, H.Y., Guo, H.W., Mockler, T., Chen, J. and Cashmore, A.R. (1998). Enhancement of blue light sensitivity of *Arabidopsis* seedlings by a blue light receptor, cryptochrome 2. *Proc. Natl. Acad. Sci. USA* **95**, 2686–2690.

Liu, H., Trieu, A.T., Blaylock, L.A. and Harrison, M. (1998). Cloning and characterization of two phosphate transporters from *Medicago truncatula* roots: regulation in response to phosphate and to colonization by arbuscular mycorrhizal (AM) fungi. *Mol. Plant–Microbe Interact.* **11**, 14–22.

Lloyd, J. and Farquhar, G.D. (1994). ^{13}C discrimination during CO_2 assimilation by the terrestrial biosphere. *Oecologia* **99**, 201–215.

Lochhead, A.G. and Rouatt, J.W. (1955). The 'rhizosphere effect' on the nutritional groups of soil bacteria. *Soil Sci. Soc. Am. Proc.* **19**, 48–49.

Logullo, M.A., Salleo, S. and Rosso, R. (1986). Drought avoidance strategy in *Ceratonia siliqua* L., a mesomorphic tree in the xeric Mediterranean area. *Ann. Bot.* **58**, 745–756.

Long, R.C. and Woltz, W.G. (1972). Depletion of nitrate reductase activity in response to soil leaching. *Agron. J.* **64**, 789–792.

Longnecker, N. and Robson, A. (1994). Leaf emergence of spring wheat receiving varying nitrogen supply at different stages of development. *Ann. Bot.* **74**, 1–7.

Loveys, B.R., Stoll, M., Dry, P.R. and McCarthy, M.G. (2000). Using plant physiology to improve the water use efficiency of horticultural crops. *Acta Hort.* **537**, 187–199.

Ludlow, M.M. (1980). Adaptive significance of stomatal responses to water stress. In: Turner, N.C. and Kramer, P.J. (eds.). *Adaptations of Plants to Water and High Temperature Stress*, pp. 123–138. Wiley-Interscience, New York.

Lüttge, U. and Smith, J.A.C. (1982). Membrane transport, osmoregulation and the control of CAM. In: Ting, I.P. and Gibbs, M. (eds.). *Crassulacean Acid Metabolism*, pp. 69–91. American Society of Plant Physiology, Rockville.

Maathuis, F.J.M. and Sanders, D. (1996). Mechanisms of potassium absorption by higher plant roots. *Physiol. Plant.* **96**, 158–168.

Mabry, J.J., Hunziker, J.H. and DiFeo, D.R. (1977). *Creosote Bush. Biology and Chemistry of* Larrea *in New World Deserts.* US/IBP synthesis Series 6. Dowden, Hutchinson and Ross Inc., Stroudsburg, Pennsylvania.

McCree, K.J. (1972). Test of current definitions of photosynthetically active radiation against leaf photosynthesis data. *Agric. Meteorol.* **10**, 443–453.

McCree, K.J. and Troughton, J.H. (1966). Prediction of growth rate at different light levels from measured photosynthesis and respiration rates. *Plant Physiol.* **41**, 559–566.

McCully, M.E. (1999). Roots in soil: unearthing the complexities of roots and their rhizospheres. *Annu. Rev. Plant Physiol. Plant Mol. Biol.* **50**, 695–718.

McGrath, S.P., Chaudri, A.M. and Giller, K.E. (1995). Long-term effects of metals in sewage sludge on soils, micro-organisms and plants. *J. Indust. Microbiol.* **14**, 94–104.

McKee, I.F., Eiblmeier, M. and Polle, A. (1997). Enhanced ozone-tolerance in wheat grown at an elevated CO_2 concentration: ozone exclusion and detoxification. *New Phytol.* **137**, 275–284.

McKey, D. (1994). Legumes and nitrogen: the evolutionary ecology of a nitrogen-demanding lifestyle. *Adv. Legume Systematics* **5**, 211–238.

McLellan, A.J., Law, R. and Fitter, A.H. (1997). Response of calcareous grassland plant species to diffuse competition: results from a removal experiment. *J. Ecol.* **85**, 479–490.

Macnair, M.R. (1993). The genetics of metal tolerance in vascular plants. *New Phytol.* **124**, 541–559.

Macnair, M.R., Cumbes, Q.J. and Meharg, A.A. (1992). The genetics of arsenate tolerance in Yorkshire fog *Holcus lanatus*. *Heredity* **69**, 325–335.

Macnair, M.R., Smith, S.E. and Cumbes, Q.J. (1993). The heritability and distribution of variation in degree of copper tolerance in *Mimulus guttatus* at Copperopolis, California. Heredity **71**, 445–455.

McNaughton, S.J. (1991). Dryland herbaceous perennials. In: Mooney, H.A., Winner, W.E. and Pell, E.J. *Responses of Plants to Multiple Stresses*, pp. 307–328. Academic Press, San Diego.

McNulty, A.K. and Cummins, W.R. (1987). The relationship between respiration and temperature in leaves of the arctic plant *Saxifraga cernua*. *Plant, Cell Env.* **10**, 319–325.

McQueen-Mason, S.J. (1995). Expansins and cell wall expansion. *J. Exp. Bot.* **46**, 1639–1650.

MacRobbie, E.A.C. (1997). Signalling in guard cells and regulation of ion channel activity. *J. Exp. Bot.* **48**, 515–528.

Mahmoud, A. and Grime, J.P. (1974). A comparison of negative relative growth rates in shaded seedlings. *New Phytol.* **73**, 1215–1220.

Mallott, P.G., Davy, A.J., Jefferies, R.L. and Hutton, M.J. (1975). Carbon dioxide exchanges in leaves of *Spartina anglica* Hubbard. *Oecologia* **20**, 351–358.

Mansfield, T.A. (1983). Movements of stomata. *Sci. Prog., Oxf.* **68**, 519–542.

Mansfield, T.A. (1999). SO_2 pollution: a bygone problem or a continuing hazard? In: Press, M.C., Scholes, J.D. and Barker, M.G. (eds.). *Physiological Plant Ecology*, pp. 219–240. Blackwell Science, Oxford.

Mansfield, T.A. and Pearson, M. (1993). Physiological basis of stress imposed by ozone pollution. In: Fowden, L., Mansfield, T.A. and Stoddart, J. (eds.). *Plant Adaptation to Environmental Stress*, pp. 155–170. Chapman and Hall, London.

Mansfield, T.A., Hetherington, A.M. and Atkinson, C.J. (1990). Some current aspects of stomatal physiology. *Annu. Rev. Plant Physiol. Plant Mol. Biol.* **41**, 55–75.

Marrs, R.H., Roberts, R.D., Skeffington, R.A. and Bradshaw, A.D. (1983). Nitrogen and the development of ecosystems. In: Lee, J.A., McNeill, S. and Rorison, I.H. (eds.). *Nitrogen as an Ecological Factor*, pp. 113–136. Blackwell Scientific Publications, Oxford.

Marschner, H. and Römheld, V. (1983). *In vivo* measurement of root-induced pH changes at the soil-root interface: effect of plant species and nutrient source. *Z. Pflanzenphysiol.* **111**, 241–251.

Mathys, W. (1973). Vergleichende Untersuchungen der Zinkaufnahme von resistenten und sensitiven Populationen von *Agrostis tenuis* Sibth. *Flora* **162**, 492–499.

Mathys, W. (1975). Enzymes of heavy-metal-resistant and non-resistant populations of *Silene cucubalus* and their interaction with some heavy metals *in vitro* and *in vivo*. *Physiol. Plant.* **33**, 161–165.

Matsumoto, H., Hirasawa, P., Morimura, S. and Takahashi, E. (1976). Localisation of aluminium in tea leaves. *Plant Cell Physiol.* **18**, 880–885.

Matsumoto, J., Muraoka, H. and Washitani, I. (2000). Ecophysiological mechanisms used by *Aster kantoensis*, an endangered species, to withstand high light and heat stresses of its gravelly floodplain habitat. *Ann. Bot.* **86**, 777–785.

Mawson, B.T., Svoboda, J. and Cummins, R.W. (1986). Thermal acclimation of photosynthesis by the arctic plant *Saxifraga cernua*. *Can. J. Bot.* **64**, 71–76.

Mayer, A.M. (1986). How do seeds sense their environment? Some biochemical aspects of the sensing of water potential, light and temperature. *Isr. J. Bot.* **35**, 3–16.

Medina, E. and Troughton, J.H. (1974). Dark CO_2 fixation and the carbon isotope ratio in Bromeliaceae. *Plant Sci. Lett.* **2**, 357–362.

Meharg, A.A. and Macnair, M.R. (1992). Polymorphism and physiology of the high affinity phosphate uptake system: a mechanism of arsenate tolerance in *Holcus lanatus* L. *J. Exp. Bot.* **43**, 519–524.

Meharg, A.A., Cumbes, Q.J. and Macnair, M.R. (1993). Pre-adaptation of Yorkshire fog *Holcus lanatus* L. (Poaceae) to arsenate tolerance. *Evolution* **47**, 313–316.

Meharg, A.A., Dennis, G.R. and Cairney, J.W.G. (1997). Biotransformation of 2,4,6-trinitrotoluene (TNT) by ectomycorrhizal basidiomycetes. *Chemosphere* **35**, 513–521.

Meidner, H. and Mansfield, T.A. (1968). *The Physiology of Stomata*. McGraw-Hill, London.

Meidner, H. and Sheriff, D.W. (1976). *Water and Plants*. Blackie, Glasgow.

Mewissen, D.J., Damblon, J. and Bacq, Z.M. (1959). Comparative sensitivity to radiation of seeds from a wild plant grown on uraniferous and non-uraniferous soil. *Nature* **183**, 1449.

Milthorpe, F.L. and Newton, P. (1963). Studies on the expansion of the leaf surface. III. The influence of radiation on cell division and expansion. *J. Exp. Bot.* **14**, 483–495.

Minchin, F.R. and Pate, J.S. (1974). Diurnal functioning of the legume root nodule. *J. Exp. Bot.* **25**, 295–308.

Mitchley, J. and Grubb, P.J. (1986). Control of relative abundance of perennials in chalk grassland in southern England. 1. Constancy of rank order and results of pot-experiments and field-experiments on the role of interference. *J. Ecol.* **74**, 1139–1166.

Miyoshi, K. and Sato, T. (1997). The effects of ethanol on the germination of seeds of japonica and indica rice (*Oryza sativa* L.) under anaerobic and aerobic conditions. *Ann. Bot.* **79**, 391–395.

Moeder, W., Anegg, S., Thomas, G., Langebartels, C. and Sandermann, H. (1999). Signal molecules in ozone activation of stress proteins in plants. In: Smallwood, M.F., Calvert, C.M. and Bowles, D.J. (eds.). *Plant Responses to Environmental Stress*, pp. 43–49. Bios Scientific Publishers, Oxford.

Molina, R., Massicotte, H. and Trappe, J.M. (1992). Specificity phenomena in mycorrhizal symbioses: community-ecological consequences and practical implications. In: Allen, M.F. (ed.). *Mycorrhizal Functioning*, pp. 357–423. Chapman and Hall, New York.

Monk, L.S., Fagerstedt, K.V. and Crawford, R.M.M. (1987). Superoxide dismutase as an anaerobic polypeptide. A key factor in the recovery from oxygen deprivation in *Iris pseudacorus*. *Plant Physiol.* **85**, 1016–1020.

Monteith, J.L. (1981). Coupling of plants to the atmosphere. In: Grace, J., Ford, E.D. and Jarvis, P.G. (eds.). *Plants and their Atmospheric Environment*, pp. 1–29. Blackwell Scientific Publications, Oxford.

Monteith, J.L. (1995). A reinterpretation of stomatal responses to humidity. *Plant, Cell Env.* **18**, 357–364.

Mooney, H.A. (1980). Seasonality and gradients in the study of stress adaptation. In: Turner, N.C. and Kramer, P.J. (eds.). *Adaptation of Plants to Water and High Temperature Stress*, pp. 279–294. Wiley-Interscience, New York.

Mooney, H.A. and Billings, W.D. (1961). Comparative physiological ecology of arctic and alpine populations of *Oxyria digyna*. *Ecol. Monogr.* **31**, 1–29.

Mooney, J. and Dunn, E. (1970). Convergent evolution of Mediterranean-climate evergreen sclerophyllous shrubs. *Evolution* **24**, 292–303.

Morecroft, M.D. and Woodward, F.I. (1996). Experiments on the causes of altitudinal differences in the leaf nutrient contents, size and $\delta^{13}C$ of *Alchemilla alpina*. *New Phytol.* **134**, 471–479.

Morgan, J.M. (1984). Osmoregulation and water stress in higher plants. *Annu. Rev. Plant Physiol.* **35**, 299–319.

Morton, J.B. and Benny, G.L. (1990). Revised classification of arbuscular mycorrhizal fungi (Zygomycetes): a new order, Glomales, two new suborders, Glomineae and Gigasporineae, and two new families, Acaulosporaceae and Gigasporaceae, with an emendation of Glomaceae. *Mycotaxon* **37**, 471–491.

Mott, J.J. (1972). Germination studies on some annual species from an arid region of Western Australia. *J. Ecol.* **60**, 293–304.

Mudd, J.B. and Kozlowski, T.T. (eds.) (1975). *Responses of Plants to Air Pollution.* Academic Press, New York and London.

Muller, B. and Touraine, B. (1992). Inhibition of NO_3^- uptake by various phloem-translocated amino acids in soybean seedlings. *J. Exp. Bot.* **43**, 617–623.

Munns, R. and Sharp, R.E. (1993). Involvement of abscisic acid in controlling plant growth in soils of low water potential. *Aust. J. Plant Physiol.* **20**, 425–437.

Munns, R., Greenway, H. and Kirst, G.O. (1983). Halotolerant eukaryotes. *Encyclopaedia of Plant Physiology* **12C**, 59–134.

Murphy, A.S. and Taiz, L. (1995). Comparison of metallothionein gene expression and non-protein thiols in ten *Arabidopsis* ecotypes. *Plant Physiol.* **109**, 1–10.

Murray, F. and Wilson, S. (1990). Growth responses of barley exposed to SO_2. *New Phytol.* **114**, 537–554.

Musgrave, A. and Walters, J. (1973). Ethylene-stimulated growth and auxin transport in *Ranunculus sceleratus* petioles. *New Phytol.* **72**, 783–789.

Nagy, E.S. (1997). Selection for native characters in hybrids between two locally adapted plant subspecies. *Evolution* **51**, 1469–1480.

Nagy, L. and Proctor, J. (1997). Plant growth and reproduction on a toxic alpine ultramafic soil: adaptation to nutrient limitation. *New Phytol.* **137**, 267–274.

Neilson, R.E., Ludlow, M.M. and Jarvis, P.G. (1972). Photosynthesis in Sitka spruce (*Picea sitchensis* (Bong.) Carr). II. Response to temperature. *J. Appl. Ecol.* **9**, 721–745.

Neumann, P.M. (1995). The role of cell wall adjustment in plant resistance to water deficits. *Crop Sci.* **35**, 1258–1266.

Newman, E.I. (1985). The rhizosphere: carbon sources and microbial populations. In: Fitter, A.H., Atkinson, D., Read, D.J. and Usher, M.B. (eds.). *Ecological Interactions in Soil*, pp. 107–121. Blackwell Scientific Publications, Oxford.

Newman, E.I. and Andrews, R.E. (1973). Uptake of P and K in relation to root growth and root density. *Plant and Soil* **38**, 49–69.

Newsham, K.K., Fitter, A.H. and Watkinson, A.R. (1995). Multi-functionality and biodiversity in arbuscular mycorrhizas. *Trends Ecol. Evol.* **10**, 407–411.

Newton, J.E. and Blackman, G.E. (1970). The penetration of solar radiation through canopies of different structure. *Ann. Bot.* **34**, 329–348.

Newton, P. (1963). Studies on the expansion of the leaf surface. II. The influence of light intensity and photoperiod. *J. Exp. Bot.* **14**, 458–482.

Niklas, K.J. (2000). The evolution of leaf form and function. In: Marshall, B. and Roberts, J.A. (eds). *Leaf Development and Canopy Growth*, pp. 1–35. Sheffield Academic Press, Sheffield.

Niu, X., Narasimhan, M.L., Salzman, R.A., Bressan, R.A. and Hasegawa, P.M. (1993). NaCl regulation of plasma membrane H^+-ATPase gene expression in a glycophyte and a halophyte. *Plant Physiol.* **103**, 713–718.

Nobel, P.S. (1974). *Introduction to Biophysical Plant Physiology.* W.H. Freeman, San Francisco.

Nobel, P.S. (1988). Principles underlying the prediction of temperature in plants, with special reference to desert succulents. In: Long, S.P. and Woodward, F.I. (eds.). *Plants and Temperature*, pp. 1–23. Company of Biologists, Cambridge, UK.

Nobel, P.S. (1989). Shoot temperatures and thermal tolerances for succulent species of *Haworthia* and *Lithops*. *Plant, Cell Env.* **12**, 643–651.

Nobel, P.S. and Jordan, P.W. (1983). Transpiration stream of desert species: resistances and capacitances for a C_3, a C_4 and a CAM plant. *J. Exp. Bot.* **34**, 1379–1391.

Nobel, P.S., Garcia-Moya, E. and Quero, E. (1992). High annual productivity of certain agaves and cacti under cultivation. *Plant, Cell Env.* **15**, 329–335.

Nye, P.H. (1966). The effect of nutrient intensity and buffering power of a soil, and the absorbing power, size and root hairs of a root, on nutrient absorption by diffusion. *Plant and Soil* **25**, 81–105.

Nye, P.H. (1969). The soil model and its application to plant nutrition. In: *Ecological Aspects of the Mineral Nutrition of Plants. Br. Ecol. Soc. Symp.* **9**, 105–114.

Nye, P.H. (1973). The relation between the radius of a root and its nutrient absorbing power (α). *J. Exp. Bot.* **24**, 783–786.

Nye, P.H. and Marriott, F.H.C. (1969). A theoretical study of the distribution of substances around roots resulting from simultaneous diffusion and mass flow. *Plant and Soil* **30**, 459–472.

Nye, P.H. and Tinker, P.B.H. (1969). The concept of root demand coefficient. *J. Appl. Ecol.* **6**, 293–300.

Nye, P.H. and Tinker, P.B.H. (1977). *Solute Movement in the Soil–Root System.* Blackwell, Oxford.

O'Dowd, N.A. and Canny, M.J. (1993). A simple method for locating the start of symplastic water flow (flumes) in leaves. *New Phytol.* **125**, 743–748.

Oechel, W.C., Cowles, S., Grulke, N., Hastings, S.J., Lawrence, B., Prudhomme, T., Riechers, G., Strain, B., Tissue, D. and Vourlitis, G. (1994). Transient nature of CO_2 fertilization in arctic tundra. *Nature* **371**, 500–503.

Ogden, J., Basher, L. and McGlone, M. (1998). Fire, forest regeneration and links with early human habitation. *Ann. Bot.* **81**, 687–696.

Ögren, E and Sundin, U. (1996). Photosynthetic responses to variable light: a comparison of species from contrasting habitats. *Oecologia* **106**, 18–27.

O'Leary, M.H. (1988). Carbon isotopes in photosynthesis. *BioScience* **38**, 328–336.

Olmsted, C.E. (1944). Growth and development in range grasses. IV. Photoperiodic responses in twelve geographic strains of side-oats grama. *Bot. Gaz.* **100**, 46–74.

Ong, C.K. (1983). Response to temperature in a stand of pearl millet (*Pennisetum typhoides* S. & T.). 4. Extension of individual leaves. *J. Exp. Bot.* **34**, 1731–1739.

Orshan, G. (1963). Seasonal dimorphism of desert and Mediterranean chamaephytes and its significance as a factor in their water economy. *Symp. Br. Ecol. Soc.* **3**, 206–222.

Osmond, C.B., Winter, K. and Ziegler, H. (1982). Functional significance of different pathways of CO_2 fixation in photosynthesis. *Encyclopaedia of Plant Physiology* **12B**, 480–547.

Osmond, C.B., Anderson, J.M., Ball, M.C. and Egerton, J.J.G. (1999). Compromising efficiency: the molecular ecology of light-resource utilization in plants. In: Press, M.C., Scholes, J.D. and Barker, M.G. (eds.). *Plant Physiological Ecology*, pp. 1–24. Blackwell Scientific Publications, Oxford.

Osmond, C.B., Ziegler, H., Stichler, W. and Trimborn, P. (1975). Carbon isotope discrimination in alpine succulent plants supposed to be capable of crassulacean acid metabolism (CAM). *Oecologia* **18**, 209–218.

Ourcival, J.-M., Berger, A. and le Floc'h, E. (1994). Absorption de l'eau atmospherique par la partie aerienne d'un chamephyte de la Tunisie presaharienne: l'*Anthyllis henoniana* (Fabaceae). *Can. J. Bot.* **72**, 1222–1227.

Owen, D.F. (1952). The relation of germination of wheat to water potential. *J. Exp. Bot.* **3**, 188–203.

Ozanne, P.G. and Shaw, T.C. (1967). Phosphate sorption by soils as a measure of the phosphate requirement for pasture growth. *Aust. J. Agric. Sci.* **18**, 601–612.

Packham, J.R. and Willis, A, J. (1977). The effects of shading on *Oxalis acetosella*. *J. Ecol.* **65**, 619–642.

Parker, J. (1972). Protoplasmic resistance to water deficits. In: Kozlowski, T.T. (ed.). *Water Deficits and Plant Growth*, vol. 3, pp. 125–176. Academic Press, New York and London.

Passioura, J.B. (1963). A mathematical model for the uptake of ions from the soil solution. *Plant and Soil* **18**, 225–238.

Pate, J.S. (1993). Structural and functional responses to fire and nutrient stress: case studies from the sandplains of South-West Australia. In: Fowden, L., Mansfield, T.A. and Stoddart, J. (eds.). *Plant Adaptation to Environmental Stress*, pp. 189–205. Chapman and Hall, London.

Pate, J.S., Casson, N.E., Rullo, J. and Kuo, J. (1985). Biology of fire ephemerals of the sandplains of the Kwongan of South-west Australia. *Aust. J. Plant Physiol.* **12**, 641–655.

Pate, J.S., Froend, R.H., Bowen, B.J., Hansen, A. and Kuo, J. (1990). Seedling growth and storage and characteristics of seeder and resprouter species of Mediterranean-type ecosystem of south-western Australia. *Ann. Bot.* **65**, 585–601.

Pate, J.S., Meney, K.A. and Dixon, K.W. (1991). Contrasting growth and morphological characteristics of fire-sensitive (obligate seeder) and fire-resistant (resprouter) species of Restionaceae (S. Hemisphere Restiads) from South-western Australia. *Aust. J. Bot.* **39**, 505–525.

Paulsen, J., Weber, U.M. and Körner, C. (2000). Tree growth near treeline: abrupt or gradual reduction with altitude? *Arc. Antarc. Alp. Res.* **32**, 14–20.

Payne, D. (1988). The behaviour of water in soil. In: Wild, A. (ed.). *Russell's Soil Conditions and Plant Growth*, pp. 315–337. Longman, London.

Pearcy, R.W. (1983). The light environment and growth of C3 and C4 tree species in the understorey of a Hawaiian forest. *Oecologia* **58**, 19–25.

Pearcy, R.W. and Ehleringer, J. (1984). Comparative ecophysiology of C3 and C4 plants. *Plant, Cell Env.* **7**, 1–13.

Pearcy, R.W. and Pfitsch, W.A. (1991). Influence of sunflecks on the δ^{13}C of *Adenocaulon bicolor* plants occurring in contrasting forest understory microsites. *Oecologia* **86**, 457–462.

Pearcy, R.W. and Valladares, F. (1999). Resource acquisition by plants: the role of crown architecture. In: Press, M.C., Scholes, J.D. and Barker, M.G. (eds.). *Plant Physiological Ecology*, pp. 45–66. Blackwell Scientific Publications, Oxford.

Pearcy, R.W. and Yang, W. (1996). A three-dimensional crown architecture model for assessment of light capture and carbon gain by understory plants. *Oecologia* **108**, 1–12.

Pearcy, R.W., Krall, J.P. and Sassenrath-Cole, G.F. (1996). Photosynthesis in fluctuating light environments. In: Baker, N.R. (ed.). *Photosynthesis and the Environment. Advances in Photosynthesis* vol. **5**, pp. 331–346. Kluwer, Dordrecht.

Pearsall, W.H. (1938). The soil complex in relation to plant communities. *J. Ecol.* **26**, 180–193.

Pearsall, W.H. (1968). *Mountains and Moorlands*, 2nd Edn, Collins, Glasgow.

Pearson, M. and Mansfield, T.A. (1993). Interacting effects of ozone and water stress on the stomatal resistance of beech (*Fagus sylvatica* L.). *New Phytol.* **123**, 351–358.

Peat, H.J. and Fitter, A.H. (1994a). Comparative analyses of ecological characteristics of British angiosperms. *Biol. Rev.* **69**, 95–115.

Peat, H.J. and Fitter, A.H. (1994b). A comparative study of the distribution and density of stomata in the British flora. *Biol. J. Linn. Soc.* **52**, 377–393.

Peoples, M.B. and Craswell, E.T. (1992). Biological nitrogen-fixation – investments, expectations and actual contributions to agriculture. *Plant and Soil* **141**, 13–39.

Percy, K.E. and Baker, E.A. (1990). Effects of simulated acid rain on epicuticular wax production, morphology, chemical composition and on cuticular membrane thickness in two clones of Sitka Spruce [*Picea sitchensis* (Bong.) Carr.]. *New Phytol.* **116**, 79–87.

Peterson, A.G., Ball, T.J., Field, C.B., Reich, P.B., Curtis, P.S., Griffin, K.L., Gunderson, C.A., Norby, R.J., Tissue, D.T., Fostreuter, M., Rey, A., Vogel, C. and CMEAL Participants (1999). The photosynthesis–leaf nitrogen relationship at ambient and elevated atmospheric carbon dioxide: a meta-analysis. *Global Change Biol.* **5**, 331–346.

Petterson, M.W. (1991). Flower herbivory and seed predation in *Silene vulgaris* (Caryophyllaceae) – effects of pollination and phenology. *Holarctic Ecol.* **14**, 45–50.

Pfitsch, W.A. and Pearcy, R.W. (1989). Daily carbon gain by *Adenocaulon bicolor*, a redwood forest understory herb, in relation to its light environment. *Oecologia* **80**, 465–470.

Pigott, C.D. and Huntley, J.P. (1981). Factors controlling the distribution of *Tilia cordata* at the northern limit of its geographical range. 3. Nature and cause of seed sterility. *New Phytol.* **87**, 817–839.

Pineros, M. and Tester, M. (1993). Plasma membrane Ca^{2+} channels in roots of higher plants and their role in aluminium toxicity. *Plant and Soil* **155/6**, 119–122.

Pirozynski, K.A. and Malloch, D.W. (1975). The origin of land plants: a matter of mycotrophism. *Biosystems* **6**, 153–164.

Pisek, A., Larcher, W., Vegis, A. and Napp-Zin, K. (1973). The normal temperature range. In: Precht, H., Christopherson, J., Hensel, H. and Larcher, W. (eds.). *Temperature and Life*, pp. 102–194. Springer-Verlag, Berlin.

Poirier, Y., Thoma, S., Somerville, C. and Schiefelbein, J. (1991). A mutant of *Arabidopsis* deficient in xylem loading of phosphate. *Plant Physiol.* **97**, 1087–1093.

Poljakoff-Mayber, A., Symon, D.E., Jones, G.P., Naidu, B.P. and Paleg, L.G. (1987). Nitrogenous compatible solutes in native South Australian plants. *Aust. J. Plant Physiol.* **14**, 34–50.

Pollard, J.A. and Baker, A.J.M. (1997). Deterrence of herbivory by zinc hyperaccumulation in *Thlaspi caerulescens* (Brassicaceae). *New Phytol.* **135**, 655–658.

Pollock, C.J. (1990). The response of plants to temperature change. *J. Agric. Sci., Cambridge* **115**, 1–5.

Pollock, C.J., Eagles, C.F., Howarth, C.J., Schunmann, P.H.D. and Stoddart, J.L. (1993). Temperature stress. In: Fowden, L., Mansfield, T.A. and Stoddart, J. (eds.). *Plant Adaptation to Environmental Stress*, pp. 109–132. Chapman and Hall, London.

Ponnamperuma, F.N. (1984). Effects of flooding in soils. In: Kozlowski, T.T. (ed.). *Flooding and Plant Growth*, pp. 9–45. Academic Press, New York.

Poorter, H., Roumet, C. and Campbell, B.D. (1996). Interspecific variation in the growth response of plants to elevated CO_2: a search for functional types. In: Körner, C. and Bazzaz, F.A. (eds.). *Carbon Dioxide, Populations and Communities*, pp. 375–412. Academic Press, San Diego.

Popp, M. (1995). Salt resistance in herbaceous halophytes and mangroves. *Prog. Bot.* **56**, 416–429.

Popp, M. and Smirnoff, N. (1995). Polyol accumulation and metabolism during water deficit. In: Smirnoff, N. (ed.). *Environment and Plant Metabolism*, pp. 119–215. Bios Scientific Publishers, Oxford.

Potvin, M.A. and Werner, P.A. (1983). Water use physiologies of co-occurring goldenrods (*Solidago juncea* and *S. canadensis*): implication for natural distributions. *Oecologia* **56**, 148–152.

Pritchard, J. (1994). The control of cell expansion in roots. *New Phytol.* **127**, 3–26.

Pritchard, J., Barlow, P.W., Adam, J.S. and Tomos, A.D. (1990). Biophysics of the inhibition of the growth of maize roots by lowered temperature. *Plant Physiol.* **93**, 222–230.

Proctor, J. (1971). The plant ecology of serpentine. III. The influence of high magnesium/calcium ratio and high nickel and chromium levels in British and Swedish serpentine soils. *J. Ecol.* **59**, 827–842.

Quail, P.H., Boylan, M.T., Parks, B.M., Short, T.W., Xu, Y. and Wagner, D. (1995). Phytochromes – photosensory perception and signal transduction. *Science* **268**, 675–680.

Queiroz, O. (1977). CAM: rhythms of enzyme capacity and activity in adaptive control mechanisms. *Encyclopaedia of Plant Physiology* **2B**, 126–137.

Raab, T.K., Lipson, D.A. and Monson, R.K. (1999). Soil amino acid utilization among species of the Cyperaceae: plant and soil processes. *Ecology* **80**, 2408–2419.

Raaimakers, D. and Lambers, H. (1996). Response to phosphorus supply of tropical tree seedlings: a comparison between a pioneer species *Tapirira obtusa* and a climax species *Lecythis corrugata*. *New Phytol.* **132**, 97–102.

Rada, F., Goldstein, G., Azocar, A. and Meinzer, F. (1985). Freezing avoidance in Andean giant rosette plants. *Plant, Cell Env.* **8**, 501–507.

Ragothama, K.G. (1999). Phosphate acquisition. *Annu. Rev. Plant Physiol. Plant Mol. Biol.* **50**, 665–693.

Raison, J.K., Berry, J.A., Armond, P.A. and Pike, C.S. (1980). Membrane properties in relation to the adaptation of plants to high and low temperature stress. In: Turner, N.C. and Kramer, P.J. (eds.). *Adaptation of Plants to Water and High Temperature Stress*, pp. 261–273. Wiley-Interscience, New York.

Raschke, K. (1976). How stomata resolve the dilemma of opposing priorities. *Phil. Trans. R. Soc. Lond.* **B273**, 551–560.

Raunkiaer, C. (1907). *Planterigets Livsformer og deres Betydning for Geografien*, Copenhagen. (Translated as *The Life-Forms of Plants and Statistical Plant Geography (1934)*, Clarendon Press, Oxford.)

Rauser, W.E. (1990). Phytochelatins. *Annu. Rev. Biochem.* **59**, 61–86.

Rayder, L. and Ting, I.P. (1983). CAM idling in *Hoya carnosa* (Asclepiadaceae). *Photosynthesis Res.* **4**, 203–211.

Read, D.J. (1991). Mycorrhizas in ecosystems. *Experientia* **47**, 376–391.

Redecker, D., Kodner, R. and Graham, L.E. (2000). Glomalean fungi from the Ordovician. *Science* **289**, 1920–1921.

Reeves, R.D., Baker, A.J.M., Borhid, A. and Berazain, R. (1999). Nickel hyperaccumulation in the serpentine flora of Cuba. *Ann. Bot.* **83**, 29–38.

Reich, P.B., Walters, M.B. and Ellsworth, D.S. (1992). Leaf life-span in relation to leaf, plant and stand characteristics among diverse ecosystems. *Ecol. Monogr.* **62**, 365–392.

Reiling, K. and Davison, A.W. (1995). Effects of ozone on stomatal conductance and photosynthesis in populations of *Plantago major* L. *New Phytol.* **129**, 587–594.

Reilly, A. and Reilly, C. (1973). Copper-induced chlorosis in *Bechium homblei*. *Plant and Soil* **38**, 671–674.

Reimann, C. (1992). Sodium exclusion by *Chenopodium* species. *J. Exp. Bot.* **43**, 503–510.

Reimann, C. and Breckle, S.-W. (1993). Sodium relations in Chenopodiaceae: a comparative approach. *Plant, Cell Env.* **16**, 323–328.

Remy, W., Taylor, T.N., Haas, H. and Kerp, H. (1994). Four hundred-million-year old vesicular arbuscular mycorrhizae. *Proc. Natl. Acad. Sci., USA* **91**, 11841–11843.

Rengel, Z. (1992). Role of calcium in aluminium toxicity. *New Phytol.* **121**, 499–513.

Rengel, Z. (1996). Uptake of aluminium by plant cells. *New Phytol.* **134**, 389–406.

Reynolds, H.L. and D'Antonio, C. (1996). The ecological significance of plasticity in root weight ratio in response to nitrogen. *Plant and Soil* **185**, 75–97.

Rhodes, D. and Hanson, A.D. (1993). Quaternary ammonium and tertiary sulphonium compounds in higher plants. *Annu. Rev. Plant Physiol. Plant Mol. Biol.* **44**, 357–384.

Rhodes, I. (1969). Yield, canopy structure and light interception in two ryegrass varieties in mixed culture and monoculture. *J. Br. Grassld Soc.* **24**, 123–127.

Rhodes, L.H. and Gerdemann, J.W. (1975). Phosphate uptake zones of mycorrhizal and non-mycorrhizal onions. *New Phytol.* **75**, 555–562.

Richards, J.H. and Caldwell, M.M. (1987). Hydraulic lift: substantial nocturnal water transport between soil layers by *Artemisia tridentata* roots. *Oecologia* **73**, 486–489.

Richter, H. (1976). The water status in the plant. Experimental evidence. In: Lange, O.L., Kappen, L., and Schulze, E.-D. (eds.). *Water and Plant Life. Ecological Studies* **19**, 42–58. Springer-Verlag, Berlin.

Riley, D. and Barber, S.A. (1971). Effect of ammonium and nitrate fertilisation on phosphorus uptake as related to root-induced pH changes at the root–soil interface. *Soil Sci. Soc. Am. Proc.* **35**, 301–306.

Robberecht, R. and Junttila, O. (1992). The freezing response of an arctic cushion plant, *Saxifraga caespitosa* L.: acclimation, freezing tolerance and ice nucleation. *Ann. Bot.* **70**, 129–135.

Robertson, K.P. and Woolhouse, H.W. (1984). Studies of the seasonal course of carbon uptake of *Eriophorum vaginatum* in a moorland habitat. *J. Ecol.* **72**, 423–435, 685–700.

Robinson, D. (1994). The responses of plants to non-uniform supplies of nutrients. *New Phytol.* **127**, 635–674.

Robinson, D. (1996). Resource capture by localised root proliferation: why do plants bother? *Ann. Bot.* **77**, 179–186.

Robinson, D. and Rorison, I.R. (1983). Relationships between root morphology and nitrogen availability in a recent theoretical model describing nitrogen uptake from soil. *Plant, Cell Env.* **6**, 641–648.

Robinson, M.F., Heath, J. and Mansfield, T.A. (1998). Disturbances in stomatal behaviour caused by air pollutants. *J. Exp. Bot.* **49**, 461–469.

Rogers, H.H., Runion, G.B. and Krupa, S.V. (1994). Plant responses to atmospheric CO_2 enrichment with emphasis on roots and the rhizosphere. *Envron. Pollut.* **83**, 155–189.

Rorison, I.H. (1965). The effect of aluminium on the uptake and incorporation of phosphate by excised sainfoin roots. *New Phytol.* **64**, 23–27.

Rorison, I.H. (1968). The response to phosphorus of some ecologically distinct plant species. I. Growth rates and phosphorus absorption. *New Phytol.* **67**, 913–923.

Rorison, I.H. (1971). The use of nutrients in the control of the floristic composition of grassland. In: *The Scientific Management of Plant and Animal Communities for Conservation. Br. Ecol Soc. Symp.* **11**, 65–77.

Römheld, V. and Marschner, H. (1990). Genotypical differences among graminaceous species in release of phytosiderophores and uptake of iron phytosiderophores. *Plant and Soil* **123**, 147–153.

Ross, J. (1981). *The Radiation Regime and Architecture of Plant Stands.* Junk, The Hague, The Netherlands.

Rousset, O. and Lepart, J. (2000). Positive and negative interactions at different life stages of a colonizing species (*Quercus humilis*). *J. Ecol.* **88**, 401–412.

Rovira, A.D. and Davey, C.B. (1974). Biology of the rhizosphere. In: Carson, E.W. (ed.). *The Plant Root and its Environment*, pp. 153–240. University Press of Virginia.

Rovira, A.D., Newman, E.I., Bowen, H.J. and Campbell, R. (1974). Quantitative assessment of the rhizoplane microflora by direct microscopy. *Soil Biol. Biochem.* **6**, 211–216.

Rozema, J. (1976). An ecophysiological study of the response to salt of four halophytic and glycophytic *Juncus* species. *Flora* **165**, 197–209.

Russell, R.S. and Clarkson, D.T. (1976). Ion transport in root systems. In: Sunderland, N. (ed.). *Perspectives in Experimental Biology*, Vol. 2, pp. 401–411. Pergamon, Oxford.

Russell, W.A., Critchley, C. and Robinson, S.A. (1995). Photosystem II regulation and dynamics of the chloroplast D1 protein in *Arabidopsis* leaves during photosynthesis and photoinhibition. *Plant Physiol.* **107**, 943–952.

Ryals, J.A., Neuenschwander, U.H., Willits, M.G., Molina, A., Steiner, H.Y. and Hunt, M.D. (1996). Systemic acquired resistance. *Plant Cell* **8**, 1809–1819.

Ryser, P. and Eek, L. (2000). Consequences of phenotypic plasticity *vs* interspecific differences in leaf and root traits for acquisition of aboveground and belowground resources. *Am. J. Bot.* **87**, 402–411.

Sage, R. (1995). Was low atmospheric CO_2 during the Pleistocene a limiting factor for the origin of agriculture? *Global Change Biol.* **1**, 93–106.

Sakai, A. (1970). Freezing resistance in willows from different climates. *Ecology* **51**, 485–491.

Sakai, A. and Larcher, W. (1987). *Frost Survival of Plants.* Springer-Verlag, Berlin.

Salisbury, E.J. (1916). The oak–hornbeam woods of Hertfordshire. *J. Ecol.* **4**, 83–117.

Salisbury, F.B. (1963). *The Flowering Process.* Pergamon, Oxford.

Salt, D.E. and Rauser, W.E. (1995). MgATP-dependent transport of phytochelatins across the tonoplast of oat roots. *Plant Physiol.* **107**, 1293–1301.

Salvucci, M.E. (1989). Regulation of Rubisco activity *in vivo*. *Physiol. Plant.* **77**, 164–171.

Sanders, F.E. and Tinker, P.B. (1973). Phosphate flow into mycorrhizal roots. *Pestic. Sci.* **4**, 385–395.

Schmidhalter, U. (1994). Measuring and modelling water uptake by roots at spatially-variable soil matric potentials. *Proceedings of the Third Congress of the European Society for Agronomy, Abano-Padova, Italy* pp. 408–409.

Schmidhalter, U., Evequoz, M., Camp, K.-H. and Studer, C. (1998). Sequence of drought response of maize seedlings in drying soil. *Physiol. Plant.* **104**, 159–168.

Scholander, P.F., van Dam, L. and Scholander, S.I. (1955). Gas exchange in the roots of mangroves. *Am. J. Bot.* **42**, 92–98.

Schulze, E.-D. (1982). Plant life forms and their carbon, water and nutrient relations. *Encyclopaedia of Plant Physiology* **12B**, 615–676.

Schulze, E.-D. (1991). Water and nutrient interactions with plant water stress. In: Mooney, H.A., Winner, W.E. and Pell, E.J. (eds.). *Response of Plants to Multiple Stresses*, pp. 89–101. Academic Press, San Diego.

Schulze, E.-D. (1993). Soil water deficits and atmospheric humidity as environmental signals. In: Smith, J.A.C. and Griffiths, H. (eds.). *Water Deficits: Plant Responses from Cell to Community*, pp. 129–145. Bios Scientific Publishers, Oxford.

Schulze, E.-D., Fuchs, M. and Fuchs, M.I. (1977). Spatial distribution of photosynthetic capacity and performance in a mountain spruce forest of Northern Germany. 3. The significance of the evergreen habit. *Oecologia* **30**, 239–248.

Schwab, K.B. and Gaff, D.F. (1990). Influence of compatible solutes on soluble enzymes from desiccation-tolerant *Sporobolus stapfianus* and desiccation resistant *S. pyramidalis*. *J. Plant Physiol.* **137**, 208–215.

Schwartz, M.W. and Hoeksema, J.D. (1998). Specialization and resource trade: biological markets as a model of mutualisms. *Ecology* **79**, 1029–1038.

Scopel, A.L., Ballaré, C.L. and Sanchez, R.A. (1991). Induction of extreme light sensitivity in buried weed seeds and its role in the perception of soil cultivations. *Plant, Cell Env.* **14**, 501–508.

Scopel, A.L., Ballaré, C.L. and Radosevich, S.R. (1994). Photostimulation of seed germination during soil tillage. *New Phytol.* **126**, 145–152.

Scott, B.J. and Fisher, J.A. (1989). Selection of genotypes tolerant of aluminium and manganese. In: Robson, A.D. (ed.). *Soil Acidity and Plant Growth*, pp. 167–203. Academic Press, Sydney.

Seymour, R.S. and Schultze-Motel, P. (1996). Thermoregulating lotus flowers. *Nature* **383**, 305.

Shantz, H.L. and Piemesal, L.N. (1927). The water requirements at Akron, Colorado. *J. Agric. Res.* **34**, 1093–1190.

Sharman, B.C. (1942). Developmental anatomy of the shoot of *Zea mays* L. *Ann. Bot.* **6**, 245–282.

Shaver, G.R., Chapin, F.S. and Gartner, B.L. (1986). Factors limiting seasonal growth and peak biomass accumulation in *Eriophorum vaginatum* in Alaskan tussock tundra. *J. Ecol.* **74**, 257–278.

Shaw, A.J. (Ed.) (1990). *Heavy Metal Tolerance in Plants: Evolutionary Aspects.* CRC Press, Boca Raton, Florida.

Sheehy, J.E. and Peacock, J.M. (1975). Canopy photosynthesis and crop growth rate of eight temperate forage grasses. *J. Exp. Bot.* **26**, 679–691.

Sheldon, J.C. (1974). The behaviour of seeds in soil. III. The influence of seed morphology and the behaviour of seedlings on the establishment of plants from surface-lying seeds. *J. Ecol.* **62**, 47–66.

Sheppard, L.J. (1994). Causal mechanisms by which sulphate, nitrate and acidity influence frost hardiness in red spruce: review and hypothesis. *New Phytol.* **127**, 69–82.

Sherwin, H.W. and Farrant, J.M. (1996). Differences in rehydration of three desiccation-tolerant Angiosperm species. *Ann. Bot.* **78**, 703–710.

Sherwin, H.W., Pammenter, N.W., February, E., Vander Willigen, C. and Farrant, J.M. (1998). Xylem hydraulic characteristics, water relations and wood anatomy of the resurrection plant *Myrothamnus flabellifolius* Welw. *Ann. Bot.* **81**, 567–575.

Shinozaki, K. and Yamaguchi-Shinozaki, K. (1997). Gene expression and signal transduction in water-stress response. *Plant Physiol.* **115**, 327–334.

Shirley, H.L. (1929). The influence of light intensity and light quality on growth of plants. *Am. J. Bot.* **16**, 354–390.

Shreve, F. and Wiggins, I.L. (1964). *Vegetation and Flora of the Sonoran Desert.* Stanford University Press.

Shropshire, W. (1971). Photoinduced parental control of seed germination and the spectral quality of solar radiation. *Solar Energy* **15**, 99–105.

Simon, L., Bousquet, J., Levesque, R.C. and Lalonde, M. (1993). Origin and diversification of endomycorrhizal fungi and coincidence with vascular land plants. *Nature* **363**, 67–69.

Skene, M. (1924). *The Biology of Flowering Plants.* Sidgwick and Jackson, London.

Skillman, J.B., Garcia, M. and Winter, K. (1999). Whole-plant consequences of crassulacean acid metabolism for a tropical forest understorey plant. *Ecology* **80**, 1584–1593.

Slatyer, R.O. (1967). *Plant–Water Relationships.* Academic Press, London and New York.

Small, E. (1972). Photosynthetic rates in relation to nitrogen recycling as an adaptation to nutrient deficiency. *Can. J. Bot.* **50**, 2227–2233.

Smirnoff, N. (1995). Antioxidant systems and plant response to the environment. In: Smirnoff, N. (ed.). *Environment and Plant Metabolism*, pp. 217–243. Bios Scientific Publishers, Oxford.

Smirnoff, N. (1996). The function and metabolism of ascorbic acid in plants. *Ann. Bot.* **78**, 661–669.

Smith, A.P. and Young, T.P. (1987). Tropical alpine ecology. *Annu. Rev. Ecol. Syst.* **18**, 137–158.

Smith, H. (1995). Physiological and ecological function within the phytochrome family. *Annu. Rev. Plant Physiol. Plant Mol. Biol.* **46**, 289–315.

Smith, H. (2000). Phytochromes and light signal perception by plants – an emerging synthesis. *Nature* **407**, 585–591.

Smith, J.A.C. and Winter, K. (1996). Taxonomic distribution of crassulacean acid metabolism. In: Winter, K. and Smith, J.A.C. (eds.). *Crassulacean Acid Metabolism: Biochemistry, Ecophysiology and Evolution*, pp. 427–436. *Ecological Studies* **114**. Springer-Verlag, Berlin.

Smith, S.D. and Osmond, C.B. (1987). Stem photosynthesis in a desert ephemeral, *Erogonum inflatum*. *Oecologia* **72**, 533–541.

Smith, S.D., Monson, R.K. and Anderson, J.E. (1997). *Physiological Ecology of North American Desert Plants.* Springer-Verlag, Berlin.

Smith, S.E. and Read, D.J. (1996). *Mycorrhizal Symbiosis.* Academic Press, London and New York.

Smucker, A.J.M. and Aiken, R.M. (1992). Dynamic root responses to water deficits. *Soil Sci.* **154**, 281–289.

Sojka, R.E. (1992). Stomatal closure in oxygen-stressed plants. *Soil Sci.* **154**, 269–280.

Somers, D.E., Devlin, P.F. and Kay, S.A. (1998). Phytochromes and cryptochromes in the entrainment of the *Arabidopsis* circadian clock. *Science* **282**, 1488–1490.

Sonesson, M. and Callaghan, T.V. (1991). Strategies of survival in plants of the Fennoscandian tundra. *Arctic* **44**, 95–105.

Sperry, J.S., Donelly, J.R. and Tyree, M.T. (1988). Seasonal occurrence of xylem embolism in sugar maple (*Acer saccharum*). *Am. J. Bot.* **75**, 1212–1218.

Spollen, W.G., Sharp, R.E., Saab, I.N. and Wu, Y. (1993). Regulation of cell expansion at low water potentials. In: Smith, J.A.C. and Griffiths, H. (eds.). *Water Deficits: Plant Responses from Cell to Community*, pp. 37–52. Bios Scientific Publishers, Oxford.

Squeo, F.A., Rada, F., Azocar, A. and Goldstein, G. (1991). Freezing tolerance and avoidance in high tropical Andean plants: is it equally represented in species with different plant height? *Oecologia* **86**, 378–382.

Squire, G.R., Gregory, P.J., Monteith, J.L., Russell, M.B. and Singh, P. (1984). Control of water use by pearl millett (*Pennisetum typhoides*). *Exp. Agric.* **20**, 135–149.

Stamp, N.E. (1984). Self-burial behaviour of *Erodium cicutarium* seeds. *J. Ecol.* **72**, 611–620.

Stanton, M.L. and Galen, C. (1993). Blue light controls solar tracking by flowers of an alpine plant. *Plant, Cell Env.* **16**, 983–989.

Stanton, M.L., Roy, B.A. and Thiede, D.A. (2000). Evolution in stressful environments. I. Phenotypic variability, phenotypic selection, and response to selection in five distinct environmental stresses. *Evolution* **54**, 93–111.

Starkey, R.L. (1929). Some influences of the development of higher plants upon the micro-organisms in the soil. *Soil Sci.* **27**, 319–334.

Steudle, E. and Henzler, T. (1995). Water channels in plants: do basic concepts of water transport change? *J. Exp. Bot.* **46**, 1067–1076.

Steudle, E. and Peterson, C.A. (1998). How does water get through roots? *J. Exp. Bot.* **49**, 775–788.

Stevanovic, B., Thu, P.T.A., Monteiro de Paula, F. and Vieira da Silva, J. (1992). Effects of dehydration and rehydration on the polar lipid and fatty acid composition of *Ramada* species. *Can. J. Bot.* **70**, 107–113.

Stevens, G.C. and Fox, J.F. (1991). The causes of treeline. *Annu. Rev. Ecol. Systemat.* **22**, 177–191.

Stewart, G.R. and Lee, J.A. (1974). The role of proline accumulation in halophytes. *Planta* **120**, 279–289.

Stewart, G.R. and Schmidt, S. (1999). Evolution and ecology of plant mineral nutrition. In: Press, M.C., Scholes, J.D. and Barker, M.G. (eds.). *Plant Physiological Ecology*, pp. 91–114. Blackwell Science, Oxford.

Stiller, V. and Sperry, J.S. (1999). Canny's compensating pressure theory fails a test. *Am. J. Bot.* **86**, 1082–1086.

Stocker, O. (1976). The water-photosynthesis syndrome and the geographical plant distribution in the Sahara deserts. In: Lange, O.L., Kappen, L. and Schulze, E.-D. (eds.). *Water and Plant Life. Ecological Studies* **19**, 506–521. Springer-Verlag, Berlin.

Storey, R., Ahmad, N. and Wyn Jones, R.G. (1977). Taxonomic and ecological aspects of the distribution of glycine-betaine and related compounds in plants. *Oecologia* **27**, 319–332.

Stoutjesdijk, P.H. (1972). Spectral transmission curves of some types of leaf canopies with a note on seed germination. *Acta Bot. Neerl.* **21**, 185–191.

Strange, J. and Macnair, M.R. (1991). Evidence for a role for the cell membrane in copper tolerance of *Mimulus guttatus* Fischer ex DC. *New Phytol.* **119**, 383–388.

Sutinen, S., Skarby, L., Wallin, G. and Sellden, G. (1990). Long term exposure of Norway Spruce *Picea abies* (L.) Karst. to ozone in open-top chambers. 2. Effects on the ultrastructure of needles. *New Phytol.* **115**, 345–355.

Szeicz, G. (1974). Solar radiation in crop canopies. *J. Appl. Ecol.* **11**, 1117–1156.

Talbott, L.D. and Zeiger, E. (1998). The role of sucrose in guard cell osmoregulation. *J. Exp. Bot.* **49**, 329–337.

Tardieu, F., Lafarge, T. and Simonneau, T. (1996). Stomatal control by fed or endogenous xylem ABA in sunflower: interpretation of correlations between leaf water potential and stomatal conductance in anisohydric species. *Plant, Cell Env.* **19**, 75–84.

Taylor, G.E. and Tingley, D.T. (1983). Sulphur dioxide flux into leaves of *Geranium carolinianum. Plant Physiol.* **72**, 237–244.

Taylor, G.J. and Foy, C.D. (1985). Mechanisms of aluminium tolerance in *Triticum aestivum* L. (wheat). 1. Differential pH induced by winter cultivars in nutrient solutions. *Am. J. Bot.* **72**, 695–701.

Taylor, G.E., Pitelka, L.F. and Clegg, M.T. (Eds.) (1991). *Ecological Genetics and Air Pollution.* Springer-Verlag, New York.

Tezara, W., Mitchell, V.J. and Lawlor, D.W. (1999). Water stress inhibits plant photosynthesis by decreasing coupling factor and ATP. *Nature* **401**, 914–917.

Thomas, T.H. (1992). Some reflections on the relationship between endogenous hormones and light-mediated seed dormancy. *Plant Growth Reg.* **11**, 239–248.

Thomashow, M.F. (1999). Plant cold acclimation: freezing tolerance genes and regulatory mechanisms. *Annu. Rev. Plant Physiol. Plant Mol. Biol.* **50**, 571–599.

Thurston, J.M. (1960). Dormancy in weed seeds. *Symp. Br. Ecol. Soc.* **1**, 69–82.

Tiedje, J.M. (1995). Approaches to the comprehensive evaluation of prokaryote diversity of a habitat. In: Allsopp, D., Colwell, R.R. and Hawksworth, D.L. (eds.). *Microbial Diversity and Ecosystem Function*, pp. 73–87. CABI, Wallingford.

Tieszen, L.L. and Wieland, N.K. (1975). Physiological ecology of arctic and alpine photosynthesis and respiration. In: Vernberg, F.J. (ed.). *Physiological Adaptation to the Environment*, pp. 157–200. Intext Educational Publishers, New York.

Tilman, D. (1987). The importance of the mechanisms of interspecific competition. *Am. Nat.* **129**, 769–774.

Tilman, E.A., Tilman, D., Crawley, M.J. and Johnston, A.E. (1999). Biological weed control *via* nutrient competition: potassium limitation of dandelions. *Ecol. Appl.* **9**, 103–111.

Ting, I.P. (1985). Crassulacean acid metabolism. *Annu. Rev. Plant Physiol.* **36**, 595–622.

Tinker, P.B.H. (1969). The transport of ions in the soil around plant roots. In: *Ecological Aspects of the Mineral Nutrition of Plants. Br. Ecol. Soc. Symp.* **9**, 135–148.

Tinker, P.B.H. and Nye, P.H. (2000). *Solute Movement in the Rhizosphere.* Oxford University Press, Oxford.

Torsvik, V., Goksøyr, J. and Daae, F.L. (1990). High diversity of DNA of soil bacteria. *Appl. Env. Microbiol.* **56**, 782–787.

Totland, O. (1996). Flower heliotropism in an alpine population of *Ranunculus acris* (Ranunculaceae): effects on flower temperature, insect visitation and seed production. *Am. J. Bot.* **83**, 452–458.

Touraine, B., Muller, B. and Grignon, C. (1992). Effect of phloem-translocated malate on NO_3^- uptake by roots of intact soybean plants. *Plant Physiol.* **93**, 1118–1123.

Tranquillini, W. (1964). The physiology of plants at high altitudes. *Annu. Rev. Plant Physiol.* **15**, 345–362.

Tranquillini, W. (1979). *Physiological Ecology of the Alpine Timberline.* Springer-Verlag, Berlin.

Treichel, S. (1975). Der Einfluss von NaCl auf die Prolinkonzentration verschiedener Halophyten. *Z. Pflanzenphysiol.* **76**, 56–68.

Trenbath, B.R. and Harper, J.L. (1973). Neighbour effects in the genus *Avena*. I. Comparison of crop species. *J. Appl. Ecol.* **10**, 379–400.

Turesson, G. (1922). The genotypical response of the plant species to the habitat. *Hereditas* **3**, 211–350.

Tyler, G. (1976). Soil factors controlling metal ion absorption in the wood anemone *Anemone nemorosa. Oikos* **27**, 71–80.

Tyler, G. (1996). Mineral nutrient limitations of calcifuge plants in phosphate sufficient limestone soil. *Ann. Bot.* **77**, 649–656.

Tyree, M.T. (1997). The cohesion-tension theory of sap ascent: current controversies. *J. Exp. Bot.* **48**, 1753–1765.

Tyree, M.T. and Ewers, F.W. (1991). The hydraulic architecture of trees and other woody plants. *New Phytol.* **119**, 345–360.

Tyree, M.T. and Sperry, J.S. (1989). Vulnerability of xylem to cavitation and embolism. *Annu. Rev. Plant Physiol. Plant Mol. Biol.* **40**, 19–38.

Tyree, M.T., Cochard, H., Cruiziet, P. and Ameglio, T. (1993). Drought-induced leaf shedding in walnut: evidence for vulnerability segmentation. *Plant, Cell Env.* **16**, 879–882.

Tyree, M.T., Fiscus, E.L., Wullschleger, S.D. and Dixon, M.A. (1986). Detection of xylem cavitation in corn under field conditions. *Plant Physiol.* **82**, 597–599.

Unsworth, M.H. and Colls, J.J. (1994). Air pollution – what's in it for crops? In: Monteith, J.L., Scott, R.K. and Unsworth, M.H. (eds.). *Resource Capture by Crops*, pp. 189–209. Nottingham University Press, Nottingham.

Vaartaja, O. (1954). Photoperiodic ecotypes of trees. *Can. J. Bot.* **32**, 392–399.

Vaganov, E.A., Hughes, M.K., Kirdyanov, A.V., Schweingruber, F.H. and Silkin, P.P. (1999). Influence of snowfall and melt timing on tree growth in subarctic Eurasia. *Nature* **400**, 149–151.

Vallack, H.W., Bakker, D.J., Brandt, I., Broström-Lundén, E., Brouwer, A., Bull, K.R., Gough, C., Guardans, R., Oloubek, I., Jansson, B., Koch, K., Kuylenstierna, J., Lecloux, A., Mackay, D., McCutcheon, P., Mocarelli, P., Scheidegger, N.M.I., Sunden-Bylehn, A. and Taalman, R.D.F. (1998). Controlling persistent organic pollutants – what next? *Environ. Toxicol. Pharmacol.* **6**, 143–175.

Vance, C.P. and Heichel, G.H. (1991). Carbon in N_2 fixation: limitation or exquisite adaptation. *Annu. Rev. Plant Physiol. Plant Mol. Biol.* **42**, 373–392.

van de Water, P.K., Leavitt, S.W. and Betancourt, J.L. (1994). Trends in stomatal density and $^{13}C/^{12}C$ ratios of *Pinus flexilis* needles during the last glacial-interglacial cycle. *Science* **264**, 239–243.

van den Bergh, J.P. (1969). Distribution of pasture plants in relation to the chemical properties of the soil. In: *Ecological Aspects of the Mineral Nutrition of Plants. Br. Ecol. Soc. Symp.* **9**, 11–23.

van der Wal, R., Egas, M., van der Veen, A. *et al.* (2000). Effects of resource competition and herbivory on plant performance along a natural productivity gradient. *J. Ecol.* **88**, 317–330.

van der Werf, A., Visser, A.J., Schieving, F. and Lambers, H. (1993). Contribution of physiological and morphological plant characteristics to a species' competitive ability at high and low nitrogen supply. A hypothesis for inherently fast- and slow-growing monocots. *Oecologia* **94**, 434–440.

van der Zaal, B.J., Neuteboom, L.W., Pinas, J.E., Chardonnens, A.N., Schat, H., Verkleij, J.A.C. and Hooykaas, P.J.J. (1999). Overexpression of a novel *Arabidopsis* gene related to putative zinc-transporter genes from animals can lead to enhanced zinc resistance and accumulation. *Plant Physiol.* **119**, 1047–1055.

van Oosten, J.-J., Wilkins, D. and Besford, R. (1994). Regulation of the expression of photosynthetic nuclear genes by CO_2 is mimicked by regulation by carbohydrates: a mechanism for the acclimation of photosynthesis to high CO_2? *Plant, Cell Env.* **17**, 913–923.

van Steveninck, R.F.M., van Steveninck, M.E., Wells, A.J. and Fernando, D.R. (1990). Zinc tolerance and the binding of zinc as zinc phytate in *Lemna minor*. X-ray microanalytical evidence. *J. Plant Physiol.* **137**, 140–146.

van Vuuren, M.M.I., Robinson, D. and Griffiths, B.S. (1996). Nutrient inflow and root proliferation during the exploitation of a temporarily and spatially discrete source of nitrogen in soil. *Plant and Soil* **178**, 185–192.

Vartapetian, B.B. and Jackson, M.B. (1997). Plant adaptations to anaerobic stress. *Ann. Bot.* **79 (Suppl. A)**, 3–20.

Vasquez, M.D., Poschenrieder, C.H., Barcelo, J., Baker, A.J. M, Hatton, P. and Cope, G.H. (1994). Compartmentation of zinc in roots and leaves of the zinc hyperaccumulator *Thlaspi caerulescens*. *Bot. Acta* **107**, 243–250.

Veenendaal, E.M., Swaine, M.D., Lecha, R.T., Walsh, M.F., Abebrese, I.K. and Owusu-Afriye, K. (1996). Response of West African forest tree seedlings to irradiance and soil fertility. *Funct. Ecol.* **10**, 501–511.

Veit, M., Bilger, W., Mühlbauer, T., Brummert, W. and Winter, K. (1996). Diurnal changes in flavonoids. *J. Plant Physiol.* **148**, 478–482.

Verkleij, J.A.C. and Schat, H. (1990). Mechanisms of metal tolerance in higher plants. In: Shaw, A.J. (ed.). *Heavy Metal Tolerance in Plants*, pp. 179–193. CRC Press, Boca Raton, Florida.

Vince-Prue, D. (1983). The perception of light-dark transitions. *Phil. Trans. R. Soc. London* **303**, 523–536.

Vogel, J.C., Fuls, A. and Danin, A. (1986). Geographical and environmental distribution of C3 and C4 grasses in the Sinai, Negev and Judean deserts. *Oecologia* **70**, 258–265.

von Arnim, A. and Deng, X.-W. (1996). Light control of seedling development. *Annu. Rev. Plant Physiol. Plant Mol. Biol.* **47**, 215–243.

von Willert, D.J. (1985). *Welwitchia mirabilis* – new aspects in the biology of an old plant. *Adv. Bot. Res.* **11**, 157–191.

von Willert, D.J., Treichel, J., Kirst, G.O. and Curdts, E. (1976). Environmentally controlled changes of phosphoenolpyruvate carboxylases in *Mesembryanthemum*. *Phytochemistry*. **15**, 1435–1436.

Wainwright, S.J. and Woolhouse, H.W. (1975). Physiological mechanisms of heavy metal tolerance in plants. In: *The Ecology of Resource Degradation and Renewal. Symp. Br. Ecol. Soc.* **15**, 231–258.

Wallender, H. (2000). Uptake of P from apatite by *Pinus sylvestris* seedlings colonised by different ectomycorrhizal fungi. *Plant and Soil* **218**, 249–256.

Walley, K.A., Khan, M.S.I. and Bradshaw, A.D. (1974). The potential for evolution of heavy metal tolerance in plants. 1. Copper and zinc tolerance in *Agrostis tenuis*. *Heredity* **32**, 309–319.

Walter, H. and Leith, H. (1960). *Klimadiagramm Weltatlas*. Fischer Verlag, Jena.

Wardlaw, I.F. (1979). The physiological effects of temperature on plant growth. *Proc. Agron. Soc. NZ* **9**, 39–48.

Warming, E. (1909). *Oecology of Plants*. Oxford University Press, Oxford.

Weaver, J.E. (1958). Classification of root systems of forbs of grassland and a consideration of their significance. *Ecology* **39**, 393–401.

Webster, G.L., Brown, W.V. and Smith, B.N. (1975) Systematics of photosynthetic carbon fixation pathways in *Euphorbia*. *Taxon* **24**, 27–34.

Weiner, J. (1990). Asymmetric competition in plant populations. *Trends Ecol. Evol.* **5**, 360–364.

Weinig, C. (2000). Plasticity versus canalization: population differences in the timing of shade-avoidance responses. *Evolution* **54**, 441–451.

Weiss, E. and Berry, J.A. (1988). Plants and high temperature stress. In: Long, S.P. and Woodward, F.I. (eds.). *Plants and Temperature*, pp. 329–346. Company of Biologists, Cambridge, UK.

Weissenhorn, I. and Leyval, C. (1995). Root colonisation of maize by a Cd-sensitive and a Cd-tolerant *Glomus mosseae* and cadmium uptake in sand culture. *Plant and Soil* **175**, 233–238.

Welbank, P.J. (1961). A study of the nitrogen and water factors in competition with *Agropyron repens* (L.) Beau. *Ann. Bot.* **25**, 116–137.

Wellburn, A. (1994). *Air Pollution and Climate Change.* Longman Scientific and Technical, Harlow, Essex.

Werk, K.S. and Ehleringer, J. R. (1984). Non-random leaf orientation in *Lactuca serriola. Plant, Cell Env.* **7**, 81–88.

Wheeler, D.M. (1995). Effect of root hair length on aluminium tolerance in white clover. *J. Plant Nutr.* **18**, 955–958.

Whitehead, F.H. (1971). Comparative autecology as a guide to plant distribution. In: *The Scientific Management of Animal and Plant Communities for Conservation. Symp. Br. Ecol. Soc.* **11**, 167–176.

Wild, A., Skarlou, V., Clement, C.R. and Snaydon, R.W. (1974). Comparison of potassium uptake by four plant species grown in sand and in flowing solution culture. *J. Appl. Ecol.* **11**, 801–802.

Wilson, G.B. and Bell, J.N.B. (1990). Studies on the tolerance to sulphur dioxide of grass populations in polluted areas. *New Phytol.* **116**, 313–317.

Wilson, J.R. and Ludlow, M.M. (1983). Time trends in osmotic adjustment and water relations of leaves of *Cenchrus ciliaris* during and after water stress. *Aust. J. Plant Physiol.* **10**, 15–24.

Wilson, S.D. and Tilman, D. (1991). Components of plant competition along an experimental gradient of nitrogen availability. *Ecology* **72**, 1050–1065.

Winding, A. (1994). Fingerprinting bacterial soil communities using Biolog microtitre plates. In: Ritz, K., Dighton, J. and Giller, K.E. (eds.). *Beyond the Biomass*, pp. 85–94. Wiley, Chichester.

Wingsle, G. and Hällgren, J.-E. (1993). Influence of SO_2 and NO_2 exposure on glutathione, superoxide dismutase and gutathione reductase activities in Scots Pine needles. *J. Exp. Bot.* **44**, 463–470.

Winner, W.E., Coleman, J.S., Gillespie, C., Mooney, H.A. and Pell, E. (1991). Consequences of evolving resistance to air pollutants. In: Taylor, G.E., Pitelka, L.F. and Clegg, M.T. (eds.). *Ecological Genetics and Air Pollution*, pp. 177–202. Springer-Verlag, New York.

Winter, K. (1974). Einfluss von Wasserstress auf die Aktivität der Phosphoenolpyruvat-Carboxylase bei *Mesembryanthemun crystallinum. Planta* **121**, 147–153.

Winter, K. and Ziegler, H. (1992). Induction of a crassulacean acid metabolism in *Mesembryanthemum crystallinum* increases reproductive success under conditions of drought and salinity stress. *Oecologia* **92**, 475–479.

Winter, K., Lüttge, U., Winter, E. and Troughton, J.H. (1978). Seasonal shift from C3 photosynthesis to crassulacean acid metabolism in *Mesembryanthemum crystallinum* growing in its natural environment. *Oecologia* **34**, 225–237.

Woledge, J. (1971). The effect of light intensity during growth on the subsequent rate of photosynthesis of leaves of tall fescue. *Ann. Bot.* **35**, 311–322.

Wolfenden, J. and Diggle, P.J. (1995). Canopy gas exchange and growth of upland pasture swards in elevated CO_2. *New Phytol.* **130**, 369–380.

Wollenweber-Ratzer, B. and Crawford, R.M.M. (1994). Enzymatic defence against post-anoxic injury in higher plants. *Proc. R. Soc. Edin.* **102B**, 381–390.

Woodward, F.I. (1987). Stomatal numbers are sensitive to increases in CO_2 from pre-industrial levels. *Nature* **327**, 617–618.

Woolhouse, H.W. (1966). The effect of bicarbonate on the uptake of iron in four different species. *New Phytol.* **65**, 372–375.

Wookey, P.A., Parsons, A.N., Welker, J.M., Potter, J.A., Callaghan, J.A., Lee, J.A. and Press, M.C. (1993). Comparative responses of phenology and reproductive development to simulated environmental change in sub-arctic and high arctic plants. *Oikos* **67**, 490–502.

Wright, W.J., Fitter, A.H. and Meharg, A.A. (2000). Reproductive biomass in *Holcus lanatus* clones that differ in their phosphate uptake kinetics and mycorrhizal colonization. *New Phytol.* **146**, 493–501.

Wu, L. and Antonovics, J. (1975). Zinc and copper uptake by *Agrostis stolonifera*, tolerant to both zinc and copper. *New Phytol.* **75**, 231–237.

Wu, L. and Bradshaw, A.D. (1972). Aerial pollution and the rapid evolution of copper tolerance. *Nature* **238**, 167.

Wuest, S.B., Albrecht, S.L. and Skirvin, K.W. (1999). Vapor transport vs. seed-soil contact in wheat germination. *Agron. J.* **91**, 783–787.

Yang, S. and Tyree, M.T. (1994). Hydraulic architecture of *Acer saccharum* and *rubrum*: comparison of branches to whole trees and the contribution of leaves to hydraulic resistance. *J. Exp. Bot.* **45**, 179–186.

Yanovsky, M.J., Casal, J.J. and Whitelam, G.C. (1995). Phytochrome A, phytochrome B and HY4 are involved in hypocotyl growth responses to natural radiation in *Arabidopsis*: weak de-etiolation of the *phyA* mutant under dense canopies. *Plant, Cell Env.* **18**, 788–794.

Young, J.P.W. and Haukka, K.E. (1996). Diversity and phylogeny of rhizobia. *New Phytol.* **133**, 87–94.

Yu, Q. and Rengel, Z. (1999). Micronutrient deficiency influences plant growth and activities of superoxide dismutase in narrow-leafed lupins. *Ann. Bot.* **83**, 175–182.

Zeiger, E. and Zhu, J. (1998). Role of zeaxanthin in blue light photoreception and the modulation of light-CO_2 interactions in guard cells. *J. Exp. Bot.* **49**, 433–442.

Zhang, H. and Forde, B.G. (1998). An *Arabidopsis* MADS box gene that controls nutrient-induced changes in root architecture. *Science* **279**, 407–409.

Ziegler, I. (1972). The effect of SO_2 on the activity of ribulose-1,5-diphosphate carboxylase in isolated spinach chloroplasts. *Planta* **103**, 155–163.

Zimmermann, M.H. and Brown, C.L. (1971). *Trees: Structure and Function.* Springer-Verlag, New York.

Zimmermann, M.H. and McDonough, J. (1978). Dysfunction in the flow of food. In: Horsfall, J.G. and Cowling, E.B. (eds.). *Plant Disease*, vol. 3, pp. 117–140. Academic Press, New York and London.

Zimmermann, U., Meinzer, F.C., Benkert, R., Zhu, J.J., Schneider, H., Goldstein, G., Kuchenbrod, E. and Haase, A. (1994). Xylem water transport: is the available evidence consistent with the cohesion theory? *Plant, Cell Env.* **17**, 1169–1181.

Zotz, G. and Thomas, V. (1999). How much water is in the tank? Model calculations for two epiphytic bromeliads. *Ann. Bot.* **83**, 183–192.

Name Index

Note: page numbers after 313 are bibliographic references

Species Index

Note: references to Plates are in *italics*

Subject Index

Note: **emboldened** page numbers indicate major topics. *All* references are to **plants**, unless otherwise indicated

Subject Index